Progress in Microalgae Research - A Path for Shaping Sustainable Futures

Edited by Leila Queiroz Zepka, Eduardo Jacob-Lopes and Mariany Costa Deprá

Published in London, United Kingdom

Progress in Microalgae Research - A Path for Shaping Sustainable Futures
http://dx.doi.org/10.5772/intechopen.98164
Edited by Leila Queiroz Zepka, Eduardo Jacob-Lopes and Mariany Costa Deprá

Contributors
Leila Zarandi Miandoab, Gerardo Eloy Suárez Álvarez, Teresita de Jesús Romero López, Yan Li, Ethan Wood, Gergely Kosa, Bushra Muzamil, Christian Vogelsang, Rune Holmstad, Gulshan Kumar Sharma, Shakeel Ahmad Khan, Amit Kumar, Ittyamkandath Rashmi, Fayaz Ahmad Malla, Gopal Lal Meena, Maria Concepción Lora Vilchis, Ramsenthil Ramadoss, Dhanasekaran Subramaniam, Durai Gunasekaran, Nasreen Alddin Mohi Abdulrahman, Ishita Bhattacharya, David de la Lama-Calvente, Juan Cubero, María José Fernández-Rodríguez, Antonia Jiménez-Rodríguez, Rafael Borja, Abhishek Saxena, Archana Tiwari, Aditi Raj, Umme Tamanna Ferdous, Zetty Norhana Balia Yusof, Nagashree Nagaraj Rao, Kausthubh Sumanth, Sanjana Subramanya, Sourav Umashankar, Supriya Gummalam, Rajeswari Mallikarjunaiah, Ashwani Sharma, María Guadalupe Figueroa-Torres, Saúl Almanza Encarnación, María Jesús Ferrara Guerrero, Aída del Rosario Malpica Sánchez, José Roberto Angeles Vázquez, Ahmed Aidan Al Hussieny Aidan Al-Hussieny, Leila Queiroz Zepka, Mariany Costa Deprá, Eduardo Jacob-Lopes, Sana Gammoudi, Neila Annabi-Trabelsi, Mariem Bidhi, Nouha Diwani, Amira Rekik, Hajer Khmekhem, Habib Ayadi, Wassim Guermazi, Jannet Elloumi, Juan C. Castro, Marianela Cobos, Carlos G. Castro, Corayma L. Cabezudo, Miguel A. Grandez, Alvaro B. Tresierra, Segundo L. Estela, Santiago Galindo, Sheyla L. Pérez, Jhon A. Vargas, Roger Ruiz, Jorge L. Marapara, Angélica V. Rios, Pedro M. Adrianzen

First published in London, United Kingdom, 2022 by IntechOpen
IntechOpen is the global imprint of INTECHOPEN LIMITED, registered in England and Wales, registration number: 11086078, 5 Princes Gate Court, London, SW7 2QJ, United Kingdom

British Library Cataloguing-in-Publication Data
A catalogue record for this book is available from the British Library

Additional hard and PDF copies can be obtained from orders@intechopen.com

Progress in Microalgae Research - A Path for Shaping Sustainable Futures
Edited by Leila Queiroz Zepka, Eduardo Jacob-Lopes and Mariany Costa Deprá
p. cm.
Print ISBN 978-1-80356-023-6
Online ISBN 978-1-80356-024-3
eBook (PDF) ISBN 978-1-80356-025-0

For EU product safety concerns:
IN TECH d.o.o., Prolaz Marije Krucifikse Kozulić 3, 51000 Rijeka, Croatia,
info@intechopen.com or visit our website at intechopen.com.

Meet the editors

Prof. Dr. Leila Queiroz Zepka is currently an associate professor at the Department of Food Technology and Science, Federal University of Santa Maria, Brazil. She has more than fifteen years of teaching and research experience. She has published more than 550 scientific publications/communications, including 15 books, 60 book chapters, 120 original research papers, 380 research communications in national and international conferences, and 12 patents. She is a member of the editorial board of five journals and acts as a reviewer for several national and international journals. Her research interests include microalgal biotechnology with an emphasis on microalgae-based products.

Dr. Mariany Coista Deprá is currently a researcher at the Department of Food Technology and Science, Federal University of Santa Maria, Brazil. She has published more than 100 scientific publications/communications, including books, book chapters, original research papers, research communications in national and international conferences, and patents. She acts as a reviewer for several international journals. Her research interests include microalgae-based processes and products with an emphasis on sustainability metrics and indicators.

Prof. Dr. Eduardo Jacob-Lopes is currently an associate professor at the Department of Food Technology and Science, Federal University of Santa Maria, Brazil. He has more than fifteen years of teaching and research experience. He has coordinated/coordinates more than fifty research projects and/or technological developments financed by public funding agencies and private initiatives. He has published more than 600 scientific publications/communications, including 15 books, 60 book chapters, 150 original research papers, 400 research communications in national and international conferences, and 12 patents. He is a member of the editorial board of ten journals and acts as a reviewer for several national and international journals. His research interests include bioprocess engineering with an emphasis on microalgal biotechnology.

Contents

Preface

The rapid pace of industrialization and expanding urbanization is a global vulnerability that requires immediate attention, particularly to the replacement of fossil resources. Anthropogenic activities continually force ecosystem remodeling for more sustainable projects. With the high interest in renewable resources, the field of biotechnology has become greatly important in recent years, and microalgal biotechnology has come to be seen as a promising biofactory of the future.

Indeed, microalgae biotechnology and the valorization of the primary and secondary metabolites of microalgae are driving forces for sustainable development. Several applications of these emerging microorganisms are becoming increasingly attractive in terms of research and the market. The close interaction of biology and process engineering is necessary to realize a "green" future.

This book provides a comprehensive overview of microalgae research to build and strengthen more sustainable pathways. It discusses technological barriers, future efforts, and microalgal challenges involving current issues of biological nature, culture bioprocessing, bioactive potential, applications beyond those for the environment, and biorefinery approaches. We hope that the information contained herein will trigger a process of critical reflection and provide future avenues for research in this area.

Dr. Leila Queiroz Zepka, Dr. Mariany Costa Deprá and Dr. Eduardo Jacob-Lopes
Federal University of Santa Maria,
Santa Maria-RS, Brazil

Chapter 1

Introductory Chapter: Microalgal Biotechnology - The Gold-Green in Modern Sustainable World

Mariany Costa Deprá, Eduardo Jacob-Lopes and Leila Queiroz Zepka

1. Introduction

As the global economy unfolds and the population increases, environmental concerns associated with greenhouse gas emissions and wastewater discharges, as well as issues involving food security, become unavoidable problems [1]. Given this devastating and worrying scenario, the challenges of sustainable development aroused interest in defining a potentially promising target that was recognized, consensually, as the future salvation.

Under this slant, the macro- and microalgae—known as the most primitive form of life on Earth—become the hype in the academic world. Over years, the discoveries surrounding these green microorganisms formed a glorious path in the biotechnological industry. So, its robust adaptability and metabolic plasticity, beyond its highly flexible capacity for genetic and environmental changes, have reiterated the microalgae, such as gold coins in the global market [2].

Without delay, macro- and microalgae were seen as a biological pump that not just could drive, balance, and maintain global ecosystems, but could also be used in many different technological fields. From this perspective, pioneering studies were traced to the application of microalgae as the main feedstock for bioenergy production [3]. However, obstacles involving low productivity, fractionation of biomass, as well as technical questions of scalability, have repercussions on technical-economic attributes, delaying the consolidation and commercialization of biofuels [4]. Later, the clean technologies concept applied to microalgal biotechnology was extended to the microalgae application as an effective tool for environmental processes, such as wastewater bioremediation and also the capacity to mitigate polluting gases. However, more than once the limitations of the process, already advanced in various aspects, have continued to haunt the microalgae consolidation [5].

Since then, the research objective has changed over time. To rectify this gap, it was necessary to understand basic aspects involving the biological fundamentals of microorganisms and extend the research to optimization of the production stages. Forthwith, the new research efforts return for advances in genetic engineering, mapping from to complex evaluation of the upstream and downstream stages until the ascertainment of new methods of biomass harvesting and extraction, gave the path to new erudition [6].

However, the current plans, at the time, to establish the consolidation of microalgal biotechnology in the fields of bioenergy and environmental bioremediation were still not sufficiently developed, and research agencies and private companies have shown interest in returning to microalgal biotechnology as a viable economy in the immediate future. Consequently, the green gold could not lose its brilliance, and the hype of microalgal biotechnology could not disappear without more, nor less.

Therefore, the most sublime strategy to intertwine the microalgae application was to establish a key connection for the hungry reduction, giving weight to health aspects, concomitant to a sustainable environment [7]. Thus, microalgae as the future food were launched as a promising game against the fulfillment of the sustainable development goals [8]. This is because its excellent nutritional content, attributing prominence to its high protein, lipid, and carbohydrate fractions, is considered a food with high value-added. Besides, the attention paid to fine chemical products, such as pigments, fatty acids, and nucleic acids contained in their biomass would attract the food and pharmaceutical industries as a platform for marketing the function of their biological properties [9].

However, although microalgae have been strategic resources for sustainability and promising market trends, the process challenges remained. Among them, the problems associated with excessive demand for nutrients and energy imputed by microalgae processes urgently need a solution [10]. In this way, it is the need to align the optimization and integration of processes, which turned into the key point for success. Thus, the triumph of microalgae-based processes would consist of a technological advance that would lead to a significant degree of self-sufficiency [11].

In this sense, in order for the microalgae processes in conjecture with biotechnology to grow larger and, in fact, become self-sufficient, it is imperative that public and private research tread exploration paths that lead to the discovery of new scientific studies that break the barriers hitherto imposed. The truth is that these jewels of nature, regardless of which way they emerge, will always play a vital role in the transformation of a sustainable modern world. This is because, in this scientific universe, there are tens of thousands of other green microscopic singularities, with a variety of shapes and functions to breathtaking, still waiting to be explored. For science, microalgae are not just a scum from lakes and seas, but they are in a universe of possibilities. From a commercial point of view, they are considered the feedstock of the future, this is, the jack of all trades.

So, perhaps, it would be appropriate to think that—in a period that is difficult to predict—all the benefits of microalgae for new processes and products are just waiting to be harvested. However, now, it is important that we become part of a revolution in the way we consider and conduct our research, ensuring that the results obtained will eventually, indeed, be used more and more for our benefit and applied to our planet. In this regard, we expect that the next chapters found in this book will serve as building blocks for the advancement of microalgal biotechnology, to build on top of a robust theoretical knowledge, an important practice contribution to the transition to a more sustainable society.

Acknowledgements

The authors would like to thank the National Council for Scientific and Technological Development (CNPq)—Brazil, and the Coordination of Improvement of Higher Education Personnel (CAPES), Finance Code 001.

Conflict of interest

The authors declare that they have no known competing financial interests or personal relationships that could have appeared to influence the work reported in this chapter.

Author details

Mariany Costa Deprá, Eduardo Jacob-Lopes and Leila Queiroz Zepka*
Federal University of Santa Maria, Brazil

*Address all correspondence to: zepkaleila@yahoo.com.br

References

[1] You X, Yang L, Zhou X, Zhang Y. Sustainability and carbon neutrality trends for microalgae-based wastewater treatment: A review. Environmental Research. 2022;**209**:112860. DOI: 10.1016/j.envres.2022.112860

[2] Islam MB, Khalekuzzaman M, Kabir SB, Hossain MR, Alam MA. Substituting microalgal biomass with faecal sludge for high-quality biocrude production through co-liquefaction: A sustainable biorefinery approach. Fuel Processing Technology. 2022;**225**:107063. DOI: 10.1016/j.fuproc.2021.107063

[3] de Carvalho Silvello MA, Gonçalves IS, Azambuja SPH, Costa SS, Silva PGP, Santos LO, et al. Microalgae-based carbohydrates: A green innovative source of bioenergy. Bioresource Technology. 2022;**344**:126304. DOI: 10.1016/j.biortech.2021.126304

[4] Min KH, Kim DH, Ki MR, Pack SP. Recent progress in flocculation, dewatering, and drying technologies for microalgae utilization: Scalable and low-cost harvesting process development. Bioresource Technology. 2022;**344**:126404. DOI: 10.1016/j.biortech.2021.126404

[5] Carraro CDFF, Loures CCA, de Castro JA. Microalgae bioremediation and CO2 fixation of industrial wastewater. Cleaner Engineering and Technology. 2022;**8**:100466. DOI: 10.1016/j.clet.2022.100466

[6] Muthukrishnan L. Bio-engineering of microalgae: Challenges and future prospects toward industrial and environmental applications. Journal of Basic Microbiology. 2022;**62**:310-329. DOI: 10.1002/jobm.202100417

[7] Liu R, Li S, Tu Y, Hao X, Qiu F. Recovery of value-added products by mining microalgae. Journal of Environmental Management. 2022;**307**:114512. DOI: 10.1016/j.jenvman.2022.114512

[8] Mishra B, Tiwari A, Mahmoud AED. Microalgal potential for sustainable aquaculture applications: Bioremediation, biocontrol, aquafeed. Clean Technologies and Environmental Policy. 2022;**1**:1-13. DOI: 10.1007/s10098-021-02254-1

[9] Chandrasekhar K, Raj T, Ramanaiah SV, Kumar G, Banu JR, Varjani S, et al. Algae biorefinery: A promising approach to promote microalgae industry and waste utilization. Journal of Biotechnology. 2022;**345**:1-16. DOI: 10.1016/j.jbiotec.2021.12.008

[10] Deprá MC, Severo IA, dos Santos AM, Zepka LQ, Jacob-Lopes E. Environmental impacts on commercial microalgae-based products: Sustainability metrics and indicators. Algal Research. 2020;**51**:102056. DOI: 10.1016/j.algal.2020.102056

[11] Cheirsilp B, Maneechote W. Insight on zero waste approach for sustainable microalgae biorefinery: Sequential fractionation, conversion and applications for high-to-low value-added products. Bioresource Technology Reports. 2022;**18**:101003. DOI: 10.1016/j.biteb.2022.101003

Chapter 2

Climate Change and Algal Communities

Umme Tamanna Ferdous and Zetty Norhana Balia Yusof

Abstract

Climate change is one of the major global concerns jeopardizing human health and wildlife. This event is considered a threat to the marine ecosystem as well. Marine algae are the leading producer in the benthic food chain. Therefore, any change in marine algal communities will disrupt the whole ecosystem. Currently, algal species face significant changes in their abundance and distribution worldwide. Toxic species are frequently invading and causing a phenomenon called the harmful algal bloom, which threatens the seafood industry and public health. This chapter will focus on the significant distribution of algal communities worldwide and the impact of climate change on these marine algal species. Besides, this chapter will shed some light on how these changes affect the marine food chain and ultimately affect human health.

Keywords: climate change, harmful algal bloom, Marine algae, public health, seafood

1. Introduction

Microalgae are unicellular photosynthetic organisms that can be eukaryotic or prokaryotic (cyanobacteria) and are ubiquitous; they can be found in fresh or marine water, soil and even extreme habitats [1]. Microalgae can transform atmospheric inorganic carbon into organic carbon biomass, and it has been estimated that half of their biomass comprises carbon (w/w) [2]. In addition, they contain approximately 50 % (w/w) carbon in their biomass, which helps them grow better [3]. Surprisingly, some species of marine microalgae can grow at high CO_2 (40 %) concentrations, which is known to help curb greenhouse gases that ultimately have a positive effect on Blue Economy [4]. One of the main benefits of using microalgae in mitigating CO_2 is that they can fix this gas up to 50 times more than terrestrial plants [5].

On the other hand, macroalgae or seaweeds are multicellular and macroscopic algae that contribute as huge biomass producers in the benthic region. Besides their consumption as nutritious food and feed, they are frequently used as fertiliser [6]. Like microalgae, seaweeds also contribute to mitigating climate change through the Blue Carbon strategy. Seaweed can sequester a large amount of CO_2 from the environment that has been predicted that ongoing seaweed farming can capture more than 6 % of CO_2 globally by 2050 [7]. Seaweed aquaculture benefits coastal habitats in different ways. For instance, a study showed that Chinese seaweed farming upgrades the coastal water quality by eliminating nitrogen and phosphate as well as by sequestering carbon, which in turn also helps in decreasing eutrophication. Interestingly, this kind of farming needs no chemical fertiliser pesticides as well saves cultivable lands [8].

Currently, changes in climate patterns give rise to several environmental hazards and the most drastic effect is evident in the marine ecosystem. Consequently, marine microalgae and seaweeds are also experiencing the negative impact of climate change. Studies revealed that changes in habitat conditions make the species migrate or to extinction [9]. Therefore, migration in seaweed species is becoming common. Range shifts in seaweed communities disturb the local distribution pattern and species richness, reported globally [10].

2. Climate change

According to NASA, climate change can be defined as persisting modification in the earth's climate, resulting from anthropogenic activities (industrialisation, burning fossil fuels, deforestation, etc.), which gives rise to greenhouse gases (GHGs) and ultimately contributes to global warming. Climate change is attributed to the increase of greenhouse gases, mainly CO_2 (418 ppm currently), which helps increase the surface temperature of our planet by about 1.01°C, and this trend is on the rise. This phenomenon also engenders warming of the oceans exceeding 0.4°F since the early 70s, a rapid decrease in ice mass (about 147 gigatons/year) in Antarctica, increasing in sea level (3.4 mm/year) and finally gives rise to different natural disasters like drought, heavy rainfall or acidification in the ocean [11].

3. Effect of climate change on living organisms

Changing climate conditions have a profound impact on human health. It can directly affect our health through extreme environmental conditions like drought, heavy rainfall or indirectly through transmissible diseases or malnutrition along with other mental health complications [12]. Severe weather condition helps spread pathogenic microorganisms yielding different waterborne or zoonotic diseases as well as increasing allergens in the air can cause respiratory infections [13, 14]. Climate change is not only impacting human life but also jeopardising wildlife. Many terrestrial and aquatic organisms are at the risk of extinction due to environmental chemical pollutants. Altered environmental conditions can make wildlife more susceptible to chemical toxicants and retard their physiological capability [15]. The impact of climate change on tree mortality is another pressing problem in the forest ecosystem. This temporal rise in tree mortality has been seen in both areas with an increased water shortfall or no such undersupply [16]. Birds, one of the main indicator groups of climate change effects, are more prone to extinction due to severe weather change. It has been predicted that about 900 land bird species will be terminated because of global warming. Moreover, loss of habitat, augmentation of invasive bird species and hunting and spreading of infectious diseases are also contributing factors to bird extinction, which is the outcome of change in climate [17]. Climate change has the most vulnerable effect on agriculture. Changing in climate pattern gives rise to land degradation, desertification or heavy rainfall and floods, which in turn can able to make the soil nutrient deficit, highly saline and less productive. All these stresses are attributable to the risk of global food security [18].

The marine ecosystem is greatly affected by climate change. The oceans absorb 90% of the extra heat of the climate, which would be increased to 5–7 times by 2100 if global warming exceeds 2 °C. Additionally, the ocean absorbs up to 30 % of anthropogenic CO_2 emissions, which renders acidification in the marine system. Ocean acidification and heat cause depletion of nutrients and O_2 supply, which eventually endanger distribution and abundance of marine fishes as well as other organisms, and along with these, climate change also affects demography, calcification and phenology of phytoplankton and zooplankton [19, 20]. It accelerates the bleaching of reef-building corals, which makes great barrier reef (GBR) vulnerable. It has been estimated that a 0.5 °C increase in local temperature may cause rapid degeneration of GBR [21].

4. Global distribution of microalgae and macroalgae

In the marine coastal area, green algae are the major contributor to photosynthesis, especially microalgae from the Mamiellophyceae class are abundant in number while Chlorophyceae are the least. Microalgae from Pyramimonadales and Chlorodendrophyceae are ubiquitous in marine coastal regions [22]. In desert soil, cyanobacterial species are mostly found than other microalgae species because of their greater desiccation tolerating capacity [23]. In the alpine region of New Zealand, microalgae from Chrysophyceae, Xanthophyceae, Chlorophyceae, Trebouxiophyceae, and Charophyceae are more prevalent [24]. A study on the southeast coast of India showed that *Isochrysis galbana* was the dominant species in that area, while *Nannochloropsis oculata* was the less common species observed. The dominance of other species found were *Chlorella marina* > *Chromulina freibergensis* > *Dicrateria inornata* > *Chaetoceros calcitrans* [25]. On the other hand, the coastal region of Indonesia is dominated by diatoms, especially *Rhizosolenia* spp., *Chaetoceros* spp. and *Coscinodiscus* spp. [26]. In the case of benthic harmful algal bloom species (BHABs), two main species, *Gambierdiscu* and *Fukuyoa,* are widespread in the Caribbean Sea, the South Pacific, Indian Oceans, North Atlantic Ocean and the Canary Islands. Additionally, they are also reported from the coastal area of Japan, Korea, New Zealand and Australia. On the other hand, *Ostreopsis* spp. is common in the Mediterranean Sea [27].

A wide variety of seaweed species are distributed on the coasts, and their similarity is dependent on climatic zones. Most of the Rhodophyta belongs to warm temperate Pacific flora. The number of Rhodophyta is more than double in these Pacific florae compared to the cold temperate floras [28]. Seaweed species are vastly available in 30–50° latitude, which covers southern Australia, Japan and the Mediterranean as well as in the Philippines [29]. In recent studies, the distribution of invasive seaweeds has been documented. The highest invasion cases belong to the Mediterranean region (132), followed by the NE Atlantic region, covering the North and Baltic Sea and the eastern Atlantic Islands. Among the Rhodophytes, the Rhodomelaceae and the Ceramiaceae accommodate more invaders, while Chlorophyta, *Caulerpa* spp. and *Codium* spp. are recorded as the most successful invader species [30].

5. Effect of climate change on microalgae globally

The growth of microalgae and the production of microalgal bioactives are significantly driven by different stress conditions. For instance, under oxidative and salinity stress, carotenoids production is upregulated [31]. Similarly, phenolics, antioxidative properties and thiamin biosynthesis are augmented under abiotic stress [32, 33]. In water bodies, microalgal growth is greatly influenced by temperature, nutrients, salinity, the direction of wind and current, light as well as other organisms present in that habitat [34]. Increased CO_2 level helps increase nutrient acquisition, photosynthetic activity and growth of microalgae in the freshwater ecosystem [35]. In a freshwater ecosystem, green microalgae, compared to diatom and cyanobacteria, can better adapt to elevated CO_2. Still, with increased temperature, cyanobacteria start to proliferate rapidly and become dominant [36]. However, the pitfalls of climate change like elevated temperature, CO_2 or UV radiation have severely affected algal growth and productivity [37].

Microalgae produce a myriad of pharmaceutically important secondary metabolites, especially antioxidant and anticancer agents [38]. They are also a reservoir of various biotoxins, and the production of these toxins is more influenced by climate change [34]. Warming in the global climate affects fresh and marine water bodies by the formation of algal blooms of harmful species, which harms other organisms in those aquatic systems as well as human health, food security and the overall economy. Mainly, harmful algal blooms (HAB) in the freshwater system are caused by cyanobacterial species, and in the marine ecosystem, dinoflagellates are responsible for this kind of blooms [39]. Dinoflagellates like *Gambierdiscus* spp., *Fukuyoa* spp. and *Ostreopsis* spp. are well-known HAB species (also known as BHABs) in the benthic region of the ocean. Though microalgae are used as fish feed in mariculture industries and also used as vectors for vaccine delivery for these fishes, HAB species are getting a major threat to aquaculture [40]. Their notorious toxins production (ciguatoxin, palytoxins, ovatoxins) leads to many health risks, such as food poisoning, irritation and respiratory illness. Climate-directed ocean warming is the key factor for the dense and prompt growth of BHABs in many tropical and subtropical marine ecosystems globally. Elevated temperature helps to flourish these toxin-producing BHABs beyond their geographic area and even in an overly populated area where toxicity is a rare concern. Surprisingly, *Gambierdiscus* sp., one of the BHABs, can propagate in the degraded coral environment due to bleaching events [27]. On benthic microalgae, sea warming exhibits a direct positive and strong correlation to whole biomass accrual and growth rate where the presence of abundant mesograzers (gastropods, crustaceans) even cannot deplete their total biomass [41]. Furthermore, pelagic HABs, including *Pseudo-nitzschia*, *Alexandrium catenella* and *Pseudochattonella,* promote toxic blooms which are highly linked to climate change and cause huge mortality in fish farms [42]. Moreover, *Alexandrium catenella* and *Pyrodinium bahamense* can form cysts that can persist as long-dormant phases in warmer conditions have short quiescence and germination phases [43].

Studies have shown that tropical seas, such as the Red, sea often face HAB from different microalgae genres, such as dinoflagellates, raphidophytes, cyanobacteria, diatoms, due to the warmer climate. Currently, five other HABs have been reported in this tropical sea area, including *Noctiluca scintillans*, *P. bahamense, Protoperidinium quinquecorne, Heterosigma akashiwo* and *Trichodesmium erythraeum*. Moreover, cysts of toxic microalgae species have been found on the Red sea coast responsible for further bloom formation [44]. Additionally, microalgae can withstand ocean acidification effects though the capability is different from species to species. A study showed that *Tetraselmis chuii* exhibited better adaptability in terms of metabolic activity in comparison to *Phaeodactylum tricornutum* [45].

6. Effect of climate change on global seaweed communities

The development and the survival of macroalgae depend on a range of environmental factors, such as temperature, CO_2 concentration, nutrients, wave peak, etc., which are also related to climate. Among these parameters, the temperature has an intense influence on their growth and distribution [46]. Elevated oceanic temperature from the result of global warming accelerates the poleward shift of macroalgal species. In a study of Australian seaweed distribution, it has been reported that continued ocean warming is responsible for the abrupt range shift of indigenous seaweed species, which may cause the extinction of 100–350 species within the next few decades. It is noteworthy that over 25% of macroalgal species solely belong to southern Australia, which infers that extinction of these species has a huge impact on total seaweed communities worldwide [47]. Along the North-Atlantic shores, canopy-forming seaweeds will show a northward shift due to the ongoing ocean warming incident. *Fucus serratus, Ascophyllum nodosum, Saccharina latissima, Laminaria hyperborea* and *Chondrus crispus* will be eradicated from the warm-temperate region of the North-Atlantic. Moreover, temperate species will be migrated to cooler Southern Arctic areas. With the absence of these seaweeds, Northwest-African and Northwest-Atlantic coasts will be transformed completely [48].

In a study of ocean acidification (OA) impact on the Baltic sea, researchers found that with elevated CO_2 levels, the photosynthetic rate of *Furcellaria lumbricalis* and *Coccotylus truncatus* has been augmented. This augmentation rate is much higher in *C. truncates* in comparison to *F. lumbricalis*. However, the effect of CO_2 on photosynthesis is also influenced by other environmental factors, especially temperature. Increased water temperature can lessen the photosynthesis activity of *F. lumbricalis* [49]. Though a positive effect is always expected regarding the growth benefit of non-calcifying seaweeds with increased CO_2 concentration in coastal water, the opposite result has been found in some seaweeds like *Fucus vesiculosus,* which is frequently found in the North Sea. With increased CO_2 in water, *F. vesiculosus* growth rate has been reported to decrease up to 15% [50]. Increased CO_2 has a drastic effect on the tissue density of *F. vesiculosus*. It lowers the breaking strength, making this seaweed vulnerable to water waves and storms. Simultaneously, phlorotannin contents may decrease in this phenomenon [51].

OA has an impact on the different developmental stages of seaweeds. *Macrocystis pyrifera* can face a slightly reduced germination rate when OA is at an extreme level, though this seaweed successfully withstands increased OA with even lowered pH [52]. Surprisingly, extract of brown seaweed *Sargassum vulgare* from the acidified ocean exhibited higher bioactivity like anti-oxidant, anti-microbial, anti-lipid peroxidation, as well as anti-cancer activities [53].

7. Impact of HAB on animal and human health

Frequent HAB events in a particular area are reported to endanger and kill marine organisms, especially fish, shellfish, birds, and other marine organisms on a large scale. HAB formation can be associated with either toxic or non-toxic phytoplanktons. Non-toxic species are responsible for discolouration sometimes for fish and shellfish mortality. In this case, oxygen depletion due to dense bloom is accountable for such killing. On the other hand, toxic species cause poisoning in marine fish and bivalves, which renders large-scale fish killing and also severe health hazards for humans [54]. Ciguatera poisoning (CP) is the most common type of HAB-related contamination in fish. Dinoflagellates, *Gambierdiscus and Fukuyoa*, are attributed to produce ciguatoxins, which ultimately cause CP. Upon consuming contaminated fish, individuals may experience gastrointestinal, cardiovascular, and neurological disturbances [55].

Other HAB-related diseases include amnesic shellfish poisoning, caused mainly by *Pseudo-nitzschia* and *Nitzschia*; azaspiracid shellfish poisoning, caused by *Azadinium* and *Amphidoma*; diarrhetic shellfish poisoning, caused by Dinophysis, Phalacroma and Prorocentrum; neurotoxic shellfish poisoning, caused by *Karenia* and *Chatonella*; paralytic shellfish poisoning, caused by *Alexandrium, Gymnodinium* and *Pyrodinium*; and palytoxicosis ovatoxin poisoning, caused by Dinoflagellate, *Ostreopsis*. All of these poisonings are mainly associated with gastrointestinal, respiratory or neurological disturbances in humans. In severe cases, it may lead to coma or even death. HAB toxins kill marine piscine species and may amalgamate in these marine creatures, which may further transfer to the food web. This may affect other animals as well [56]. For example, birds, snakes and turtles found in HAB affected wetlands are reported to face sub-lithal effects. HAB toxins like microcystin can elevate physiological stress in these organisms, resulting in immune function or reproduction anomalies [57].

Besides direct consumption, exposure to HAB related toxins may also affect human health. Aerosolised brevetoxin from *Karenia brevis* is responsible for impaired respiratory function. This brevetoxin can spread quickly to the entire water column and maybe end up in the diet of the biota of that area. Not only that, this toxin has a detrimental impact on the larval stages of aquatic creatures [58].

8. HAB mitigation strategies

HABs are the major threats to public health and the economy in benthic areas. HAB formation is not only responsible for destroying aquaculture industries but also severely affecting the tourism industries. Besides fish poisoning, HAB affected beaches become vulnerable due to fouling of water, which may cause skin diseases and other illnesses [59].

Several biological, chemical or physical measures are taken worldwide to mitigate these HAB events. The most popular and effective action is modified clay (MC). By using inorganic and organic substances clay surface is reformed and used to remove the blooms. MC does not damage the water quality and can decrease the eutrophication by absorbing nutrients. Furthermore, MC can quickly reduce algal toxins with a faster degradation rate without affecting other marine organisms [60]. Biological control measures like using lactic acid bacteria and other marine bacteria (e.g. Paracoccus spp.) also effectively reduce HAB toxin [61]. Macroalgae cultivation is another potent way to control HAB events. A study showed macroalgae *Saccharina latissima*, *Chondrus crispus*, and *Ulva* spp. can hinder the growth and toxin production of *Alexandrium catenella* in aquaculture settings [62].

Besides these control measures, early risk assessment for eutrophication around the fish farming areas may help the related authorities to take precautionary measures. Climate change risk maps also can play a vital role in predicting upcoming HAB events in risk areas [63]. To further minimise economic loss and emotional stress, income diversification and skill development are highly recommended [64].

9. Conclusion

Though seaweed and microalgae farming help in curbing greenhouse gases, these algal communities are now observing threats of extinction and migration. Moreover, toxic microalgae species are becoming a major public health hazard in coastal areas. A proper mitigation plan would help slow down the negative impact of climate change. More research is needed to take necessary and effective mitigation action in terms of the current status of the indigenous seaweed species globally. The actual effect of global warming on the macroalgal communities should be scrutinised first. Moreover, the harmful microalgae bloom events in the coastal regions should be documented and studied profusely, which, in turn, will help the authorities to take actions for preserving the Blue economy.

Acknowledgements

This work is supported by the Higher Institution Centre of Excellence (HICOE) Research Grant (Innovative Vaccines and Therapeutics against Fish Diseases) (Project No. 6369100), and SATREPS (JICA-JST): COSMOS-MOHE G4-B Research Grant (Microalgae for Sustainable Aquaculture Health: Microalgae Vaccine Delivery System) (Project No. 6300866).

Author details

Umme Tamanna Ferdous[1] and Zetty Norhana Balia Yusof[1,2,3]*

1 Aquatic Animal Health and Therapeutics Laboratory (AquaHealth), Institute of Bioscience, Universiti Putra Malaysia, Serdang, Selangor, Malaysia

2 Faculty of Biotechnology and Biomolecular Sciences, Department of Biochemistry, Universiti Putra Malaysia, Serdang, Selangor, Malaysia

3 Bioprocessing and Biomanufacturing Research Center, Universiti Putra Malaysia, Serdang, Selangor, Malaysia

*Address all correspondence to: zettynorhana@upm.edu.my

References

[1] Venkatesan J, Manivasagan P, Kim S-K. Marine microalgae biotechnology: Present trends and future advances. In: Se-Kwon K, editor. Handbook of Marine Microalgae. Boston: Academic Press; 2015. pp. 1-9

[2] Yen HW, Chen Hu I, Chen CY, Ho SH, Lee DJ, Chang JS. Microalgae-based biorefinery: From biofuels to natural products. Bioresource Technology. 2013;**135**:166-174

[3] de Morais MG, de Morais EG, Duarte JH, Deamici KM, Mitchell BG, Costa JAV. Biological CO_2 mitigation by microalgae: Technological trends, future prospects and challenges. World Journal of Microbiology and Biotechnology. 2019;**35**(5):1-7

[4] Costa JAV, de Freitas BCB, Lisboa CR, Santos TD, de Fraga Brusch LR, de Morais MG. Microalgal biorefinery from CO_2 and the effects under the blue economy. Renewable and Sustainable Energy Reviews. 2019;**99**(October 2017): 58-65

[5] Cheng J, Huang Y, Feng J, Sun J, Zhou J, Cen K. Improving CO_2 fixation efficiency by optimizing chlorella PY-ZU1 culture conditions in sequential bioreactors. Bioresource Technology. 2013;**144**:321-327

[6] Salehi B, Sharifi-rad J, Seca AML, Pinto DCGA. Current trends on seaweeds : Looking at chemical composition, phytopharmacology, and cosmetic applications. Molecules. 2019;**24**(22):4182

[7] Duarte CM, Jiaping W, Xiao X, Bruhn A, Krause-Jensen D. Can seaweed farming play a role in climate change mitigation and adaptation? Frontiers in Marine Science. 2017;**4**:100

[8] Zheng Y, Jin R, Zhang X, Wang Q, Jiaping W. The considerable environmental benefits of seaweed aquaculture in China. Stochastic Environmental Research and Risk Assessment. 2019;**33**(4-6):1203-1221

[9] Henson SA, Beaulieu C, Ilyina T, John JG, Long M, Séférian R, et al. Rapid emergence of climate change in environmental drivers of marine ecosystems. Nature Communications. 2017;**8**(5020):1-9

[10] Fernández C, Hoz D, Ramos E, Puente A, Juanes JA. Climate change induced range shifts in seaweeds distributions in Europe. Marine Environmental Research. 2019;**148**(January):1-11

[11] NASA. "Global Climate Change." https://climate.nasa.gov/evidence/. 2022

[12] Wheeler N, Watts N. Climate change: From science to practice. Current Environmental Health Reports. 2018;**5**(1):170-178

[13] Demain JG. Climate change and the impact on respiratory and allergic disease: 2018. Current Allergy and Asthma Reports. 2018;**18**(4):22

[14] Funari E, Manganelli M, Sinisi L. Impact of climate change on waterborne diseases. Ann Ist Super SAnItà. 2012;**48**(4):473-487

[15] Noyes PD, Lema SC. Forecasting the impacts of chemical pollution and climate change interactions on the health of wildlife. Current Zoology. 2015;**61**(4):669-689

[16] Luo Y, Chen HYH. Climate change-associated tree mortality increases

without decreasing water availability. Ecology Letters. 2015;**18**(11):1207-1215

[17] Şekercioĝlu CH, Primack RB, Wormworth J. The effects of climate change on tropical birds. Biological Conservation. 2012;**148**(1):1-18

[18] Arora NK. Impact of climate change on agriculture production and its sustainable solutions. Environmental Sustainability. 2019;**2**(2):95-96

[19] IPCC. "Global Warming of 1.5°C." An IPCC Special Report on the Impacts of Global Warming of 1.5°C above Pre-Industrial Levels and Related Global Greenhouse Gas Emission Pathways, in the Context of Strengthening the Global Response to the Threat of Climate Change, Sustainable Development. Vol. 85. 2018

[20] Poloczanska ES, Burrows MT, Brown CJ, Molinos JG, Halpern BS, Hoegh-Guldberg O, et al. Responses of marine organisms to climate change across oceans. Frontiers in Marine Science. 2016;**3**(May):1-21

[21] Ainsworth TD, Heron SF, Ortiz JC, Mumby PJ, Grech A, Daisie Ogawa C, et al. Climate change disables coral bleaching protection on the Great Barrier Reef. Science. 2016;**352**(6283):338-342

[22] Tragin M, Vaulot D. Green microalgae in marine coastal waters: The Ocean Sampling Day (OSD) dataset. Scientific Reports. 2018;**8**(1):1-12

[23] Zhang B, Zhang Y, Downing A, Niu Y. Distribution and composition of cyanobacteria and microalgae associated with biological soil crusts in the Gurbantunggut Desert, China. Arid Land Research and Management. 2011;**25**(3):275-293

[24] Novis PM, Beer T, Vallance J. New records of microalgae from the New Zealand Alpine Zone, and their distribution and dispersal. New Zealand Journal of Botany. 2008;**46**(3):347-366

[25] Srinivasakumar KP, Rajasekha M. The population abundance, distribution pattern and culture studies of isolated microalgal strains from selective sampling sites along the South East Coast of India. African Journal of Biotechnology. 2009;**8**(16):3814-3826

[26] Kadim MK, Arsad S. Distribution and abundance of microalgae based on coastal characteristic and ecology in Bone Bolango Coastal Region, Indonesia. Asian Journal of Microbiology, Biotechnology and Environmental Sciences. 2016;**18**(2):395-401

[27] Tester PA, Wayne Litaker R, Berdalet E. Climate change and harmful benthic microalgae. Harmful Algae. 2020;**91**(January):101655

[28] van den Hoek C. World-wide latitudinal and longitudinal seaweed distribution patterns and their possible causes, as illustrated by the distribution of Rhodophytan Genera. Helgoländer Meeresuntersuchungen. 1984;**38**(2):227-257

[29] Bolton JJ. Global seaweed diversity: Patterns and anomalies. Botanica Marina. 1994;**37**(3):241-246

[30] Williams SL, Smith JE, Biological Centre JJ, Bolton MH, Zawawi MH, Idris AHM, et al. Climate change induced range shifts in seaweeds distributions in Europe. Botanica Marina. 2006;**7145**(8):1064-1078

[31] Abidin AA, Zainal NA, Othman FY, Yusof ZNB. Determination of transgene stability in nannochloropsis Sp . transformed with immunogenic peptide for oral vaccination against

vibriosis. Aquaculture International. 2021a;**29**:477-486

[32] Azim NH, Subki A, Balia ZN, Yusof. Abiotic stresses induce total phenolic, total flavonoid and antioxidant properties in Malaysian indigenous microalgae and cyanobacterium. Malaysian Journal of Microbiology. 2018;**14**(1):25-33

[33] Fern LL, Abidin AAZ, Yusof ZNB. Upregulation of thiamine (Vitamin B1) biosynthesis gene upon stress application in Anabaena Sp. and Nannochloropsis Oculata. Journal of Plant Biotechnology. 2017;**44**(4):462-471

[34] Estevez P, Castro D, Pequeño-Valtierra A, Giraldez J, Gago-Martinez A. Emerging marine biotoxins in seafood from European Coasts: Incidence and analytical challenges. Foods. 2019;**8**(5):1-13

[35] Brown TR, Lajeunesse MJ, Kathleen M. Strong effects of elevated CO_2 on freshwater microalgae and ecosystem chemistry. Limnology and Oceanography. 2020;**65**(2):304-313

[36] Li W, Xiaoguang X, Fujibayashi M, Niu Q, Tanaka N, Nishimura O. Response of microalgae to elevated CO_2 and temperature: Impact of climate change on freshwater ecosystems. Environmental Science and Pollution Research. 2016;**23**(19):19847-19860

[37] Beardall J, Stojkovic S. Microalgae under global environmental change : Implications for growth and productivity , populations and trophic flow. Scienceasia. 2006;**1**(s1):1-10

[38] Ferdous UT, Yusof ZNB. Medicinal prospects of antioxidants from algal sources in cancer therapy. Frontiers in Pharmacology. 2021;**12**(March):1-22

[39] Gobler CJ. Climate change and harmful algal blooms: Insights and perspective. Harmful Algae. 2020;**91**:101731

[40] Abidin AA, Zainal CY, Yusof ZNB. Carotenogenesis in nannochloropsis oculata under oxidative and salinity stress. Sains Malaysiana. 2021b;**50**(2):327-337

[41] Werner FJ, Matthiessen B. Warming has stronger direct than indirect effects on benthic microalgae in a seaweed system in spring. Marine Biology. 2017;**164**(4):1-10

[42] Trainer VL, Moore SK, Hallegraeff G, Kudela RM, Clement A, Mardones JI, et al. Pelagic harmful algal blooms and climate change: Lessons from nature's experiments with extremes. Harmful Algae. 2020;**91**(May 2019):101591

[43] Brosnahan ML, Fischer AD, Lopez CB, Moore SK, Anderson DM. Cyst-forming dinoflagellates in a warming climate. Harmful Algae. 2020;**91**:101728

[44] Mohamed ZA. Potentially harmful microalgae and algal blooms in the Red Sea: Current knowledge and research needs. Marine Environmental Research. 2018;**140**(October 2017):234-242

[45] Bautista-Chamizo E, Sendra M, Cid Á, Seoane M, Romano M, de Orte, and Inmaculada Riba. Will temperature and salinity changes exacerbate the effects of seawater acidification on the marine microalga phaeodactylum tricornutum? Science of the Total Environment. 2018;**634**:87-94

[46] Harley CDG, Anderson KM, Demes KW, Jorve JP, Kordas RL, Coyle TA, et al. Effect of climate change on global seaweed communities. Journal of Phycology. 2012;**1078**:1064-1078

[47] Wernberg T, Russell BD, Thomsen MS, Frederico C, Gurgel D, Bradshaw CJA, et al. Seaweed communities in retreat from ocean warming. Current Biology. 2011;**21**(21):1828-1832

[48] Jueterbock A, Tyberghein L, Verbruggen H, Coyer JA, Olsen JL, Hoarau G. Climate change impact on Seaweed Meadow Distribution in the North Atlantic Rocky Intertidal. Ecology and Evolution. 2013;**2013**:1-18

[49] Pajusalu L, Albert G, Fachon E, Hepburn CD, Kotta J, Liversage K, et al. Ocean acidification may threaten a unique seaweed community and associated industry in the Baltic Sea. Journal of Applied Phycology. 2019;**32**:2469-2478

[50] Gutow L, Mo M, Bartl K, Saborowski R, Bartsch I, Wiencke C. Ocean acidification affects growth but not nutritional quality of the Seaweed Fucus Vesiculosus (Phaeophyceae , Fucales). Journal of Experimental Marine Biology and Ecology. 2014;**453**:84-90

[51] Kinnby A, White JCB, Toth GB, Pavia H. Ocean acidification decreases grazing pressure but alters morphological structure in a dominant Coastal Seaweed. PLoS ONE. 2021;**16** (1 January):1-14

[52] Roleda MY, Morris JN, McGraw CM, Hurd CL. Ocean Acidification and Seaweed Reproduction : Increased CO_2 Ameliorates the Negative Effect of Lowered PH on Meiospore Germination in the Giant Kelp Macrocystis Pyrifera (Laminariales , Phaeophyceae). Global Change Biology. 2012;**18**:854-864

[53] Kumar A, Buiac MC, Palumbod A, Mohamed Mohany AM, Wadaan WN, Beemster GTS, et al. Ocean acidification affects biological activities of seaweeds: A Case Study of Sargassum Vulgare from Ischia Volcanic CO_2 Vents. Environmental Pollution. 2020;**259**:113765

[54] Sidabutar T, Cappenberg H, SrimarianaES,MuawanahA,WouthuyzenS. Harmful algal blooms and their impact on fish mortalities in Lampung Bay: An overview. In: IOP Conference Series: Earth and Environmental Science. 2021

[55] Chinain M, Gatti CMI, Darius HT, Quod JP, Tester PA. Ciguatera poisonings: A global review of occurrences and trends. Harmful Algae. 2021;**102**(March):101873

[56] Turner AD, Lewis AM, Bradley K, Maskrey BH. Marine invertebrate interactions with harmful algal blooms – Implications for one health. Journal of Invertebrate Pathology. 2021;**186**(February):107555

[57] Refsnider JM, Garcia JA, Holliker B, Hulbert AC, Nunez A, Streby HM. Effects of harmful algal blooms on stress levels and immune functioning in Wetland-associated Songbirds and Reptiles. Science of the Total Environment. 2021;**788**:147790

[58] Heil CA, Muni-Morgan AL. Florida's Harmful Algal Bloom (HAB) problem: Escalating risks to human, environmental and economic health with climate change. Frontiers in Ecology and Evolution. 2021;**9**:646080

[59] Karlson B, Andersen P, Arneborg L, Cembella A, Eikrem W, John U, et al. Harmful algal blooms and their effects in Coastal Seas of Northern Europe. Harmful Algae. 2021;**102**:101989

[60] Song X, Zhang Y, Zhiming Y. An eco-environmental assessment of harmful algal bloom mitigation using modified clay. Harmful Algae. 2021;**107**(May):102067

[61] Martelli F, Cirlini M, Dellafiora L, Neviani E, Asta CD. Mitigation of marine toxins by interactions with bacteria : The case of okadaic acid and tetrodotoxin. Food Control. 2022;**131**(April 2021): 108428

[62] Sylvers PH, Gobler CJ. Mitigation of harmful algal blooms caused by Alexandrium Catenella and reduction in saxitoxin accumulation in bivalves using cultivable seaweeds. Harmful Algae. 2021;**105**:102056

[63] Soto D, León-Muñoz J, Garreaud R, Quiñones RA, Morey F. Scientific warnings could help to reduce farmed salmon mortality due to harmful algal blooms. Marine Policy. 2021;**132**(May):1-5

[64] Moore KM, Allison EH, Dreyer SJ, Ekstrom JA, Jardine SL, Klinger T, et al. Harmful algal blooms: Identifying effective adaptive actions used in fishery-dependent communities in response to a protracted event. Frontiers in Marine Science. 2020;**6**(January):1-12

Chapter 3

Cryptophyte: Biology, Culture, and Biotechnological Applications

María Concepción Lora Vilchis

Abstract

Cryptophytes are single-cell biflagellate algae, with extrusive organelles called ejectosomes. They live in fresh and marine water, mainly in shaded environments where light levels are reduced relative to the surface. They are the product of a secondary endosymbiosis of a red alga, which still retains the endosymbiont nucleus's reminiscences and has four membranes around its plastids. Cryptophytes have a metabolic diversity that makes them very interesting from a nutritional point of view since they present a balance of fatty acids w3/w6, sterols, carotenoid pigments, and phycobiliproteins, these last also have antioxidant effects. Their composition makes them attractive for food in aquaculture and human consumption, pharmaceuticals and cosmetics; their fluorescent potential has attracted the attention of researchers in genomics, neuroscience and molecular biology. The biochemical composition of the cells is modulated by illumination, available nutrients, and its growth phase. This work reviews the general biology of cryptophytes, emphasizing the photosynthetic ones, culture properties and its biotechnological potential.

Keywords: endosymbiosis, phycoerythrin, phycocyanin, stress, culture, biotechnology

1. Introduction

Cryptophytes or cryptomonads are eukaryote algae that are biflagellate and unicellular, with sizes between 3 and 50 μm, most are photosynthetic and motile, and a few are palmelloid and form colonies surrounded by mucilaginous sheaths [1]. They are classified into the kingdom Chromista, phylum Cryptophyta, class Cryptophyceae, and order Cryptomonadales. They live in environments from fresh to marine water usually of good quality [2, 3]; also they can be found at varying light conditions and at different temperatures, including those that are extreme such as the Antarctic [4], blooms of cryptophytes have been reported in fresh [5] and marine waters [6]. Cryptophyte's genus *Goniomonad* lacks plastids; they are heterotrophic and feed on bacteria and small organic particles. Due to their small size and biochemical composition, Cryptophytes are essential contributors to the food chains of a diversity of organisms [7–10]. Some cryptophytes are plastid donors for dinoflagellates like *Dinophysis acuminata* [11] and ciliates like *Mesodinium rubrum* and *M. major* [12]. Their importance has been underestimated mainly due to their delicate structure, which can be easily altered or broken by common fixatives such as Lugol and formalin. However, cryptophyte inter-species morphology is not that different allowing for species-level taxonomy by light microscopy [13, 14].

The abundance of cryptophytes is increasing in places like the Antarctic Peninsula [6] and Chesapeake Bay [12], where a notable change has been observed in species composition and size distribution, significantly influencing local ecosystems.

There is general agreement that cryptophytes evolved from a secondary endosymbiosis, which occurred by the engulfment of a red alga by an unknown eukaryote [15, 16]; this event resulted in a cell with two nuclei, two cytoplasms, one of each is in the chloroplast, which is covered with four membranes, and with unique content and distribution among algae of harvesting-light pigments, they have chlorophylls *a* and c_2, phycocyanin (PCY) or phycoerythrin (PER) [17, 18]. The cryptophytes are complex cells with specific movements and unique structures that allow easy recognition. The main morphological characteristics are slightly ovoid asymmetric cells, two asymmetric flagella with mastigonemes bipartite, an internal and an external periplast plate that surrounds the cell membrane, the structure of the furrow/gullet (groove/throat) [19], ejecti- or ejectosomes that allows them to suddenly alter its swimming direction in the opposite direction [1, 3, 8, 20].

The composition of the cryptophytes, especially in fatty acid, phycobiliproteins (PBPs), and carbohydrates, has attracted the attention of aquaculture and diverse business sectors, such as pharmaceutical, nutraceutical, chemical, and cosmetic industries [21, 22]. The culture of these cells and the production of these substances have some difficulties to overcome, as these delicate cells have a low growth rate compared to other cells in the market.

This chapter will address the Cryptophytes, mainly focusing on those that are photosynthetic, observing their biology, biochemical composition, culture systems, and some of their products with antioxidant potential, as well as fluorescent pigments which are of increasing interest as a marker in biotechnology applications.

2. Morphology, biology, and function

The cell morphology is flatter on the ventral side and concaves on the dorsal side (**Figure 1a**); the cell shape is like a bean or a drop of water; the form is mainly influenced by the furrow/gullet complex located at the anterior part. A gullet, or a furrow, or some combination of both (**Figure 1b**) is one of the main diagnostic features of the genera [23]. The complex can be localized by light microscopy in big cells, by the presence of big ejectisomes surrounding it; in small cells, an electron microscope is necessary [9]. The furrow is a ventral groove of variable length that begins in the vestibular region and extends posteriorly to half of the cell. The gullet is an invagination that extends to the posterior side; it originates from surrounding the furrow when it is present, but when the furrow is absent it is found on the ventral side near the vestibulum [8, 20, 23]. The vestibulum is a structure at the anterior side, present in all cells, an outwardly facing depression, from which two asymmetric flagella have a subapical origin on the ventral right side; it is connected to the beginning of the furrow/gullet complex.

2.1 Evasion and defense

Ejectisomes are present in all cryptophytes and are located on the furrow/gullet complex underlying the periplast around the cell, the size of these organelles can be

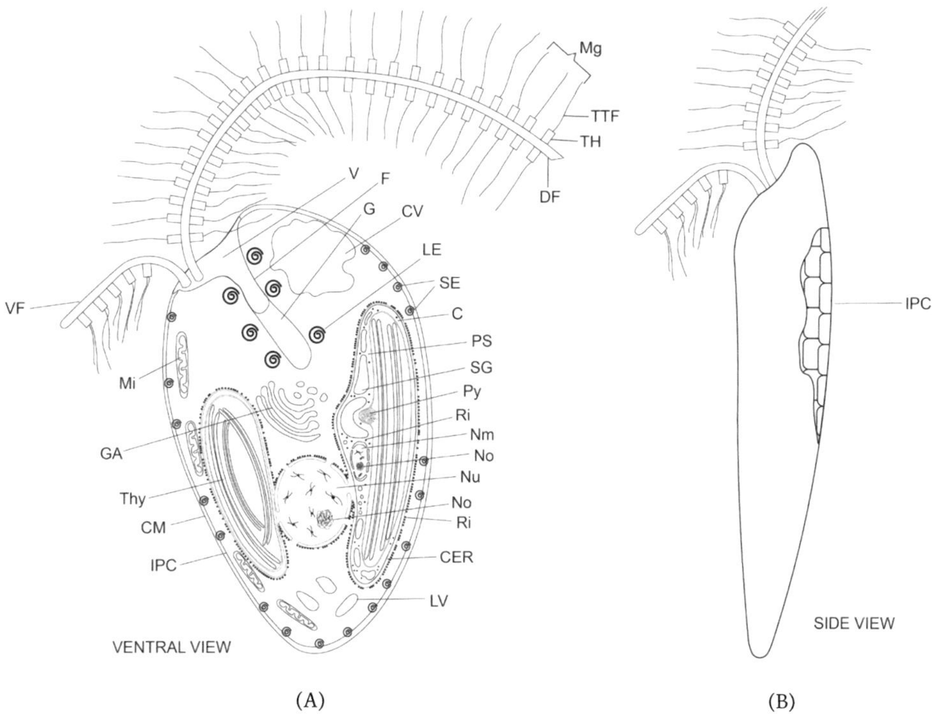

Figure 1.
General drawing of a cryptomonad cell. A. Ventral view. B. Side view showing part of the inner periplast component. C—chloroplastid, CER—chloroplastid endoplasmic reticulum, CM—cell membrane, CV—contractile vacuole, DF—dorsal flagellum, F—furrow, G—gullet, GA—Golgi apparatus, IPC—inner periplast component, LE—large ejectisome, LV—lipid vesicle, Mg—mastigonemes, Mi—mitochondrion, Nm—nucleomorph, No—nucleolus, Nu—nucleus, PS—periplastid space, Py—pyrenoid, Ri—ribosome, SE—small ejectisomes, SG—starch granule, TH—tubular hair, Thy—thylakoid, TTF—thin terminal filament, V—vestibulum, VF—ventral flagellum.

big (500–700 nm) or small (250–250 nm) [24]. Ejectisomes are cylinders-like structures formed by two tightly spiraled taper tapes of unequal size, coiled together and surrounded by a membrane. When the organisms are stressed by sudden changes in pH, osmolarity, or light intensity, ejectisomes are discharged, and the cell can be pushed backward, showing jerky movements [8, 9, 19, 25], probably by the impact of the ejectisomes with an object, it is a defense mechanism that allows the cell to escape from predators [26]. The ejected ejectisomes are tapes unrolled that look like long ribbons with ruffled edges; the big ones can be 7 μm × 228 nm and small ones 3.6 μm × 50 nm [24]. Ejectisomes are synthesized in the Golgi apparatus and are degraded in cytolysosomes during nutrient starvation [27].

2.2 Protection and support

Cryptophytes do not have a cell wall, but the cell membrane is covered and sandwiched by both an inner and surface periplast layer (**Figure 1A** and **B**), observed only by electron microscopy. The inner periplast component (IPC) or epiplast can be flat or consist of several plates of different forms, among the shapes identified are polygonal, rectangular, and hexagonal [9, 14, 28, 29]. The surface periplast layer is formed by plates and rosulate scales; however, in some a fibrous coat can be present instead.

The epiplast is made of epiplastin, a proteinaceous substance that provides flexibility and protection to the cell membrane [30]. The periplast plates decrease in size toward the rear of the cell and disappear at the furrow/gullet complex and vestibulum [9, 31].

2.3 Motility

The motility of these cells is due to two asymmetric flagella and its cellular form. Both flagellums have rows of mastigonemes similar to those of stramenopiles [32]. The longest flagellum directed to the front has two rows of bipartite hairs, and the shortest flagellum directed to the back only has one, they can be observed by electron microscopy, and structure variations have been described [28]. The bipartite hairs are composed of a tubular attached to the axoneme, it terminates with a single non-tubular filament in the longest flagellum, but in the case of the shortest, it ends with two unequal terminal filaments [9, 20, 28]. In addition, delicate seven-sided scales measuring 140–170 nm in diameter are commonly attached to the hairs [33]. Other structures that are part of the mechanism for motility are the rhizostyle and a compound rootlet system. The rhizostyle is a peculiar microtubular flagellar root that originates near one basal body and extends toward the posterior extreme of the cell; in some species until the nucleus, without a physical connection, but in others, just to the first third of the cell and has a wing-shaped lamellar projection [29, 34]. The cryptophytes have a phototaxis response that is mediated by a two-rhodopsin-based photosensory mechanism, similar to what is observed in green flagellated algae *Chlamydomonas reinhardtii* [35]. The structure of this mechanism includes an integral membrane protein, with a seven transmembrane alpha-helices covalently bonded to the retinal chromophore to make a channel structure, it is the anion channelrhodopsin, which is light-gated, and initially discovered in chlorophyte algae, which serve as photoreceptors to guide phototactic orientation [36, 37], this structure has been utilized in optogenetic applications [38].

2.4 Control of volume and osmolarity

The Cryptophytes do not have a cell wall, so variations in osmolarity could induce turgidity causing it to burst or plasmolysis causing the cell to compress. However, they possess a remarkable organelle, a contractile vacuole (CV), which has a rhythmic activity with diastolic and systolic cycles allowing for filling and emptying of CV, respectively [39]. This rhythmic mechanism maintains the cell volume and osmolarity, the cycle last either10s in freshwater or 40s in marine water [40]. The CV is near where the flagella structure originates and functions usually by discharging excess water and ions to the vestibulum [41]. Cryptophytes and red algae also employ another mechanism for osmolarity and volume control, synthesis of floridoside (2-O-D-glycerol-α-D-galactoside); which is a low molecular weight carbohydrate that functions as an osmolyte [41–43]; a similar mechanism is used by red algae algae's what signals an endosymbiotic inheritance [41].

2.5 Plastid of the cryptophytes

The plastids in alga and plants evolved from the endosymbiosis of a cyanobacterium, which means the incorporation of one cyanobacterium in a heterotrophic cell [44]; this primary endosymbiosis explains the plastid origin in chlorophytes, glaucophytes, and red algae. Other algae stramenopiles, haptophytes, and cryptophytes

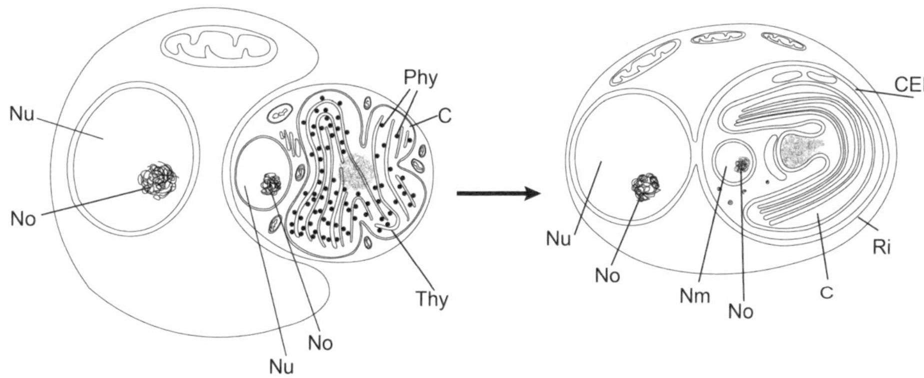

Figure 2.
Secondary endosymbiosis of a red algae that evolved in a cryptomonad cell. The cryptomonads have four membranes plastids; the outer is the plastid endoplasmic reticulum which surrounds plastid and nucleus. The phycobilisomes disappeared, and only one pigment not organized in a structure that is in the thylakoid lumen. Nu—nucleus, No—nucleolus, Thy—thylakoid, Nm—nucleomorph, C—chloroplastid, Ri—ribosome, CER—chloroplastid endoplasmic reticulum.

are the product of a secondary endosymbiosis by an unknown eukaryotic host and a red algal symbiont (**Figure 2**). Many scientific works confirm this hypothesis [45, 46]. These organisms possess four membranes around the plastid; the outermost membrane is thought to be the phagocytotic vacuole membrane that endocytosed the red algae and evolved to become the chloroplastic endoplasmic reticulum (CER) [47]. The CER is contiguous to the exterior nuclear membrane [9, 19, 48] (**Figure 1**), it involves the two outer membranes of the plastid, and has ribosomes on its outer surface (**Figure 1**) [49]. However, in contrast to other algae, the plastid of cryptophytes is more complex [48, 49]; between its two outer and two inner membranes, there is a space that is thought to correspond to the remains of the endocytosed red algae cytoplasm, it is called periplastid compartment (PC). One of the significant adaptations for endocytosis was the loss of genetic information of the endocytosed cell (from the nucleus and chloroplast of the red algae). However, not all nuclear information disappeared in Cryptophytes and Chlorarachniophytes, as occurred in all other secondary endosymbioses, which were sent to the host nucleus by endosymbiotic gene transfer [50, 51]. The remanent of the nuclear information is harbored in the PC and constitutes consists of a small nucleus or nucleomorph (NM). The PC also harbors eukaryotic ribosomes, and numerous starch globules produced over the pyrenoid are visible in the PC (**Figure 1A**) [49]. The plastids of the cryptophytes require that most of their proteins be nucleus-encoded, and are synthesized as precursors in the cytosol, and subsequently imported through the four membranes surrounding the plastid [47, 52]. This is possible because there is a mechanism for importing proteins that allows crossing of 2–5 membranes when the information is sent to the PC, stroma, or the thylakoid lumen. There should also be a mechanism for the retrograde pathway, from plastid to other organelles [47]; the CER functions are associated with this pathway [53, 54]. Other CER functions are related to bidirectional lipid and metabolite transfer and division [55]. The proteins directed to the plastid are synthesized as preproteins with a bipartite N-terminal signal sequence, which is used for a co-translational translocation of them across the outermost membrane, and after passing this membrane, the signal sequence is cleaved off [53]. The mechanism for passing the second membrane is possibly like the other four membrane plastids. The cryptophytes possess a nuclear-encoded symbiont-specific ERAD machinery (endoplasmic-reticulum-associated protein degradation) and also SELMA (symbiont-derived ERAD-like machinery); the origin of these mechanisms is unclear but is being studied [47]. To reach the stroma, Toc-Tic machinery (translocon of the outer and inner membrane of chloroplasts) similar to that of chlorophytes and diatoms may need to be present; this machinery existed in the common ancestor of all Archaeplastida, organisms with primary plastids [47, 51, 53].

All cryptophytes have one NM in each PC (**Figure 1**) with a double membrane with pores similar to those in the nucleus and three chromosomes that NM replicates in coordination with the nucleus [9, 44, 56]. The understanding of the presence of NM could resolve some fundamental questions such as the phylogeny of other algae with secondary plastids that also lack this vestigial structure [48]. The position of the NM is characteristic of the Cryptophyte species [9, 49].

2.6 Pigments and light harvesting by red algae endosymbiont and cryptophytes

Like all phytoplankton, cryptophytes have chlorophyll *a* as the primary light-harvesting pigment [57], and other accessory pigments ∝-carotene, alloxanthin, chlorophyll c2, and the PBPs for the capture of low light intensity in wavelengths not well absorbed by chlorophyll *a* (500–650 nm) [17, 58, 59].

The red algae and cyanobacteria have a color that depends on the predominance of PBPs, the orange-red PER or the blue PCY, which are in several hundred and are highly organized in supramolecular complexes, the phycobilisomes (PBS) (**Figure 2**). The PBS are their main light-harvesting antennae and cover the stromal surface of thylakoids [60]; the PBSs have mobility that lets them distribute the absorbed energy between photosystems (PSI and PSII) [61]. PBPs are composed of two kinds of α and β protein subunits and are more stable in trimer (3α + 3β) or hexamer (6α + 6β). The protein part of apoprotein is covalently bonded to a chromophore or phycobilin [32, 60, 62], these chomoproteins are united to colorless linker polypeptides [63], and constitute the light-harvesting antenna for transferring energy to chlorophyll *a* to PSII and possibly to PSI [32, 60, 62].

Like red algae, cryptophytes have PBPs pigments, but they do not have PBSs, and, they only produce one kind of PBP pigment per cell (Cr-PE or Cr-PC), packed into the thylakoid lumen [17, 64] without any arrangement [32, 65, 66], it gives the cell a red or blue color [9, 18, 19], but cells possess other accessory pigments, allowing them to display a great diversity of colors [67]. The ratio of Chlorophyll a: PBPs of cryptophytes can be several times higher than that of non-PBPs pigments [18]. The endosymbiosis provided the cryptophytes with new machinery that allowed diversification of light capture [65]; the PBPs are an auxiliary or second light-harvesting system, allowing them to occupy light spectra niches for more efficient light capture [65].

The PBPs of the cryptophytes are composed of two α and two β protein subunits and four linear tetrapyrrole chromophores or phycobilins covalently bonded by one or two thioether bonds to specific cysteine residues on the protein [68]; these PBPs provide unique spectral properties of absorption and emission fluorescence (**Table 1**).

The number and location of phycobilins within the protein are the primary factors that determine the visible absorption, the fluorescence spectrum, and the energy transfer pathway for any given PBP [66]. The complex of chromophores and protein subunits is a complete light-capturing unit; the α subunits of PBPs are encoded in the nuclear genome (derived from the ancestral host), whereas the β subunits are encoded in the plastid as in red algae, so the PBPs are unique chromoprotein

PBPs	Color	Absorption (nm)	Emission (nm)	Cryptomonad genera	References
Cr-PE 545	Red	538–551	580–587	*Teleaulax*, *Plagioselmis*, *Geminigera*, *Hanusia* *Guillardia*, *Rhinomonas*, *Pyrenomonas*/*Rhodomonas*, *Storeatula*, *Proteomonas*, *Cryptomonas*, *Baffinella*	[13, 18, 67, 69–71]
Cr-PE 555	Red	553–556	578	*Hemiselmis*	[69]
Cr-PE 566	Red	563–567	600–619	*Cryptomonas*, *Baffinella*, *Chilomonas Campylomonas*, *Falcomonas*	[13, 18, 65, 70]
Cr-PC 564	Blue	557–566	N. R.	*Hemiselmis*	[72]
Cr-PC 569	Blue	568–569	650, 656	*Falcomonas*, *Hemiselmis*	[13, 69, 73]
Cr-PC 577	Blue	576–578	634–641	*Hemiselmis*	[18, 74]
Cr-PC 615	Blue	612–615	580–589	*Hemiselmis subgen. Plagiomonas*	[18, 69, 70]
Cr-PC 630	Blue	625–630	648, 649	*Chroomonas*	[17]
Cr-PC 645	Blue	641–650	654–662	*Chroomonas*, *Komma*	[69, 70, 73]

Adapted from table VI [67, 68]. The PBPs are named, including the wavelengths of maximum absorption [9, 17].N.R. = no reported.

Table 1.
Different classes of phycobiliprotein (PBPs) in cryptophytes, spectral ranges of the main absorption and fluorescence emission maxima at visible wavelengths (nm) (Cr, cryptophytes; PE, phycoerythrin; PC, phycocyanin); and the actual Cryptomonad genera.

complexes that originated from secondary endosymbiosis [17, 65]. Another function of PBPs in cryptophytes is to help them with photoacclimation; this process involves changes in PBPs concentrations and shifts in the PBPs absorbance peaks when they are grown under red, blue, or green light [58]. The anterior means the photosynthetic system of cryptophytes is very different from other algae [75, 76], the location of PBPs in thylakoid lumen, the presence of dimeric PBPs forms, and the principal connection to PSII, but the precise mechanism remains to be discovered.

Another structure in the chloroplast is the pyrenoid, where the enzyme RUBISCO responsibly for CO2 fixing is located, there is one pyrenoid per plastid, and its position is an identifying characteristic of each species (**Figure 1**) [31].

3. Isolation, maintenance, and availability of cryptophytes

Cryptophytes are inhabitants of still waters with a low trophic state, they reach maximal biomass near the summer chemocline [77], and vertically migrate as a mechanism for harvesting inorganic nutrients; they can grow in turbid waters with low light due to their efficient greenlight harvesting PBPs [64]. Nets of different kinds and sampling bottles can be used for collecting cryptomonads from the water column; sediments or mucilage should be placed in the water, so the cells can be motile while transporting them to the laboratory, and all samples should be placed at a low temperature [20, 77]. In the lab, the samples should be gently filtered (>30 μm–<200 μm) and cultured by enriching the water with one standard culture media diluted (1:5–1:10), WARIS, and BBM (freshwater); ASP-12, ESM, f/2, and ASP (marine; check recipes on the CCAC, CCAP, CCMP, NIES, and SAG websites) [9, 78]. Cryptophytes do not grow in agar (only the palmella-forming taxa). Fluorescent light (20–50 $\mu mol/m^2/s$), 16:8 or 12:12 light:dark cycle, and 16–25 °**C.** It is recommended that culture conditions should be the same or as close as possible to that of the sampling site. When cells establish a population, they can be isolated, and the method of choice is using a micropipette [9, 79]; another way is by dilution in a sterile 24-well microplate. The delicate cells do not resist the fixation with Lugol or formol, but glutaraldehyde 2% can be an option [9, 80, 81]. The cells can be identified as cryptophytes using a standard optical microscope, with respect to size, form, and movement. Photographs are difficult to obtain because of cell movement, but dark field, phase contrast, differential interference contrast (DIC), and fluorescence microscopy help identify some features of cells [9, 20]. For maintaining the isolated cells, low temperature and low light are recommended. With heterotrophic cryptophytes like *Goniomonas* and *Chilomonas*, an organic carbon source, like a sterilized wheat seed, pea, or lentil, in a soil-water media are typically provided. Some also grow in WARIS added with soil extract [9, 20]. Different strains of cryptomonads are available in collections around the world.

4. Taxonomy

Although distinguishing cryptophyte cells from other flagellated cells is relatively easy by their movement, gross morphology, and epifluorescence; classifying by genus and species is difficult [9, 82]. Initial investigations proposed the morphology and color as a basis for this definition [20, 23, 25], but color changes depending on the cell stage and culture conditions, mainly light and nitrogen [83, 84]. Characteristics such as the groove/gullet complex, or ultrastructural features such as flagella, the position of NM, pyrenoid, and IPC, can only observed with an electron microscope, but could be characters of taxonomic value. However, this is view has changed after discovery that some species with haploid and diploid phases can have very different morphologies, suggesting that there is an intermediary stage sexual reproduction that could be confounding taxonomic certainty of species [9, 31, 81].

Among the characteristics used in Cryptophyte systematics are the presence or absence and the kind of accessory pigment (PER or PCY) (**Table 1**) [23, 28, 33]. The ultrastructural traits such as the arrangement of flagellar hairs [28], the number and location of the NM in the PC [1, 23], the presence and location of the eyespot in Chroomonas species, the type of scales comprising the outer periplast component, number location, and type of pyrenoids [8, 23], number of chloroplast per cell [85] are some of the taxonomic keys still used. Advances in molecular tools improved the phylogeny of the groups. There is a correspondence between molecular data with biliprotein, but it has not found with other morphological traits like IPC.

Life histories may be another character. Although initially it was thought that all cryptophytes divided only vegetatively, now species such as *Proteomonas sulcata* have been identified to have a dimorphic life cycle [86], *Cryptomonas/Campylomonas* [83], *Teleaulax/Plagioselmis* [13], are known to have complex life cycles. In some cases, those species have shown alternation of generations, with very different forms [10] and haploid/diploid phases. They had even been classified as different species. The aforementioned shows how complicated the taxonomic classification of cryptophytes can be.

	References
Class Goniomonadophyceae	
Order: Goniomonadida	
Family: Goniomonadaceae; Genus: *Goniomonas*	
Class Cryptophyceae	
Order: Cryptophyceae	
Family: Cryptomonadaceae; Genus: *Cryptomonas* (includes *Campylomonas/Chilomonas*)	[82, 88]
Order: Pyrenomonadales	
Family: Pyrenomonadaceae	
Genus: *Rhodomonas/Storeatula/Rhinomonas*	[89]
Family: Geminigeraceae	
Genus: *Geminigera/Teleaulax/Plagioselmis, Hanusia, Guillardia, Proteomonas*	[13]
Family: Chroomonadaceae	
Genus: Chroomonas, Falcomonas, Komma	[88]
Family: Hemiselmidaceae	
Genus: Hemiselmis	

Clay, Kugrens and Lee [23, 90]. The color of the genus indicates the kind of pigment it contains. Red = Cr-phycoerythrin, Blue = Cr-phycocyanin, Black = No color = no photosynthetic, no chloroplasts present.

Table 2.
Cryptomonad classification of the phylum Cryptophyta Cavalier-Smith emend.

Few of the genes more employed in the cryptophytes phylogeny are nuclear, 18S, ITS1, 5.8S, ITS2, 28S, SSU, LSU rDNA [13, 70, 87]; the nucleomorph SSU rDNA and 18S rDNA [83], and the chloroplast psbA [32].

The Cryptomonad classification of the Phylum Cryptophyta is shown in **Table 2**; however, this table will change in the future to reflect advances in knowledge, culturing, and electron microscopy.

5. Culturing of Cryptophyta

Due to their biochemistry, Cryptophytes have applications in aquaculture and a wide variety of biotechnology applications. The most studied species are Pyrenomonadales of the Pyrenomonadaceae family, Rhodomonas/Storeatula/Rhinomonas, and from the family of Geminigeraceae, Geminigera, Teleaulax, Guillardia, Proteomonas.

5.1 For aquaculture

At the experimental level, cryptophytes have been cultivated using batch systems to determine the effect on the growth of environmental parameters, temperature (12–32°C) [91–94], light in quality (white, blue, green, and red) [94, 95] and quantity (11–600 μM m^{-2} s^{-1}) [91, 93, 94, 96–98], nitrogen sources (nitrates, ammonium, urea) [90, 99, 100] and quantities [92, 96], all of this has been carried out to optimize the culture under laboratory conditions (**Figure 3**).

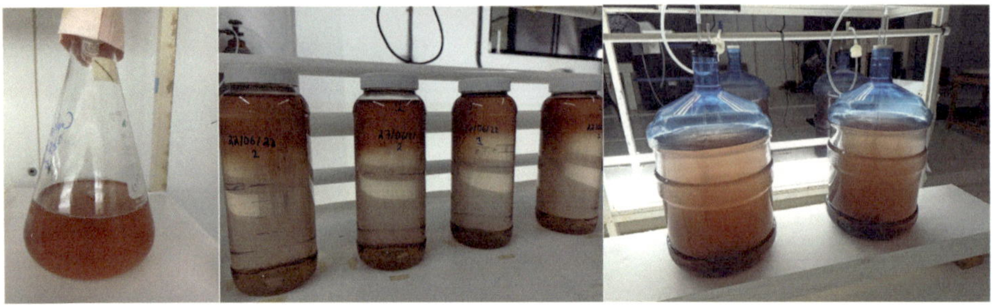

Figure 3.
Photographs of Proteomonas sulcata cultures in a batch system.

Cells are easily cultured in small volumes and can reach cell densities of 4–6 × 10^6 cells ml^{-1}; they are considered an important food source for use in aquaculture since the biomass has a high content of proteins, lipids, fatty acids (PUFA, HUFA), and sterols; with an EPA/DHA ratio close to two [22, 101, 102]. The high nutritional value was especially recognized for copepods [98, 102–105] and mollusks [105, 106]. The cryptophytes *Rhodomonas* sp., *R. salina*, *R. baltica*, and *R. reticulata* have been among the most widely used in aquaculture, and their size (5–20 μm) allows them to be ingested by copepods (medium and adult sizes) that prefer these to other algal groups (Diatoms or Chlorophytes) (personal observation), and mollusks in seed stage, juveniles, and adults. Cryptophytes do not have a cell wall, therefore allows for easier digestion and absorption. The most employed medium is f/2- Si [22, 90, 92, 93, 96, 99, 100, 107], followed by B1 [98, 108], L1 [91], Z8 [109], and 2f [94], with average specific growth rates (μ) between 0.48 and 0.88, indicted that growth is slow for these organisms. Optimal temperatures of 19–24°C for cultivation make using these cells less feasible in temperate zones.

For aquaculture, *Rhodomonas* has been cultured in carboys (20 L) as a batch system; it has been reported that bigger scale cultures have presented difficulties, with unexpected plateau phases or even death compared to *Cryptomonas* sp., more stable and predictable [102]. Semicontinuous cultures of *Rhodomonas* sp. (20 L) have been used for aquaculture, with exchange rates of 0.33 every third day [102]. The main limitations for scaling cultures in nutrient sufficiency are the maintaining light conditions and culture in suspension without damaging cells. Continuous cultures of *Rhodomonas salina* and *Rhodomonas* sp. in column photobioreactors (94 L) have been carried out for feeding copepods [108], with exchange rates of 0.46 d^{-1} and mean cell densities of 2.40 × 10^6 cells ml^{-1}. The main changes in those systems were: a smaller column diameter (0.2 m), agitation by supplying air mixed with CO_2 from the bottom, and in wider columns (0.8 m and 500 L), the central and external illumination of indoor cultures with different strains (*Teleaulax amphioxeia*—TA, *Rhinomonas* sp., *Chroomonas* sp.) [90]. In this last, cell densities were lower than previously reported (4.6–6.4 × 10^5 cells ml^{-1}). However, the percentages of EPA and DHA (of TFA) were very stable during the stationary phase, higher for TA (14.6–11%) with slight variations depending on the culture medium (f/2 or urea compound fertilizer), which confirms its high nutritional potential.

The first to work on continuous cultures of *Rhodomonas* sp. using a photobioreactor tubular (200 L) in a greenhouse with natural light [109] showed biomass concentrations up to 1.5 g L^{-1}, 2–5 times higher than previous reports [108]. These results suggest that outdoors cultures can be a good strategy for improving the productivity of cryptophytes and could be a field to open these cells to more biotechnological applications.

5.2 Pigments production (Cr-PE and Cr-PC)

There are advantages to the production of PBPs from Cryptophytes. The first is that they are water-soluble compounds; furthermore, compared to cyanobacteria and red algae, each species of Cryptophyte produces only one type of PBP pigment (Cr-PE or Cr-PC) (**Table 1**), and the cells are easily broken to free PBPs by freezing and thawing. An inversely proportional relationship relates the PBPs production to irradiance [107, 110]; it is contrary to the biomass, which is directly related [95]. The Cr-PE production (pg cell^{-1}) is also related to wavelength, and has been observed that red light induces a greater production, followed by green light [110] during the exponential stage of culture when there are sufficient nutrients. The cryptophytes studied were *Rhodomonas* sp., *Proteomonas sulcata*, [92, 98, 110–112]; *Guillardia theta*, *R. salina*, *P. sulcata*, *Storeatula* sp., and *Chroomonas mesostigmatica* [69]. The content of PBPs in Cryptophytes can be up to 20% of dry cell weight, which contrasts with other photosynthetic pigments (chlorophyll and carotenoids), which are up to 10 times lower [69]; these PBPs percentages can be reached under the sufficiency of nutrients and low irradiance (20–40 μM m^{-2} s^{-1}). It is important to mention, when cells lack nutrients, especially nitrogen, the decrease in pigment content is proportional to the nutrient deficit, this fact has suggested that pigments are a nitrogen reserve [84] like in cyanobacteria. Understanding the role of PBPs in the cryptophytes would help solve the problem of high biomass with low pigment or low biomass with high pigment.

The best method for extracting PBPs from biomass, which is obtained by centrifugation, is to suspend biomass in buffer phosphate (0.1 mol L^{-1}) (pH 6, 7–7.2) in accord to [69, 113]. Then cycles of frozen/thaw, homogenize the suspension and centrifugate (11,000 g) to remove the cellular debris, obtain the supernatant; measure the absorbance by a scan (450–750 nm), determine the maximal absorption employing 1 cm quartz cuvette, and use buffer phosphate as blank. Samples not analyzed immediately can remain at −20°C or − 80°C (until six months). For determining the PBPs content (*C* pg cell^{-1}), Eq. (1) is proposed by [69]:

$$C = \frac{A}{\varepsilon * d} \times MW \times \frac{Vb}{Vs} \times \frac{10^{12}}{N} \tag{1}$$

Where A = absorbance of sample, ε = extinction coefficient, for Cr-PE (5.67 × 10^5 L mol^{-1} cm^{-1}) [114] and for Cr-PC (5.7 × 10^5 L mol^{-1} cm^{-1}) [115], d = path length, MW = molecular weight (Cr-PE =45,000; Cr-PC = 50,000 Da) V*b* and V*s* = volume of buffer and sample, N = number of cells L^{-1}.

The extraction mentioned produces a crude extract, which has to be purified for higher quality and value; depending on the purity, PBPs can reach a price of 130 USD – to 15–33 × 10^3 USD per gram [116, 117]. In general, the purification procedures consist of removing impurities by precipitation with ammonium sulfate, dialyzation, and separation by gel filtration chromatography [118]; in [119], **Table 1** lists various procedures employed for PBPs purification.

6. Biotechnological applications

Cryptophytes are cells with unique biochemistry; for example, it has a high and balanced content of PUFA (ALA, SDA, EPA, DHA) [21, 22], phytosterols, carbohydrates [21], and fluorescent pigments PBPs. Each of these compounds has effects on health [21, 22, 83, 120], giving them a wide variety of potential applications in food, cosmetics, pharmaceuticals, medicine, immunology, as well as for scientific research [116]. There are many patents for these applications, but most are related to fluorescent properties of PBPs [121].

The food and beverage industries have increased attention to natural pigments from microorganisms, especially aquatic products from macro and microalgae [21]. This trend is predicted to continue in the future due to increasing consumer health consciousness of potential and known harmful effects of synthetic compounds. Other advantages are stability, considering temperature, pH, osmolarity, low reactivity [122–124], and the health-promoting effects from another point of view [121]. An advantage is that the market for cyanobacteria and their PBPs (mainly *Arthrospira*, *Anabaena*) and red alga (*Porphyridium* and *Galdieria*) has been established. For PBPs and their products, in 2010, it was estimated to be around US$60 million [125]. Another critical factor for industries is the stability of PBPs, they showed good range of stability between pH 5–10 and between −20 and 80°C, and it could be prolonged by adding preservatives [124]. The blue (PCY) and red (PER) pigments that Cryptophytes provide is attractive to the food and cosmetic industries [21]. Some of their uses are in chewing gum, desserts, candies, dairy products, ice cream, soft drinks in different presentations (aqueous and alcoholic), and cosmetics like lipstick and eyeliners [116, 119, 126, 127]. Some applications like food and cosmetics do not need high purity, thus semi-purified pigments are an attractive option since they are less expensive than highly purified pigments, that are needed for using them as molecular markers. The purity is measured by the ratio of the maximum pigment absorbance divided by its absorbance at 280 nm; for food, a ratio of ≥0.7 is accepted, a reagent grade is ~3.9, and analytical grade ≥4.0 [117], that is the reason why PER is expensive and sold anywhere from US$200 to US$100,000/kg.

In regards to the effects of PBPs, some are related to their antioxidant potential [119, 128, 129] as they are natural ROS scavengers, have anti-inflammatory [130, 131], and anti-aging properties [119, 128, 129]. Other protective effects of PBPs are located in mitochondria membrane from ROS stress, which could maintain cellular viability and proliferation [131, 132] in different health problems. PCY also has a pro-apoptotic effect in different cancer cell lines and is an inhibitor of COX-2 enzyme, which converts arachidonic acid to prostaglandins and plays a crucial role in tumor progression and chemical resistance, and the PGE2, which participates in promoting angiogenesis [132]. PCY has been observed as an inhibitor of viral protease activities (i. e. SARS-CoV-2) [133, 134], which could mean antiviral protection. The anti-diabetes activity of phycocyanobilin could be due to the inhibition of NADPH oxidase and the protective effects against human lymphatic endothelial cells apoptosis, which also explains its neuroprotective effects [135].

PBPs play an essential role in fluorescent-based detection systems, like with flow cytometry, epifluorescence, and confocal microscopy for fluorescent immunoassays like protein electrophoresis [136], immunophenotyping; they can also be used as selective markers of specific biological structures, i.e., arterial wall thickness, atherosclerotic plaque, luminal boundaries and to better delineate tumors mass outlines and such other fluorescent studies [121, 137]. Spectral properties, such as excitation and emission at the red end of the spectrum and diminished interference from biological matrices give it considerable stability for quenching compared to other biological compounds. In addition, PBPs have high water solubility with minimal interaction with other substances, and the ease of binding to antibodies by conventional cross-linking reagents, have made these fluorochromes unique and superior to other products, i.e. for medical tagging [137–139]. All of this is motive for developing new biotechnological processes and products, and the commercialization of new patents is still a goal in the near future.

7. Conclusion and prospects

The cryptophytes are secondary endosymbionts that inherited many red algae characteristics, but differences involving morphology and physiology make them unique and not completely understood. The functioning of PBPs as a secondary antenna into the thylakoid and its connection to photosynthesis, the PBPs as an energy reserve, to understand why these cells still have a vestigial nucleus NM with three chromosomes, the way the nucleus coordinates the activities of NM and plastid are some of the mysteries for the scientist to solve. From an applications point of view, cryptophytes have high nutritive value that includes a balanced PUFA profile, high protein content, and the possibility to induce a high content of PCY or PER, antioxidant pigments, as well as having a very stable fluorescence making it attractive for use in research, thus its future looks promising and applications using cryptophytes should grow. From a production perspective, these cells have some advantages compared to other producers of PBPs, they have only one type of pigment. Cryptophytes lack of a cell wall facilitating digestion and nutrient absorption by organism and the extraction of products, and in addition have high content of PBPs with a lower molecular weight. The low growth rates and how to achieve the scaling necessary for high biomass with high pigments content will continue to be a challenge in the near future. Improvements will come probably by improved understanding of its physiology. Also, needed is to increase the variety of organisms in collections, improve the culture procedures including photobioreactors, and finally make more accessible purification methods.

Acknowledgements

The author thanks Gerardo Hernández García for the **Figures 1** and **2** and Dr. Miguel Victor Cordova Matson for his support with edition. This work was supported by institutional projects CIBNOR-20427 and CIBNOR-20437.

Author details

María Concepción Lora Vilchis
Biological Research Center of the Northwest S.C. CIBNOR, La Paz, B. C. S., Mexico

*Address all correspondence to: cony04@cibnor.mx

References

[1] Hoef-Emden K. Revision of the genus Cryptomonas (Cryptophyceae) II: Incongruences between the classical morphospecies concept and molecular phylogeny in smaller pyrenoid-less cells. Phycologia. 2007;**46**(4):402-428. DOI: 10.2216/06-83.1

[2] Luo W, Bock C, Li HR, et al. Molecular and microscopic diversity of planktonic eukaryotes in the oligotrophic Lake Stechlin (Germany). Hydrobiologia. 2011;**661**:133-143

[3] Zheng BH, Chen ZJ, Li YY, et al. Structural characteristics and driving factors of the planktonic eukaryotic community in the Danjiangkou reservoir, China. Water (Switzerland). 2020;**12**:1-22

[4] Borges Mendes CR, Tavano VM, Segabinazzi Dotto T, et al. New insights on the dominance of cryptophytes in Antarctic coastal waters: A case study in Gerlache Strait. Deep-Sea Research Part II: Topical Studies in Oceanography. 2018;**149**:161-170

[5] Egerton TA, Marshall H, Morse RE, et al. Emergence of algal blooms: The effects of short-term variability in water quality on phytoplankton abundance, diversity, and community composition in a tidal estuary. Microorganisms. 2014;**2**:33-57

[6] Mukhanov V, Sakhon E, Polukhin A, et al. Cryptophyte and photosynthetic Picoeukaryote abundances in the Bransfield Strait during austral summer. Water (Switzerland). 2022;**14**:1-15

[7] Robinson C, Archer SD, Williams PJLB. Microbial dynamics in coastal waters of East Antarctica: Plankton production and respiration. Marine Ecology Progress Series. 1999;**180**:23-36

[8] Clay BL. Cryptomonads. In: Wehr JD, Sheath RG, Kociolek JP, editors. Freshwater Algae of North America. Ecology and Classification. 2nd ed. New York: Elsevier Inc.; 2015. pp. 809-850. DOI: 10.1016/B978-0-12-385876-4.00018-9

[9] Hoef-Emden K, Archibald JM. Cryptophyta (Cryptomonads). In: Archibald J, Simpson A, Slamovits C, editors. Handbook of the Protists. 2nd ed. Vol. 2. Cham, Switzerland: Springer International Publishing; 2017. pp. 851-891. DOI: 10.1007/978-3-319-28149-0_35

[10] Cerino F, Zingone A. European journal of phycology a survey of cryptomonad diversity and seasonality at a coastal Mediterranean site a survey of cryptomonad diversity and seasonality at a coastal Mediterranean site. European Journal of Phycology. 2006;**41**:363-378

[11] Wisecaver JH, Hackett JD. Transcriptome analysis reveals nuclear-encoded proteins for the maintenance of temporary plastids in the dinoflagellate Dinophysis acuminata. BMC Genomics. 2010;**11**:366. DOI: 10.1186/1471-2164

[12] Johnson MD, Beaudoin DJ, Frada MJ, et al. High grazing rates on Cryptophyte algae in Chesapeake Bay. Frontiers in Marine Science. 2018;**5**:1-13

[13] Altenburger A, Blossom HE, Garcia-Cuetos L, et al. Dimorphism in cryptophytes—The case of Teleaulax amphioxeia/Plagioselmis prolonga and its ecological implications. Science Advances. 2020;**6**:1-9

[14] Kugrens P, Lee RE, Andersen RA. Cell form and surface patterns in Chroomonas and Cryptomonas cells (Cryptophyta) as revelled by scanning electron microscopy. Journal of Phycology. 1986;**25**:512-522

[15] McFadden GI. Primary and secondary endosymbiosis and the origin of plastids. Journal of Phycology. 2001;**37**:951-959

[16] Keeling PJ. The endosymbiotic origin, diversification and fate of plastids. Philosophical Transactions of the Royal Society B: Biological Sciences. 2010;**365**:729-748

[17] Hill DRA, Rowan KS. The biliproteins of the Cryptophyceae. Phycologia. 1989;**28**:455-463

[18] Cunningham BR, Greenwold MJ, Lachenmyer EM, et al. Light capture and pigment diversity in marine and freshwater cryptophytes. Journal of Phycology. 2019;**55**:552-564

[19] Hoef-Emden K. Revision of the genus Chroomonas HANSGIRG: The benefits of DNA-containing specimens. Protist. 2018;**169**:662-681

[20] Kugrens P, Clay BL. Cryptomonads. In: Wehr JD, Sheath RG, editors. Freshwater Algae of North America: Ecology and Classification. San Diego: Academic Press; 2003. pp. 715-755

[21] Abidizadegan M, Peltomaa E, Blomster J. The potential of cryptophyte algae in biomedical and pharmaceutical applications. Frontiers in Pharmacology. 2021;**11**:1-17

[22] Peltomaa E, Johnson MD, Taipale SJ. Marine cryptophytes are great sources of EPA and DHA. Marine Drugs. 2018;**16**:1-11

[23] Clay BL, Kugrens P, Lee RE. A revised classification of Cryptophyta. Botanical Journal of the Linnean Society. 1999;**131**:131-151

[24] Rhiel E, Westermann M. Isolation, purification and some ultrastructural details of discharged ejectisomes of cryptophytes. Protoplasma. 2012;**249**:107-115

[25] Pringsheim EG. Some aspects of taxonomy in the Cryptophyceae. The New Phytologist. 1944;**43**:143-150

[26] Ammermann S, Schneider T, Westermann M, et al. Ejectisins: Tough and tiny polypeptides are a major component of cryptophycean ejectisomes. Protoplasma. 2012;**250**:551-563

[27] Rhiel E, Ammermann S, Hillebrand H. Effects of nutrient depletion on ejectisomes of cryptophytes. Nova Hedwigia. 2016;**102**:77-87

[28] Kugrens P, Lee RE, Andersen RA. Ultrastructural variations in Cryptomonad flagella. Journal of Phycology. 1987;**23**:511-518

[29] Gillott MA, Gibbs SP. Comparison of the flagellar rootlets and periplast in two marine cryptomonads. Canadian Journal of Botany. 1983;**61**:1964-1978

[30] Goodenough U, Roth R, Kariyawasam T, et al. Epiplasts: Membrane skeletons and epiplastin proteins in euglenids, glaucophytes, cryptophytes, ciliates, dinoflagellates, and apicomplexans. Mol Biol Physiol. 2018;**9**(1):e02020-18. DOI: 10.1128/mBio.02020-18

[31] Cerino F, Zingone A. Decrypting cryptomonads: A challenge for molecular taxonomy. In: Brodie J, Lewis J, editors. Unraveling the Algae the Past, Present and Future of Algal Systematics. Boca Raton, FL 33487-2742: CRC Press. Taylor and Francis Group; 2007. pp. 197-214

[32] Lee RE. Phycology. New York, NY, USA. Available from: https://www.booktopia.com.au/phycology-robert-edward-lee/book/9781107555655.html; 2018 [Accessed: March 17, 2022]

[33] Lee RE, Kugrens P. The occurrence and structure of flagellar scales in some freshwater Cryptophytes1. Journal of Phycology. 1986;**22**:549-552

[34] Nam SW, Shin W. Ultrastructure of the flagellar apparatus in cryptomorphic Cryptomonas curvata (Cryptophyceae) with an emphasis on taxonomic and phylogenetic implications. Algae. 2016;**31**:117-128

[35] Sineshchekov OA, Govorunova EG, Jung KH, et al. Rhodopsin-mediated photoreception in cryptophyte flagellates. Biophysical Journal. 2005;**89**:4310-4319

[36] Govorunova EG, Sineshchekov OA, Li H, et al. Cation and anion channelrhodopsins: Sequence motifs and taxonomic distribution. MBio. 2021;**12**(4):e01656-21. DOI: 10.1128/mBio.01656-21

[37] Govorunova EG, Sineshchekov OA, Spudich JL. Emerging diversity of channel rhodopsins and their structure-function relationships. Frontiers in Cellular Neuroscience. 2022;**15**:1-12

[38] Joshi J, Rubart M, Zhu W. Optogenetics: Background, methodological advances and potential applications for cardiovascular research and medicine. Front Bioeng Biotechnol. 2020;7:466. DOI: 10.3389/fbioe.2019.00466

[39] Komsic-Buchmann K, Becker B. 6 Contractile vacuoles in green algae – structure and function. In: Heimann K, Katsaros C, editors. Advances in Algal Cell Biology. Berlin, Boston: De Gruyter; 2012. pp. 123-142. DOI: 10.1515/9783110229615.123

[40] Patterson DJ, Hausmann K. The behaviour of contractile vacuole complexes of cryptophycean flagellates. British Phycological Journal. 1981;**16**:429-439

[41] Hoef-Emden K. Osmotolerance in the cryptophyceae: Jacks-of-all-trades in the chroomonas clade. Protist. 2014;**165**:123-143

[42] Eggert A, Karsten U. Low molecular weight carbohydrates in red algae – an ecophysiological and biochemical perspective. In: Seckbach D, Chapman J, editors. Red Algae in the Genomic Age. Cellullar Origin, Life Extreme Habitats and Astrobiology. 1st ed. Vol. 13. Dordrecht: Springer; 2010. pp. 443-456. DOI: 10.1007/978-90-481-3795-4

[43] Yu Y, Jia X, Wang W, et al. Floridean starch and floridoside metabolic pathways of neoporphyra haitanensis and their regulatory mechanism under continuous darkness. Marine Drugs. 2021;**19**(12):664. DOI: 10.3390/md19120664

[44] Archibald JM. Endosymbiosis and eukaryotic cell evolution. Current Biology. 2015;**25**:R911-R921

[45] Cavalier-Smith T, Couch JA, Thorsteinsen KE, et al. Cryptomonad nuclear and nucleomorph 18s rrna phylogeny. European Journal of Phycology. 1996;**31**:315-328

[46] Douglas SE, Penny SL. The plastid genome of the cryptophyte alga, Guillardia theta: Complete sequence and conserved synteny groups confirm its common ancestry with red algae. Journal of Molecular Evolution. 1999;**48**:236-244

[47] Gould SB, Maier UG, Martin WF. Protein import and the origin of red complex plastids. Current Biology. 2015;**25**:R515-R521

[48] Tanifuji G, Onodera NT. Cryptomonads: A model organism sheds light on the evolutionary history of genome reorganization in secondary endosymbioses. In: Hirakawa Y, editor. Avances in Botanical Research. 1st ed.

London: Elsevier Ltd; 2017. pp. 263-320. DOI: 10.1016/bs.abr.2017.06.005

[49] Gibbs SP. The chloroplast endoplasmic reticulum: Structure, function, and evolutionary significance. International Review of Cytology. 1981;**72**:49-99

[50] Timmis JN, Ayliff MA, Huang CY, et al. Endosymbiotic gene transfer: Organelle genomes forge eukaryotic chromosomes. Nature Reviews. Genetics. 2004;**5**:123-135

[51] Gentil J, Hempel F, Moog D, et al. Review: Origin of complex algae by secondary endosymbiosis: A journey through time. Protoplasma. 2017;**254**:1835-1843

[52] Felsner G, Sommer MS, Gruenheit N, et al. ERAD components in organisms with complex red plastids suggest recruitment of a preexisting protein transport pathway for the periplastid membrane. Genome Biology and Evolution. 2011;**3**:140-150

[53] Wastl J, Maier UG. Transport of proteins into cryptomonads complex plastids. The Journal of Biological Chemistry. 2000;**275**:23194-23198

[54] Hernández-Verdeja T, Strand Å. Update on chloroplast development retrograde signals navigate the path to chloroplast development. Plant Physiology. 2018;**176**:967-976

[55] Cavalier-Smith T. Kingdom Chromista and its eight phyla: A new synthesis emphasising periplastid protein targeting, cytoskeletal and periplastid evolution, and ancient divergences. Protoplasma. 2018;**255**:297-357. DOI: 10.1007/s00709-017-1147-3

[56] Onuma R, Mishra N, Miyagishima S-Y. Regulation of chloroplast and nucleomorph replication by the cell cycle in the cryptophyte Guillardia theta. Scientific Reports. 2017;**7**:2345. DOI: 10.1038/s41598-017-02668-2

[57] Falkowski PG, Raven JA. Aquatic Photosynthesis. STU-Stud. Princeton University Press. Available from: http://www.jstor.org/stable/j.ctt4cgbxs

[58] Spangler LC, Yu M, Jeffrey PD, et al. Controllable phycobilin modification: An alternative photoacclimation response in cryptophyte algae. ACS Central Science. 2022;**8**:340-350

[59] García N, López-Elías JA, Miranda A, et al. Effect of salinity on growth and chemical composition of the diatom Thalassiosira weissflogii at three culture phases. Latin American Journal of Aquatic Research. 2012;**40**:435-440

[60] Gantt E, Grabowski B, Cunningham FX. Antenna systems of red algae: Phycobilisomes with photosystem ll and chlorophyll complexes with photosystem I. In: Green BR, Parson WW, editors. Light-Harvesting Antennas in Photosynthesis. Advances in Photosynthesis and Respiration. 1st ed. Vol. 13. Dordrecht: Springer; 2003. pp. 307-322. DOI: 10.1007/978-94-017-2087-8_10

[61] Kaňa R, Kotabová E, Lukeš M, et al. Phycobilisome mobility and its role in the regulation of light harvesting in red algae. Plant Physiology. 2014;**165**:1618-1631

[62] Zhang J, Ma J, Liu D, et al. Structure of phycobilisome from the red alga Griffithsia pacifica. Nature. 2017;**551**:57-63

[63] Adir N. Elucidation of the molecular structures of components of the phycobilisome: Reconstructing a giant. Photosynthesis Research. 2005;**85**:15-32

[64] Spear-Bernstein L, Miller KR. Unique location of the phycobiliprotein

light-harvesting pigment in the Cryptophyceae. Journal of Phycology. 1989;**25**:412-419

[65] Greenwold MJ, Cunningham BR, Lachenmyer EM, et al. Diversification of light capture ability was accompanied by the evolution of phycobiliproteins in cryptophyte algae. Proc Biol Sci. 2019;**286**:20190655. DOI: 10.1098/rspb.2019.0655

[66] Toole CM, Allnut FCT. Red, cryptomonad and glaucocystophyte algal phycobiliproteins. In: Larkum, A, Douglas S, Raven JA, editors. Photosynthesis in Algae. Advances in Photosynthesis and Respiration. 1st ed. Vol. 14. Dordrecht: Springer; 2003. pp. 305-334. DOI: 10.1007/978-94-007-1038-2_14

[67] Greenwold MJ, Cunningham BR, Lachenmyer EM, et al. Diversification of light capture ability was accompanied by the evolution of phycobiliproteins in cryptophyte algae. Proceedings of the Royal Society B: Biological Sciences. 286. 2019. DOI: 10.1098/rspb.2019.0655

[68] Wedemayer GJ, Kidd DG, Wemmer DE, et al. Phycobilins of cryptophycean algae. Occurrence of dihydrobiliverdin and mesobiliverdin in cryptomonad biliproteins. The Journal of Biological Chemistry. 1992;**267**:7315-7331

[69] Heidenreich KM, Richardson TL. Photopigment, absorption, and growth responses of marine cryptophytes to varying spectral irradiance. Journal of Phycology. 2020;**56**:507-520

[70] Daugbjerg N, Norlin A, Lovejoy C. Baffinella frigidus gen. et sp. nov. (Baffinellaceae fam. nov., Cryptophyceae) from Baffin Bay: Morphology, pigment profile, phylogeny, and growth rate response to three abiotic factors. Journal of Phycology. 2018;**54**:665-680

[71] Deane JA, Strachan IM, Saunders GW, et al. Cryptomonad evolution: Nuclear 18S rDNA phylogeny versus cell morphology and pigmentation. Journal of Phycology. 2002;**38**:1236-1244

[72] Magalhães K, Lopes Santos A, Vaulot D, et al. Hemiselmis aquamarina sp. nov. (Cryptomonadales, Cryptophyceae), a cryptophyte with a novel phycobiliprotein type (Cr-PC 564). Protist. 2021;**172**(4):125832. DOI: 10.1016/j.protis. 2021.125832

[73] Hill DRA. Chroomonas and other blue-green cryptomonads. Journal of Phycology. 1991;**27**:133-145

[74] Overkamp KE, Langklotz S, Aras M, et al. Chromophore composition of the phycobiliprotein Cr-PC577 from the cryptophyte *Hemiselmis pacifica*. Photosynthesis Research. 2014;**122**:293-304

[75] Stadnichuk IN, Novikova TM, Miniuk GS, et al. Phycoerythrin association with photosystem II in the cryptophyte alga Rhodomonas salina. The Biochemist. 2020;**85**:679-688

[76] Stadnichuk IN. Role of phycobiliprotein antenna of Cyanobacteria, Red and Cryptophyte Algae in association with photosystems I and II. Journal of Plant Biochemistry & Physiology. 2020;**8**:1-4

[77] Camacho A. On the occurrence and ecological features of deep chlorophyll maxima (DCM) in Spanish stratified lakes. Limnetica. 2006;**25**:453-478

[78] Andersen RA. Recipes for freshwater and seawater media. In: Andersen R, editor. Algal Culturing Techniques. San Francisco: Elsevier Academic Press; 2005. pp. 429-538

[79] Andersen RA, Kawachi M. Traditional microalgae isolation

techniques. In: Andersen RA, editor. Algal Culturing Techniques. San Diego: Elsevier Academic Press; 2005. pp. 83-100

[80] Xia S, Cheng Y, Zhu H, et al. Improved methodology for identification of cryptomonads: Combining light microscopy and PCR amplification. Journal of Microbiology and Biotechnology. 2013;**23**:289-296

[81] Cerino F, Zingone A. A survey of cryptomonad diversity and seasonality at a coastal Mediterranean site. European Journal of Phycology. 2006;**41**:363-378

[82] Okamoto N, Inouye I. The katablepharids are a distant sister group of the Cryptophyta: A proposal for Katablepharidophyta divisio nova/ Kathablepharida phylum novum based on SSU rDNA and beta-tubulin phylogeny. Protist. 2005;**156**:163-179

[83] Hoef-Emden K, Melkonian M. Revision of the genus Cryptomonas (Cryptophyceae): A combination of molecular phylogeny and morphology provides insights into a long-hidden dimorphism. Protist. 2003;**154**:371-409

[84] Lewitus A, Caron D. Relative effects of nitrogen or phosphorus depletion and light intensity on the pigmentation, chemical composition, and volume of Pyrenomonas salina (Cryptophyceae). Marine Ecology Progress Series. 1990;**61**:171-181

[85] Choi B, Son M, Kim JI, et al. Taxonomy and phylogeny of the genus Cryptomonas (Cryptophyceae, Cryptophyta) from Korea. Algae. 2013;**28**:307-330

[86] Hill DRA, Wetherbee R. Proteomonas sulcata gen. et sp. nov. (Cryptophyceae), a cryptomonad with two morphologically distinct and alternating forms. Phycologia. 1986;**25**:521-543

[87] Majaneva M, Remonenc I, Rintala J-M, et al. Rhinomonas nottbecki n. sp. (Cryptomonadales) and molecular phylogeny of the family Pyrenomonadaceae. The Journal of Eukaryotic Microbiology. 2014;**61**:480-492

[88] Cavalier-Smith T. Kingdom protozoa and its 18 phyla. Microbiological Reviews. 1993;**57**:953-994

[89] Hoef-Emden K. Molecular phylogeny of phycocyanin-containing cryptophytes: Evolution of biliproteins and geographical distribution. Journal of Phycology. 2008;**44**:985-993

[90] Lee BI, Kim SK, Kim JH, et al. Intraspecific variations in macronutrient, amino acid, and fatty acid composition of mass-cultured Teleaulax amphioxeia (Cryptophyceae) strains. Algae. 2019;**34**:163-175

[91] Oostlander PC, van Houcke J, Wijffels RH, et al. Optimization of Rhodomonas sp. under continuous cultivation for industrial applications in aquaculture. Algal Research. 2020;**47**. DOI: 10.1016/j.algal.2020.101889

[92] Chaloub RM, Motta NMS, de Araujo SP, et al. Combined effects of irradiance, temperature and nitrate concentration on phycoerythrin content in the microalga rhodomonas sp. (cryptophyceae). Algal Research. 2015;**8**:89-94

[93] Guevara M, Arredondo-Vega BO, Palacios Y, et al. Comparison of growth and biochemical parameters of two strains of Rhodomonas salina (Cryptophyceae) cultivated under different combinations of irradiance, temperature, and nutrients. Journal of Applied Phycology. 2016;**28**:2651-2660

[94] Yamamoto S, Yamato R, Aritaki Y, et al. Development of a culture protocol

for Rhodomonas sp. Hf-1 strain through laboratory trials. Fisheries Science. 2019;**85**:695-703

[95] Chen B, Yeh HY, Huang CH, et al. Effects of indoor culture conditions on growth and phycoerythrin content of *Proteomonas sulcata* (Cryptophyta) assessed by flow cytometry. Journal of Applied Phycology. 2022;**34**:1201-1213

[96] Lafarga-De la Cruz F, Valenzuela-Espinoza E, Millán-Núñez R, et al. Nutrient uptake, chlorophyll a and carbon fixation by Rhodomonas sp. (Cryptophyceae) cultured at different irradiance and nutrient concentrations. Aquacultural Engineering. 2006;**35**:51-60

[97] Yamamoto S, Yamato R, Yoshimatsu T. Optimum culture conditions of Rhodomonas sp. Hf-1 strain as a live food for aquatic animals. Fisheries Science. 2018;**84**:691-697

[98] Vu MTT, Douëtte C, Rayner TA, et al. Optimization of photosynthesis, growth, and biochemical composition of the microalga Rhodomonas salina—An established diet for live feed copepods in aquaculture. Journal of Applied Phycology. 2016;**28**:1485-1500

[99] Thoisen C, Vu MTT, Carron-Cabaret T, et al. Small-scale experiments aimed at optimization of large-scale production of the microalga Rhodomonas salina. Journal of Applied Phycology. 2018;**30**:2193-2202

[100] Valenzuela-Espinoza E, Lafarga-De-La-Cruz F, Millán-Nuñez R, et al. Growth, nutrient uptake and proximate composition of Rhodomonas sp. cultured using f/2 medium and agricultural fertilizers. Ciencias Marinas. 2005;**31**:79-89

[101] Seixas P, Coutinho P, Ferreira M, et al. Nutritional value of the cryptophyte Rhodomonas lens for Artemia sp. Journal of Experimental Marine Biology and Ecology. 2009;**381**:1-9

[102] Knuckey RM, Semmens GL, Mayer RJ, et al. Development of an optimal microalgal diet for the culture of the calanoid copepod *Acartia sinjiensis*: Effect of algal species and feed concentration on copepod development. Aquaculture. 2005;**249**:339-351

[103] Drillet G, Jørgensen NOG, Sørensen TF, et al. Biochemical and technical observations supporting the use of copepods as live feed organisms in marine larviculture. Aquaculture Research. 2006;**37**:756-772

[104] Rasdi NW, Qin JG. Improvement of copepod nutritional quality as live food for aquaculture: A review. Aquaculture Research. 2016;**47**:1-20

[105] Stottrup JG. Production and nutritional value of copepods. In: Stottrup JG, McEvoy LA, editors. Live Feeds in Marine Aquaculture. Oxford OX4 2DQ, UK: Blackwell Science Ltd.; 2003. pp. 145-205

[106] Brown MR. Nutritional value and use of microalgae in aquaculture. In: Cruz-Suárez LE, Ricque-Marie D, Tapia-Salazar M, Gaxiola-Cortés MG, Simoes N, editors. Avances en Nutrición Acuícola. Cancún, Quintana Roo, Mexico: Simposium Internacional de Nutrición Acuícola; 2002. pp. 281-292

[107] Bartual A, Lubián L, Gálvez J, et al. Effect of irradiance on growth, photosynthesis, pigment content and nutrient consumption in dense cultures of Rhodomonas salina (Wislouch) (Cryptophyceae). Ciencias Marinas. 2002;**28**:381-392

[108] Vu MTT, Jepsen PM, Jørgensen NOG, et al. Testing the yield

of a pilot-scale bubble column photobioreactor for cultivation of the microalga Rhodomonas salina as feed for intensive calanoid copepod cultures. Aquaculture Research. 2018;**50**:63-71

[109] Patil V, Källqvist T, Olsen E, et al. Fatty acid composition of 12 microalgae for possible use in aquaculture feed. Aquaculture International. 2007;**15**:1-9

[110] Latsos C, van Houcke J, Blommaert L, et al. Effect of light quality and quantity on productivity and phycoerythrin concentration in the cryptophyte Rhodomonas sp. Journal of Applied Phycology. 2021;**33**:729-741

[111] Mercier L, Peltomaa E, Ojala A. Comparative analysis of phycoerythrin production in cryptophytes. Journal of Applied Phycology. 2022;**34**:789-797

[112] Derbel H, Elleuch J, Tounsi L, et al. Improvement of biomass and phycoerythrin production by a strain of Rhodomonas sp. isolated from the Tunisian coast of Sidi Mansour. Biomolecules. 2022;**12**:885

[113] Lawrenz E, Fedewa EJ, Richardson TL. Extraction protocols for the quantification of phycobilins in aqueous phytoplankton extracts. Journal of Applied Phycology. 2011;**23**:865-871

[114] MacColl R, Berns DS, Gibbons O. Characterization of cryptomonad phycoerythrin and phycocyanin. Archives of Biochemistry and Biophysics. 1976;**177**:265-275

[115] MacColl R, Habig W, Berns DS. Characterization of phycocyanin from Chroomonas species. The Journal of Biological Chemistry. 1973;**248**:7080-7086

[116] Mysliwa-Kurdziel B, Solymosi K. Phycobilins and phycobiliproteins used in food industry and medicine. Mini-Reviews Med Chem. 2017;**17**(13):1173-1193. DOI: 10.2174/1389557516666160912180155

[117] Rito-Palomares M, Nuez L, Amador D. Practical application of aqueous two-phase systems for the development of a prototype process for c-phycocyanin recovery from *Spirulina maxima*. Journal of Chemical Technology and Biotechnology. 2001;**76**:1273-1280

[118] Sathuvan M, Thangam R, Venkateshbabu G, et al. Single-step purified R-phycoerythrin transmits cellular imaging functionalities in vitro. International Journal of Biological Macromolecules. 2022;**194**:563-570

[119] Sonani RR. Recent advances in production, purification and applications of phycobiliproteins. World Journal of Biological Chemistry. 2016;**7**:100

[120] Eriksen NT. Production of phycocyanin—A pigment with applications in biology, biotechnology, foods and medicine. Applied Microbiology and Biotechnology. 2008;**80**:1-14

[121] Sekar S. Phycobiliproteins as a commodity: Trends in applied research , patents and commercialization. Journal of Applied Phycology. 2008;**20**:113-136

[122] Sen T, Barrow CJ, Deshmukh SK. Microbial pigments in the food industry—Challenges and the way forward. Frontiers in Nutrition. 2019;**6**:1-14

[123] Cardozo KHM, Guaratini T, Barros MP, et al. Metabolites from algae with economical impact. Comparative Biochemistry and Physiology, Part C. 2007;**146**:60-78

[124] Nowruzi B, Konur O, Amir S, et al. The Stability of the phycobiliproteins in the adverse environmental conditions relevant to the food storage. Food

Bioprocess Technol. 2022. DOI: 10.1007/s11947-022-02855-8

[125] Wang A, Yan K, Chu D, et al. Microalgae as a mainstream food ingredient: Demand and supply perspective. In: Alam M, Xu JL, Wang Z, editors. Microalgae Biotechnology for Food, Health and High Value Products. Singapore: Springer Nature Singapore Pte Ltd; 2020. pp. 29-79

[126] Chakdar H, Pabbi S. Cyanobacterial phycobilins: Production, purification, and regulation. In: Shukla P, editor. Frontier Discoveries and Innovations in Interdisciplinary Microbiology. 1st ed. New Delhi: Springer; 2016. pp. 45-70. DOI: 10.1007/978-81-322-2610-9_4

[127] Matos ÂP. The impact of microalgae in food science and technology. JAOCS, Journal of the American Oil Chemists' Society. 2017;**94**:1333-1350

[128] Sonani RR, Singh KN, Kumar J, et al. Concurrent purification and antioxidant activity of phycobiliproteins from Lyngbya sp . A09DM : An antioxidant and anti-aging potential of phycoerythrin in *Caenorhabditis elegans*. Process Biochemistry. 2014;**49**:1757-1766

[129] Sonani RR, Rastogi RP, Madamwar D. Antioxidant potential of phycobiliproteins: Role in anti-aging research. Biochemistry and Analytical Biochemistry. 2015;**4**:4-11

[130] Díaz-Domínguez G, Marsán-Suárez V, Del Valle-Pérez L. Principales propiedades inmunomoduladoras y antinflamatorias de la ficobiliproteína C-ficocianina main immunomodulatory and anti-inflamatory properties of phycobiliproteins C-phycocyanin. Revista Cubana de Hematologia, Inmunologia y Hemoterapia. 2016;**32**:447-454

[131] Romay C, González R, Ledón N, et al. C-phycocyanin: A biliprotein with antioxidant, anti-inflammatory and neuroprotective effects. Current Protein & Peptide Science. 2003;**4**:207-216

[132] Jiang L, Wang Y, Yin Q, et al. Phycocyanin: A potential drug for cancer treatment. Journal of Cancer. 2017;**8**:3416-3429

[133] Sansone C, Brunet C, Noonan DM, et al. Marine algal antioxidants as potential vectors for controlling viral diseases. Antioxidants. 2020;**9**:1-12

[134] Pendyala B, Patras A, Dash C. Phycobilins as potent food bioactive broad-spectrum inhibitors against proteases of SARS-CoV-2 and other coronaviruses: A preliminary study. Frontiers in Microbiology. 2021;**12**:1-10

[135] Dagnino-Leone J, Figueroa CP, Castañeda ML, et al. Phycobiliproteins: Structural aspects, functional characteristics, and biotechnological perspectives. Computational and Structural Biotechnology Journal. 2022;**20**:1506-1527

[136] Aráoz R, Lebert M, Hader D. Electrophoretic applications of phycobiliproteins. Electrophoresis. 1998;**19**:215-219

[137] Glazer AN. Phycobiliproteins—A family of valuable, widely used fluorophores. Journal of Applied Phycology. 1994;**6**:105-112

[138] Kronick MN, Grossman PD. Immunoassay techniques with fluorescent phycobiliprotein conjugates. Clinical Chemistry. 1983;**29**:1582-1586

[139] Sun L, Wang S, Chen L, et al. Promising fluorescent probes from phycobiliproteins. IEEE Journal of Selected Topics in Quantum Electronics. 2003;**9**:177-188

Chapter 4

Transcription Flexibility of *Dunaliella* Chloroplast Genome

Leila Zarandi Miandoab

Abstract

When a *Dunaliella salina* cell is stressed, a series of adaptive changes occur, including gene expression regulation, acclimating to new conditions, and maintaining survival. Due to the natural habitat and the high adaptability of this extraordinary organism to the intolerable environment for other photosynthetic organisms, the plasticity of metabolic pathways has been proven. In this regard, it seems that manipulating the amount and activity of enzymes involved in these pathways is inevitable. Therefore, both nuclear and organelles genomes must sense environmental fluctuation quickly and accurately to respond appropriately to those changes during transcription or post-transcriptional stages. In addition to the nuclear genome, *D. salina* has an autonomous chloroplast genome, consisting of 66, and a mitochondria genome consisting of seven genes encoding proteins. The mystery of *D. salina* survival in harsh environments, from 5 M salinity salt lakes to the Atacama Desert Caves, lies in this flexibility and adaptability from molecular levels to the metabolic pathway of *D. salina* cells. Therefore, who can say prudently that the prosperity of *D. salina* depends on flexibility in the regulation of plastid gene expression?

Keywords: acclimation, transcription, *Dunaliella salina*, chloroplast genome, survival

1. Introduction

Microalgae are worldwide microscopic organisms capable of producing valuable bioactive components from biomass to molecules with drug properties. The rapid growth, utilization of a wide variety of water sources, and photosynthetic activity are the reason for microalgae's success in producing such compounds. High photosynthesis efficiency in microalgae and acclimation power in various ecosystems make microalgae attractive model organisms for the investigation of tolerance mechanisms.

Dunaliella is one of the important microalgae Genus with special characteristics in the dominated territory. *Dunaliella* spices also have wonderful ecological attributes, for example, *Dunaliella acidophila* can survive in pH 1 and *D. salina can live* in hypersaline even crystallizer ponds.

Though the genus and its species have been studied for over a century and a half, there are still a lot of unanswered questions about its magic tools for such behavior.

The interest in *D. salina* and the hired methods and used strategies by it to survive in intolerable environments for other photosynthetic organisms dates back to the 1870s. Till now the research on those strategies has reached molecular levels, and regulating the expression of genes involved in cellular metabolic pathways leads to acclimation and adaptation to diverse environments. It is thought that the flexibility of nuclear and organellar gene expression is one of these mysterious tools.

So, in this chapter, an attempt has been made to be paid a slight aspect of the flexibility of *D. salina,* this particular cell, in the regulation of chloroplast gene expression focusing on **transcription factors**.

2. Endosymbiotic theory and the origin of the organelle genome

Endosymbiosis proposes that the origin of today's eukaryotic organelles evolved through a symbiosis process between two primitive free-living cells. One prokaryote cell was swallowed by the precursor of modern eukaryotes. According to Martin et al. [1], the primary nucleus probably consists of pressing a piece of the cytoplasmic membrane around chromosomes. The nucleated primary cell, which moved like an amoeba by creating false feet, then swallowed the primary prokaryotic cells during phagocytosis, and, for some unknown reason, a number of the ingested prokaryotes survived inside the amoebic cell and became the basis for the symbiosis of above-mentioned cells [1].

According to endosymbiont theory, ingested early prokaryotes retained their specific traits while surviving. They benefited their nucleated host from the advantages of their specific metabolic abilities. It is believed that a bacterium capable of oxidative metabolism is the ancestor of primary mitochondria and a photosynthetic bacterium is the ancestor of primary chloroplast [2].

They eventually lost their cell wall and much of their DNA because they were useless inside the host cell. Thus, both mitochondria and chloroplasts have their DNA, but both also depend on nuclear genes for some functions [2]. Some organellar protein genes (as whole or partial) are present in the nucleus.

Three issues mentioned by Mereschkowsky [3] about plastid insertion into eucaryotic cells are as follows:

i. plastids are indubitably diminished cyanobacteria that entered into a symbiosis with a heterotrophic nucleated host cell in early evolution,

ii. heterotrophic nucleated host cell was itself the product of an earlier symbiosis between a larger, heterotrophic, amoeboid host cell and a smaller "micrococcal" endosymbiont that make the nucleus, which gained plastids,

iii. the plant's autotrophy and self-sufficiency are completely beholden to cyanobacteria [1].

The following provides further evidence for the Endosymbiotic Theory:

- Chloroplast's size, division method (fission), and the existence of Fts proteins at their dividing surface is similar to prokaryotic cells.

- Mitochondria size, division method (dual fission), and the existence of Fts homologs at their dividing surface are similar to prokaryotic cells.

- Mitochondria and chloroplasts have their DNA, which is circular, not linear.

- Mitochondria and chloroplasts have their organellar ribosomes (the 30S and 50S subunits), not 40S and 60S [1].

The possibility of the presence of some genes outside the nucleus – which were originally known as extra-chromosomal genes –was first proposed in the 1950s to justify the unusual inherited pattern of some genes in the fungus *Neurospora crassa*, the yeast *Saccharomyces cerevisiae*, and the photosynthetic alga *Chlamydomonas reinhardtii*.

Simultaneous biochemical and electron microscopy studies have increased the possibility of the presence of DNA in mitochondria and chloroplasts. As a result, in the early 1960s, different sets of data were put together and the existence of chloroplasts and mitochondria genomes was accepted independently of the eukaryotic nucleus genome.

The theory of endosymbiotic is corroborated by observations in which the processes of gene expression in organelles are similar to those processes in bacteria. In addition, when the nucleotide sequences were compared, the genes of the organelles were very similar to their counterparts in the bacteria rather than to the genes in the eukaryotic nucleus.

The theory of endosymbiotic was confirmed by the discovery of organisms that show more primitive stages of endosymbiosis than mitochondria and chloroplasts. For example, the early stages of endosymbiosis have been observed in the single-called *Cyanophora paradox*, whose photosynthetic structures are different from those of chloroplasts and consist of a community of ingested cyanobacteria. In a similar vein, Rickettsia, which lives inside eukaryotic cells, is likely to be an advanced type of bacteria that makes up mitochondria.

According to the theory of endosymbiotic, after the primary cyanobacteria eaten by the primary eukaryotic cell are not digested for unknown reasons, the newcomer's behavior must be controlled by the host and transform from a self-sufficient organism to a semi-self-sufficient employee. They exchanged genetic material and somehow divided tasks. Regarding the information required for the biosynthesis of the important photosynthetic enzyme Rubisco, this division of tasks has been done in such a way that the genes related to the large subunit remain in the genome of the old cyanobacterium and the new organ that had been the feature of oxygen photosynthesis. However, small subunit genes that were responsible for regulating enzyme function and activity were transferred to the host cell nucleus. It seems that the original genome thus wanted and was able to control and initiate the function and status and activity of the enzyme within the primary chloroplast. Another group of genes in the chloroplast organelle genome is related to the proteins and nucleic acids of the organellar ribosomes.

2.1 Physical properties of the organelle genome

Almost all eukaryotes have a mitochondrial genome and all photosynthetic eukaryotes have a chloroplast genome in addition to the mitochondrial ones. All organelle genomes were initially thought to be circular DNA molecules. Electron microscopy showed that in some organelles, DNA was present in both circular and linear shapes. But linear molecules were assumed to be simple fragments of circular genomes created by breaking circular genomes during sample preparation for electron microscopy.

The genomes of most mitochondria and chloroplasts are now believed to be circular, but it has recently been discovered that there are many different forms of genomes in different organisms. In many eukaryotes, circular genomes are present along with linear types in the organelles, and in chloroplasts, there are small circular fragments that make up the entire subset of the genome. A recent pattern culminates in seaweed Dinoflagellate, whose chloroplast genome is divided into many small rings, each carrying only one gene. We now find that the mitochondrial genomes of some microbial eukaryotes, such as Paramecium, Chlamydomonas, and the types of yeasts, are always linear.

The number of organelles genome copies is not well defined. Each mitochondrion of a human cell has approximately 10 identical molecules, reaching about 8000 copies per cell, but in *S. cerevisiae*, even if there are more than 100 genomic copies in each mitochondrion, the total number of genomes per cell is less (less than 6500) will be. Photosynthetic microorganisms, such as Chlamydomonas, have approximately 1000 chloroplast genomes per cell, which is about one-fifth of the number in the plant cell.

The size of the mitochondrial genome varies and does not depend on the complexity of the organism. Most multicellular organisms have small mitochondrial genomes with a compact genetic organization in which the genes are close together and slightly apart. Most lower eukaryotes, such as *S. cerevisiae* and flowering plants, have larger, less compact mitochondrial genomes, some of which have introns. The genome of chloroplasts is less varied in size and most of them have the same structure as the genome of rice chloroplast.

2.2 Genetic content of an organelle genome

The genome of organelles is much smaller than the genome of the cell nucleus, so their gene content is expected to be very limited. In terms of genetic content, the mitochondrial genome shows more diversity. Their gene content varies from five genes in the malaria parasite Plasmodium falciparum to 92 genes in the *Reclinomonas americana* unicellular. All mitochondrial genomes have genes for noncoding rRNAs and some respiratory chain protein components that are linked to the main biochemical characteristics of mitochondria.

In genomes with higher gene content, there are genes for tRNA, ribosomal proteins, and proteins involved in the transcription, translation, and transfer of other proteins from the cell cytoplasm into the mitochondria. Most chloroplast genomes have a similar set of 200 genes or more than encode rRNA, tRNA, ribosomal proteins, and photosynthetic proteins.

An important principle of the endosymbiotic theory is the preservation of organelles genomes. Why have organelles preserved their DNA? John F. Allen's CoRR hypothesis (co-location for redox regulation) described the best answer to that question: It proposes that organelles have protected genomes to be independent in the expression of the respiratory and photosynthetic electron transport chains elements. This independence is essential to maintain **Redox Balance** in the bioenergetic membrane. Hence the CoRR hypothesis states that plastids and mitochondria have focused on genes encoded electron transport chain components, and organellar ribosome rRNA and proteins as organelle translation machine tools. The ribosome biogenesis and assembly process require that some proteins need to be co-expressed in the same compartment as their nascent rRNAs. The convergence observed in gene content in plastid and mitochondrial genomes is striking [4].

For the explanation of the redox balance phrase, can be said it refers to the smooth flow of electrons through the electron transport chain in mitochondria and chloroplasts. These two organelles have electron transport chains that generate proton gradients and produce ATP. Quinols and quinones are essential components in both electron transport chains [5].

If the flow of electrons through the inner mitochondrial membrane or the thylakoid as a bioenergetic membrane is disrupted, the steady-state quinol (reduced form of the quinones) concentration increases and the quinols can transfer electrons non-enzymatically to O_2 and generate the superoxide radical (O^{2-}), the start point of ROS dissemination. Electron flow disturbance occurs when, downstream components are in insufficient amounts, or upstream components are too active. Without retaining the genome, the electron transport chain and the redox state of the organelle will be abandoned, leading to the destruction of the organelle [6].

3. Why Dunaliella

Given the title of this book, the author intends to clarify the importance and role of adaptation and flexibility of transcription regulation of all proteins encoding proteins in organelles genomes, especially chloroplasts, by focusing on popular microalgae *D. salina*.

D. salina is considered for high ability in production and accumulation of massive amounts of β-carotene. Because of its spatial properties, various fields of scientists and researchers like *D. salins*. From 1870 decades of investigation about mechanisms and strategies for optimization of production, extraction and application started and followed now. But it is not only about orange pigment and other subjects, such as genetics, proteins, bioactive compounds, and phytoremediation properties, also attractive [7–13]. Among all *Dunaliella* spices, *D. salina* is a typic organism for the magic and strong power of survival in harsh environments.

D. salina lives in saline rivers, saline lakes exposed to intense light and dryness, and out-of-mind places on earth. Azúa-Bustos and his collage in 2010 reported a novel subaerial *Dunaliella* species growing on cave spiderwebs in the Atacama Desert, which was very surprising. In ancient Atacameño culture and the original language of the Atacameños "Kunza," there is no word equivalent to "rain," and the growth and survival of a photosynthetic organism in such conditions are very wonderful.

Easiness of cultivation, diversity of known strains (such as CCAP19/18, CCAP 19/20, and CCAP 19/30) and geographical isolates, lack of disturbing rigid cell wall for DNA extraction, being unicellular and having only one cup shape chloroplast (that means only a plastid genome that facilitates the develop homoplasmic lines of plastid transformants versus multicellular species) and relativity with *C. reinhardtii* and *Volvox carteri,* has made *D. salina* marvelous algae for organelle genome research and plastome engineering [14].

Such a large area and habitation in a diverse environment in terms of physical and chemical conditions can only indicate and confirm the fact that "*D. salina* has the solution to deal with any environmental fluctuations." *D. salina* can easily and quickly understand the changes in its living environment and select and implement the best response leading to survival.

Environmental changes can include light intensity, temperature, acidity (pH), the amount and concentration of nutrients, the amount of water, salinity, heavy metals, and even the presence of other organisms for which they may appear as pathogens or pests.

Understanding such a variety of physicochemical and biological factors requires highly sensitive and efficient sensors and receivers that can transmit environmental messages to the cell control room scilicet NUCLEUS.

In the next step, the nucleus genome modulates the biosynthesis of some metabolites and overproduces some other metabolites, including glycerol and beta-carotene, by regulating the transcription of specific genes, especially those involved in specific metabolic pathways.

The presence of some protein-coding genes in organelle genomes inevitably regulates their transcription and coordination with the transcription process of nuclear genes. Therefore, the nucleus sends representatives, including transcription factors, to the organelles to control transcription. Each TF is affected by one or more environmental factors. It carries the message to genes that have the corresponding transcription elements and the TF binding site above the initial codon.

One or more TF may be located in the regulatory and promoter region of a gene or gene clusters (some organelles genes, especially in chloroplasts, are operated under the control of a promoter), the result of which can accelerate gene recognition by RNA polymerase and start transcription, or vice versa, prevent the establishment of RNA polymerase in its area and do not allow transcription.

3.1 *Dunaliella* organellar genome

The *D. salina* mitochondrial and plastid genomes are 28.3 and 269 kb, respectively, and assemble as circular molecules The mitochondrial genome (mtDNA) of *D. salina* is average, 51.5 kb; the size of the *D. salina* plastid genome (ptDNA) is more pronounced than its mitochondrial counterpart, being the largest ptDNA sequenced thus far, complete mitochondrial DNA named mtDNA and plastid DNA as ptDNA. A pair of inverted repeats (14.4 kb), in the *D. salina* ptDNA, divide into a large (127.3 kb) and a small single-copy region (112.9 kb), named the LSC and SSC regions.

The GC content of the *D. salina* mitochondria DNAs is 34.4% and plastid 32.1%, which regarding other Archaeplastida organelle genomes is common.

D. salina organelles in the members of the *Chlamydomonadales* are poor in GC or rich in AT, which is important because the *Chlamydomonadales* contain species with GC-rich mitochondrial genomes. The different regions of the *D. salina* mitochondrial and plastid genomes have relatively constant GC content. As:

- Coding DNA: 33%(mtDNA) and 34%(ptDNA);
- Introns and intronic open reading frames (ORFs): 34%(mtDNA) and 32%(ptDNA);
- Intergenic regions: 37%(mtDNA) and 31%(ptDNA).

It is better to know the GC content for the different codon-site positions of the mtDNA and ptDNA protein-coding regions, is approximately.

- 1st position: 38%(mtDNA) and 42%(ptDNA);
- 2nd position: 38% (mtDNA) and 52% (ptDNA);
- 3rd position: 19% (mtDNA) and 13% (ptDNA).

The *D. salina* organelle genomes are large, circular-mapping molecules with ~60% noncoding DNA, this amount of noncoding DNA led to placing them among the most inflated organelle DNAs sampled from the Chlorophyta. The *D. salina* plastid genome, about 269 kb, is the largest complete plastid DNA sequence currently deposited in GenBank. *D. salina* organelle genomes have uniquely high intron densities. For mitochondria DNA ~1.5 and plastid DNA ~0.4 introns per gene [14].

3.2 Transfer of genetic material between the *Dunaliella* chloroplast and nucleus

The CoRR theory seems to explain well the presence of independent genomes of organelles comes from the Endosymbiotic theory. But the grade of independence of organelle genomes has changed over time.

In this way, several genes have been transferred to the nucleus genome, and some have been intelligently conserved in the organelle's genome. But the same genes located in the organelle's genome can be controlled and regulated by the nucleus genome.

Numerous regulatory regions and sites can be identified above the origin codon organelle's genome. Cis-regulatory elements are known to be controllable by organ-specific transcription factors. These transcription factors originate from the nuclear genome and enter the organelle to regulate the transcription of the organelle's genome.

4. The importance of collaboration and coordination between *Dunaliella* genomes

Nuclear monitoring of chloroplast transcription is essential for harmony. The expression of genes encoded by the chloroplast genome is highly dependent on a wide range of factors of nuclear origin. In turn, these factors regulate the expression of plastid genes in response to various environmental and developmental signals. Regulatory factors are widely present at various stages of plastid gene expression, including transcription, RNA editing, post-transcriptional RNA modification, RNA binding, and translation [15].

The transcription system of prokaryotes is different and simplest than that of eukaryotes. It is believed that the prokaryotic gene transcription features have been hired for genome transcription in chloroplasts. Although, at the plastome whole-genome level, the polycistronic operon transcription model cannot account for all the chloroplast transcription products, especially regarding various RNA isoforms. Analysis of algal and higher plants plastids and cyanobacteria transcriptomes revealed that the entire plastome is transcribed and that this attribute is inherited from prokaryotic cyanobacteria, the ancestor of the chloroplast genomes that separated about 1 billion years ago. A multiple arrangement transcription model was proposed by Shi and Wang that multiple transcription initiations and terminations combine randomly to execute the genome transcription followed by subsequent RNA processing events, which elucidates the full chloroplast genome transcription phenomenon and numerous functional and/or aberrant pre-RNAs [16].

Despite living in eukaryotic host cells for almost one billion years since their coexistence event, plastids still retain their prokaryotic properties. Previous studies have shown that plastids preserved some prokaryotic properties, such as prokaryotic gene promoters and terminators, and clustered transcripts of the gene. At first, it was thought that some chloroplast functional genes are transcribed as polystyrene transcripts and then processed into small, mature RNAs. There are almost 20 large transcription units and most of these areas are not transcribed, such as areas between two transcription units. Under such a polystyrene operon transcription model, plastome genes can be transcribed from intrinsic true promoters and later constituted constant-size transcripts. However, this model cannot consider all transcription products across the genome, including massive plastid-encoded RNA output, gene-like transcription, multiple or multiple alternative promoters and terminators, overlapping isoforms, and gene transcription binding in the same polystyrene. This transcriptional and heterogeneous dynamics suggest that an additional overall transcriptional mechanism causes transcription of the entire plastom.

Transcription of chloroplast genome genes is controlled by various factors of nuclear origin. Primary factors affecting the transcription of genes in the chloroplast genome are NEP and additional and non-nuclear PEP subunits. In a group of additional PEP subunits, additional nuclear-encoded protein factors (PAPs) and transcription initiation factors (sigma factors) can be detected. It is well known that PAPs are essential in transcription regulation, however, some of their exact functions have not been proven.

5. Transcription machine and transcription regulation methods in chloroplasts

Chloroplasts are believed to have emerged as a result of the coexistence of photosynthetic cyanobacteria and the ancestors of modern eukaryotic plant cells following various genomic rearrangements. Compared to the genome of cyanobacteria, which is 3Mpz, the genome of terrestrial chloroplasts is 20 times smaller. Despite such a size difference between the two genomes, the expression of chloroplast genes is regulated by more complex systems than cyanobacterial genes. Most importantly, the expression of chloroplast genes strongly depends on post-transcriptional regulation, which includes polystyrene mRNA processing, intron binding, and RNA editing. Chloroplast genes are transcribed in flowering plants by two types of RNA polymerase. Multi-subunit bacterial species (PEP) RNA is encoded by the chloroplast genome and Phage Polymerase T3/T7 RNA (NEP) is encoded by the nuclear genome. In adult chloroplasts, PEP represents the primary transcription machine, which transcribes more than 80% of the original chloroplast transcripts. NEP, on the other hand, transcribes chloroplast housekeeping genes. NEP is a phage-type RNA polymerase enzyme with a single subunit [15].

Both PEP and NEP are essential for the transcription of chloroplast proteins. Even though NEP and PEP identify different promoters, many chloroplast genes have promoters that are detected by both PEP and NEP. Promoters detected by NEP can be divided into three groups—2,1a, 1b. All promoters belonging to species 1a are identified by a protected nuclear motif called YRTa. Promoters of this type are several nucleotides higher than the transcription initiation sequence. Type 1b promoters, in addition to the protected motif in their structure called the GAAbox, are located between the 18th and 20th nucleotides above the YRTa motif.

Experiments on mutant tobacco have shown an essential role of the GAA motif for proper recognition of the promoter performed by NEP.

In contrast, type 2 promoters all consist of NEP promoters without the YRTa motif. Many promoters have been identified by PEP similar to bacterial G70 promoters and are identified by two motifs of normal sequences spaced from the transcription start site by −10 and −35 nucleotides. The first motif of 10 nucleotides that is farther from the transcription site is the TATAAT sequence. Whereas, the second motif of 35 nucleotides that is farther away from the transcription site is the TTG ACT sequence. Due to the great diversity among plants, the position of conventional sequences of specific PEE promoters may be different. For example, in the barley chloroplast genome, the TATAAT 3–9 nucleotide sequence is located above the transcription start site, while the TTG ACT 15–21 nucleotide sequence is located upstream of the transcription start site.

PEP is mainly responsible for the transcription of chloroplast genes, which are protein products that are related to photosynthesis in various ways. However, some genes encoding proteins involved in photosynthesis are encoded by NEP. There is a small group of chloroplast genes that are not related to photosynthesis and are specifically transcribed by NEP. It includes the ACCD gene encoding acetyl-CoA carboxylase subunit in dicotyledonous plants, RPL23 gene encoding ribosomal protein L23, CLPP gene encoding ATP-dependent proteolytic subunit in monocotyledonous plants, and RPOB gene encoding all major subunits in PE. Therefore, chloroplast genes with their promoters can be divided into three categories, which are as follows:

1. Transcribed genes only with the participation of NEP.

2. Genes transcribed only with the participation of PEP.

3. Genes transcribed with both NEP and PEP.

In the case of dicotyledonous plants where two different NEPs have been detected, an extended portion of the promoters can be suggested. The activity of RPOTp and RPOTmp in tissues is different at different stages of plant development. In *Arabidopsis taliana*, increased RPOTmp activity has been observed mainly in young dividing cells and photosynthetic inactive tissues, whereas enhanced RPOTp activity has been observed in photosynthetic green tissues. Differences in the structure of known promoters were identified by the two types of NEP. Interestingly, both NEP and PEP are active at all stages of plant development in plastids of unbleached tissues, including roots, fruits, and seeds. The persistent activity of these two types of polymerases is related to their involvement in the transcription of housekeeping genes, such as tRNA-encoding genes [17].

Regulation of transcription of chloroplast genes is essential for the proper functioning of chloroplasts and overall plant growth under normal and adverse conditions.

5.1 RNA polymerase types

It is an enzyme that consists of several subunits. Although most of the genes in the PEP subunit have been transferred to the nuclear genome, the genes encoding the primary and nuclear PEP subunits (α, β, β', and β'') have been preserved in the chloroplast genome. One of the main differences between the central subunits of PEP is their molecular weight. The alpha subunit has a molecular weight of 38 kDa, the beta subunit has a molecular weight of 120 kDa, the beta subunit has a molecular weight of 85 kDa, and the beta subunit has a molecular weight of 185 kDa. Similarly, in bacteria and most land plants, the RPOA gene encodes the alpha major subunit and with the ribosomal protein genes, organizes into an operon under the control of the same promoter. In contrast, the RPOC2, RPOC1, and RPOB genes encode the major and central subunits of β, β', and β'', respectively, and form a separate operon designated as RPOBC. In addition to the major subunits, PEP is composed of additional protein factors encoded by the nucleus genome, including sigma factors (SIGs) and PAPs, polymerase-related proteins. Sigma chloroplast factors, which are stations for the bacterial transcription initiation factor, play an essential role in the transcription of the chloroplast encoded gene. Sigma factors regulate transcription at different developmental stages by identifying different promoters and allowing a complete set of PEP to begin to polymerize. The highlighted secondary PEP factor appears to be PAPs that are involved in almost every stage of transcription [17].

PEP has a promoter-identifier subunit called the sigma factor. The major PEP enzyme subunits are encoded by a set of genes located in the plastid genome: rpoA, rpoB, rpoC1, and rpoC2. Conversely, during evolution, the sigma factor genes, which specifically provide the promoter needed for PEP, have been transferred to the nucleus genome to possibly allow the nucleus to regulate the expression of the chloroplast gene in response to environmental and developmental signals. PEP and a set of proteins associated with polymerase PAPs constitute a massive protein complex required for transcription. All PAPs are encoded by genes in the nucleus, and most are components of the active pTAC transcription chromosome. These proteins are predicted to be involved in DNA and RNA metabolism, regulate redox from photosynthesis, and protect the PEP complex from reactive oxygen species (ROS).

5.2 Sigma factors

Chloroplasts are the cytoplasmic organs in which photosynthesis takes place in plants and algae. Due to their cyanobacterial strain, chloroplasts contain a small transcriptional active genome and a bacterial gene expression machine.

Sigma factors are separable subunits of bacterial RNA polymerases that ensure effective transcription initiation of gene promoters. Chloroplasts together have a type of bacterial RNA polymerase with a sigma factor subunit due to their prokaryotic origin. The excellent plant *A. taliana* contains six sigma factors (SIGs) for hundreds of its chloroplast genes. The role of this relatively large number of transcription initiation factors for the small chloroplast genome is not fully understood.

Sigma factors are bacterial RNA polymerase subunits. They are capable of effective transcription of bacterial genes with their three distinct activities—transferring the promoter recognition property to RNA polymerase, melting two strands of the promoter region into single single-stranded open complexes capable of transcription, and interacting with other DNA-binding transcription factors that are regulated for gene expression.

The genome of chloroplasts usually contains 100–300 genes that are mostly organized in polystyrene operons, such as bacteria. Most chloroplast genes contain promoter elements of bacterial types 10 and 35 that are identified and transcribed by a multi-subunit bacterial RNA polymerase. The major subunits of the bacterial RNA polymerase are encoded in the plastid genome and are called PEP. Like bacterial polymerase, the original PEP enzyme requires reversible binding to the sigma factor subunit encoded in the nucleus for effective transcription [18].

Is a single subunit enzyme that performs a single transcription protein; from the identification of the promoter to the end of the process, regardless of the structural pattern of DNA. This enzyme evolved through replication of the mitochondrial RNA polymerase encoding nuclear gene and bears a strong resemblance to phage species RNA polymerase and is also made of a subunit. Three types of NEP are detectable, and all three are encoded by RPOT genes. RPOTp occurs in monocotyledonous plants, while RPOTp and RPOTmp have been identified in dicotyledonous plants. In addition, the third form of the NEP family enzyme, RPOTm, has occurred in mitochondria [17]. In *A. taliana*, NEP is encoded by the two nuclear genes rpoTp, and romp [15].

In flowering plants and mosses, one or more single subunits of phage RNA polymerase, known as NEP, transcribe a small subset of chloroplast genes from distinct promoter elements. Genes transcribed by NEP include rpoB, the PEP beta subunit encoder, and several tRNA genes. Some chloroplast genes include NEP and PEP promoters, which are transcribed by two RNA polymerases at all stages of plastid development and in all plant tissues [18].

Conversely, SIGs (sigma factors) are important for PEP binding to promoters of relevant genes. Six different SIGs are involved in the transcription of *A. thaliana* genes, bacterial-derived sigma factors that show great similarity to their ancestors only through the conservative region at the end of their molecules. The nonconservative region seems to be crucial for the functioning of certain sigma factors. SIGs sigma factors are regulated by phosphorylation of specific sequences in the aforementioned nonconservative region. Research on the activation of sigma factors has often been conducted exclusively in SIG1 and SIG6. Phosphorylation of SIGs appears to be a complex process performed by various enzymes. In 1996, Baginsky and his team proposed that SIG6 phosphorylation was mediated by a PEP associated with the serine–threonine protein kinase called plastid transcription kinase. In addition, in the literature, PTK has been abbreviated as cpck2 due to the high similarity between the catalytic components of PTK and the casein kinase2 (ck2) subunit. However, cpck2 has been shown to use SIG6 as a substrate for regulatory phosphorylation and is not the most sensitive site for SIG6 phosphorylation for conventional cpck2. Therefore, this hypothesis requires the presence of other kinases involved in this process. Phosphorylation of SIGs can both initiate and stop the transcription of genes identified by the PEP complex. A type of promoter known to be determined by a specific factor in this process appears to be determined, for example, SIG6 factor phosphorylation is necessary for transcription of the ATPB gene, but does not affect transcription of the PSBA gene. In turn, the lack of phosphorylation of SIG1 reduces the transcription of two DOGA and PSBA genes. The transcription of genes encoded by the chloroplast genome can be affected by smaller molecules, such as rare nucleotides. Recent studies show an interest in these molecules, they are considered new signal molecules that stimulate the response of plants to various types of biotic and abiotic stresses, and they are called Alarmones. Laboratory studies have shown that under different stress conditions, guanosine tetraphosphate is produced in plastids and then binds to the beta subunit of the PEP enzyme and inhibits RNA synthesis [19].

Large-scale prediction and analysis of primary operons in plastids reveal unique genetic features in the evolution of chloroplasts. While bacterial operons have been thoroughly studied, therefore, there is little analysis of chloroplast operons (limited ability to study the basic elements of these structures and apply them to synthetic biology).

Plastids are cellular organelles found mainly in a diverse group of photosynthetic organisms. The endosymbiotic theory explains the origin of plastids (Section 2). Since the beginning of this common evolutionary interaction, most cyanobacterial genes have been lost or transferred horizontally to the host nucleus genome, while the plastid genome has largely retained photosynthesis-related genes and conservative genes. As a result, the plastid is heavily dependent on encoded nuclear proteins for basic operations, making it a non-autonomous organelle. Nevertheless, it retains many of its ancestral characteristics and genomic traits, such as the cyclic structure of the genome, bacterial 70S ribosomes, and PEP or the organization of genes in operon transcription units such as bacteria. Operons are DNA units made up of several genes controlled by a single promoter that often share a common function.

Unlike bacteria, plastid operons are not available in databases and are only examined by a small number of studies focused on higher plant model organisms, in which the entire operon map was revealed in the atmosphere by a differential RNA sequence. In tobacco, part of the polystyrene transcripts was detected using the northern stain. In spinach, psbB and rpoBC operons were detected using Northern blot, while ATP synthase operons were proposed in Escherichia coli by comparing their gene content and ordering their homology to gene clusters.

In algae, part of the operons of *C. reinhardtii* was studied, and several operons were detected using the northern spot. Whereas, two recent reports identified 16 and 22 polystyrene subunits by research for compatibility in RNA sequence overlap, called intergenic regions of adjacent genes. However, there is no extensive analysis of chloroplast operons. The ability to identify them, identify their features, and use this data for synthetic biology purposes remained limited.

The expression of the plastid gene differs from the bacterial model in terms of different characteristics. Chloroplast transcripts are often edited and bound by RNA. The role of transcription termination is significantly reduced, many noncoding RNAs are replicated repeatedly, and the plastid genome is suggested to be completely transcribed. In addition, gene expression often relies on RNA-binding proteins (Pentatricopeptide repeat family) that bind Cis elements upstream of the starting codon. Thus, it inhibits the activity of exoribonucleases and stimulates translation by suppressing the stem-loop, which inhibits ribosome binding. In addition, polystyrenes are regulated by several promoters and are widely processed, thus contributing to the formation of various transcript isoforms derived from a single primary transcription unit. Thus, the structure of plastid operons has evolved significantly compared to classical bacterial operons. This difference likely affected the composition and properties of chloroplast operons and gave rise to unique properties. Consequently, the ability to convert synthetic and synthetic genes into plastids has had a major impact on plant biotechnology, when it points to significant advantages over nuclear deformation. These benefits include uniform composition based on specific coalition location, no gene silencing, relatively high expression of dissimilar genes, and long-term deformation in most crops due to maternal inheritance.

One obvious advantage specifically for this study is the use of the ability of natural plastids to express polystyrenes and to design vectors with multiple genes under the control of a single promoter, thus minimizing plasmid sizes and allowing the introduction of multiple metabolites to the cells. Associated with transgenes in a single deformation. Because both basic scientific questions and biological ideas are hampered by the lack of extensive information on plastid operons.

5.3 Transcription factors

Transcription factors are proteins that bind to DNA regulatory sequences (amplifiers, suppressors, or extinguishers), which are usually located in the 5′ region of target genes, to adjust the speed of transcription and the number of transcripts. This may lead to enhancing or reducing gene transcription and protein synthesis, consequently altering cell function. There are several families of transcription factors, and members of each family may have common structural characteristics. These families include:

- Helix-turn-helix
- Helix–loop–helix
- Zinc finger
- Basic protein-leucine zipper
- β-sheet motifs.

Many transcription factors are common to several cell types (ubiquitous), while others are cell-specific and may specify the phenotypic characteristics of a cell. Transcription factors may be activated directly by ligands, such as glucocorticoids and vitamins A and D. Also, stimulation of cell surface receptors launches multiple intracellular signal transduction pathways, including MAPK, PKA, JAK, and PKC that lead to indirect activation of transcription factors. Transcription factors may act as the nucleus messenger and translate transient peripheral signals at the cell surface into long-term changes in gene transcription. Transcription factors may be activated within the nucleus, often with a transcription factor already attached to DNA, or within the cytoplasm, leading to exposure to nuclear localization signals and targeting the nucleus. Phosphorylation, acetylation, and nitration in transcription factors as post-translational changes can affect the DNA binding quality or transcriptional activity [20].

Membrane-transcription factors are transcription factors that are anchored in the membranes in a passive state and are activated by external or internal stimuli. These transcription factors are released from the maternal membranes and transported to the nucleus. Research shows that some proteins attached to the cytoplasmic membrane (PM) and some proteins attached to the endoplasmic reticulum can enter the nucleus. Based on specific signal recognition signals, some transcription factors attached to membrane-bound proteins undergo proteolytic cleavage to release intracellular fragments that enter the nucleus to control gene transcription. In addition, some transcription factors bind to membrane proteins as integral proteins in the cell nucleus through smuggling into the Golgi and endoplasmic reticulum, where membrane-releasing mechanisms rely on endocytosis. In contrast, transcription factors attached to the membrane of the endoplasmic reticulum are transmitted directly to the nucleus or by transfer to the Golgi. In both pathways, only fragments of transcription factors attached to the membrane of the endoplasmic reticulum are transported to the nucleus. Most transcription factors are located in the cytoplasm. After receiving a signal from the transmembrane signal transduction, the transcription factors are activated and transported to the nucleus after the cytoplasm, where they interact with the corresponding DNA framework (cis active elements) [21]. But keep in mind that transcription factors originate from the nucleus and target the promoter of nuclear and chloroplast genes.

Transcription factors are proteins involved in the process of converting DNA to RNA. They contain a large number of proteins, except RNA polymerase, which initiates and regulates gene transcription. One of the hallmarks of transcription factors is that they have DNA-binding domains that give them the ability to bind to specific DNA sequences called amplifier sequences or promoter sequences. Other transcription factors bind to regulatory sequences, such as activating and suppressing sequences, which can stimulate or suppress transcription of the relevant gene [22].

The task of transcription factors is to regulate and turn off genes to ensure that they are expressed in the cell at the right time and in the right amount throughout the life of the cell. The transcription factor group acts in concert to guide cell division, cell growth, and cell death throughout life. Transcription factors are members of the proteome as well as the regulome. Transcription factors alone or with other proteins in a complex act as activators or inhibitors of RNA polymerase affinity to specific genes [23]. They are classified into different classes based on their DNA-binding domains [24].

5.3.1 Family classification of transcription factors belonging to D. salina species

D. salina transcription factors include a total of 31 different transcription factors, presented in **Table 1**, each of which is classified into specific families of transcription factors. In *D. salina CCAP 19/18* TF involved in nuclear and chloroplast gene transcription is a few differences.

Out of a total of 31 types of transcription factors related to *D. salina*, 10 types of transcription factors are involved in the regulation of the nucleus genome and are not involved in the regulation of the chloroplast genome. The 21 other types are involved in the regulation of the chloroplast genome and are located on the promoter regions of 66 chloroplastic proteins encoding gens.

In this way can be arranged *D. salina* 31TFs in 23 families according to **Table 2**.

The GARP transcription factor family (made up of G2-like and ARR-B) (family number 11 in **Table 2**) has a structural but distant relationship with the MYB transcription factor (family number 12 in **Table 2**).

The SBP transcription factor family (family number 20 in **Table 2**) interacts with the C3H transcription factor (family number 7 in **Table 2**).

Homeobox encodes a DNA helix-turn-helix binding motif called the homeodomain. The second DNA binding is a second independent folded protein that contains at least one structural motif that recognizes dual or single-stranded DNA. A second DNA binding can identify a specific DNA sequence (a recognition sequence) or have a general tendency for DNA [7].

Transcription-activating domains are regions of transcription factors that, in conjunction with a DNA-binding domain, can activate transcription from the promoter by direct contact with the transcription machine (general transcription factor and

TF involve in nuclear gene transcription (10 No)	TF involve in chloroplast gene transcription (21 No)
RWP-RK, C2C2-CO-like, C2C2-LSD, DDT, TUB, NF-X1, C3H, Whirly, PLATZ, Nin-like.	AP2, ERF, HSF, CSD, B3, WRKY, CPP, MADS box, ARR-B, MYB, MYB-related, C2C2-YABBY, C2C2-GATA, SBP, GARP-G2-like, C2H2, NF-YB, NF-YC, bZIP, Homeodomain, bHLH.

Table 1.
D. salina CCAP 19/18 TF involved in nuclear and chloroplast gene transcription.

No	*D. salina* 31TFs	TF families
1	1. AP2 (APETALA2) 2. ERF (ETHYLENE-RESPONSIVE FACTOR)	AP2/ERF Domain
2	3. B3	B3 DNA-binding Domain
3	4. bHLH (Basic Helix–Loop–Helix)	HLH DNA-binding domain
4	5. bZIP (Basic Region/Leucine Zipper Motif)	bZIP Domain
5	6. C2C2-LSD 7. C2C2-CO-like (CONSTANS LIKE) 8. C2C2-YABBY 9. C2C2-GATA	C2C2 Zinc Finger
6	10. C2H2 (Cysteine2Histidine2)	C2H2 Zinc Finger
7	11. C3H	CCCH Zinc Finger
8	12. CPP (Cysteine-rich Polycomb-Like Protein)	CxC Domain
9	13. CSD (Cold Shock Domain)	Cold Shock Domain, β-Sheet
10	14. DDT	DDT Zinc Finger (NEW)*
11	15. GARP-G2-like (Golden 2-Like) GLKP 16. ARR-B (*Arabidopsis* Response Regulator-B type) or ARRM	MYB-like DNA Binding Domain
12	17. MYB (myeloblastosis Oncogene) 18. MYB-related	MYB-like DNA-binding Domain
13.	19. HB-other (Homeobox)	Homeodomain (helix-turn-helix)
14	20. MADS box (M-type, ...) (Maintenance of minichromosome1 Agamous deficiency Serum)	MADS Domain
15	21. HSF (Heat Shock Factor)	Heat Shock Domain
16	22. NF-X1	NF-X1 type zinc finger
17	23. NF-YB (Nuclear Factor Y Subunit β) 24. NF-YC (Nuclear Factor Y Subunit γ or Gamma)	CCAAT motif, CBF (NF-Y, CP1), HisFold Domain
18	25. PLATZ (Plant AT-rich sequence and zinc binding Protein 1)	PLAT zinc binding Domain
19	26. RWP-RK27. Nin-like (Nodule inception) NLP	RWP-RK Domain or RKD
20	28. SBP (Squamosa Promoter Binding Protein-like) SPL	SBP Domain
21	29. TUB (TUBBY-like) or TUB Bipartite TF	-tubulin, TUB Familyβ
22	30. Whirly (Why) ssDNA-binding transcriptional regulator or Protein (TF)	single-stranded DNA-binding Domain
23	31. WRKY (WRKY DNA-binding Protein)	WRKY Domain or β-Sheet DBD (ZF-Like)

Table 2.
Classification of D. salina transcription factors.

RNA polymerase) or via other proteins that are known as co-activators. Transcription suppressor domains are regions of transcription factors that, in conjunction with a DNA binding domain, can suppress transcription from the promoter by contact with transcription machines or through other proteins known as Co. repressors.

5.3.2 Stresses affecting transcription factors of D. salina

Plant stress is a condition in which the plant grows in non-ideal conditions, which increases the demand for it. The effects of stress can lead to stunted growth, crop yield, permanent damage, or death if the stress is too much for the plant. Plant stress factors are mainly classified into two main groups: Biotic and abiotic factors. Abiotic factors include various environmental factors that affect plant growth (such as light, water, and temperature), while biotic factors are other organisms that share the environment with plants (such as pathogens, pests, and weeds). Stress response usually involves complex molecular mechanisms, including changes in gene expression and regulatory networks [19].

Stress-responsive transcription factors play a key role in responding to abiotic stresses and stress tolerance [25]. Therefore, these stress-responsive transcription factors may be important targets for product development by increasing abiotic stress tolerance. Plant stress hormones, such as abscisic acid and jasmonic acid, regulate plant abiotic stress responses. The abscisic acid signaling pathways activate target transcription factors. For example, bZIP, ABF, and Jasmonic acid signaling pathways activate MIC bHLH transcription factors. This abscisic acid and jasmonic acid-dependent transcription factors control the expression of stress-responsive genes, as demonstrated by overexpression and deletion systems. In addition, computational and experimental approaches have identified other transcription factors belonging to the WRKY, MYB, AP2/ERF, and NAC families that are not direct components of the abscisic acid and jasmonic acid signaling pathways but are essential to responding plants to Abiotic stress [26].

6. Conclusion

Plants are constantly exposed to environmental fluctuations, including biotic and abiotic stresses that cause metabolic, morphological, physiological, and molecular changes in plants/algae and all photosynthetic organisms that affect the growth, development, and final production of the plant. In response to these stresses, algae-like plants have evolved various defense systems and have mechanisms to deal with abiotic and biotic stresses. As a result, the stress response in cells begins with perception. Stress clues to living or nonliving factors in cell walls or membranes that are perceived by secondary messengers (such as calcium ions, reactive oxygen species, and hormones) and as intracellular signals. They are transmitted to signal transduction or transduction pathways of downstream signals, such as kinases or phosphatases. Transduction pathways regulate the expression of transcription factors, which in turn modulate the expression of stress-responsive genes in photosynthetic cells.

When a plant is stressed, several adaptive changes occur in plant cells to maintain growth, including over-regulation or under-regulation of various genes. Regulatory proteins act in stress signal transduction by influencing the expression of downstream target genes (functional genes). These regulatory proteins include protein kinases, protein phosphatases, and transcription factors. Transcription factors bind to specific sequences (Cis elements) in the promoter of target genes (stress-related genes), thereby regulating gene expression and affecting biological phenotypes. Transcription factors are key regulatory components of the biotic and abiotic stress signaling pathway.

The chloroplast genome like the nuclear genome is a potential binding site for transcription factors. It can be concluded that environmental stresses affect the regulation of expression and transcription of *D. salina* chloroplast genes.

Author details

Leila Zarandi Miandoab
Faculty of Basic Sciences, Department of Biology, Azarbaijan Shahid Madani University, Tabriz, Iran

*Address all correspondence to: leilazarandym@gmail.com

References

[1] Martin WF, Garg S, Zimorski V. Endosymbiotic theories for eukaryote origin. Philosophical Transactions of the Royal Society B: Biological Sciences. 2015;**370**(1678):20140330

[2] Raven J, Mason L. Chapter 4 cell structure. In: Biology. New York: Singer McGraw-Hill Education; 2017. p. 75

[3] Martin W, Kowallik KV. Annotated English translation of Mereschkowsky's 1905 paper 'Über Natur und Ursprung der Chromatophoren im Pflanzenreiche.' European Journal of Phycology. 1999; **34**(3):287-295

[4] Maier U-G et al. Massively convergent evolution for ribosomal protein gene content in plastid and mitochondrial genomes. Genome Biology and Evolution. 2013;**5**(12):2318-2329

[5] Van Noorden R. The vital question: Why is life the way it is? Nature. 2015;**523**(7562):SB1-SB1

[6] Müller M et al. Biochemistry and evolution of anaerobic energy metabolism in eukaryotes. Microbiology and Molecular Biology Reviews. 2012;**76**(2):444-495

[7] Yesudhas D, Batool M, Anwar MA, Panneerselvam S, Choi S. Proteins recognizing DNA: Structural uniqueness and versatility of DNA-binding domains in stem cell transcription factors. Genes. 2017;**8**(8):192

[8] Borowitzka MA, Borowitzka LJ, Kessly D. Effects of salinity increase on carotenoid accumulation in the green alga Dunaliella salina. Journal of Applied Phycology. 1990;**2**(2):111-119

[9] Dhanam DS, Dhandayuthapani K. Optimization of β-carotene production by marine microalga Dunaliella salina. International Journal of Current Microbiology and Applied Sciences. 2013;**3**:37-43

[10] Zarandi-Miandoab L, Hejazi M-A, Bagherieh-Najjar M-B, Chaparzadeh N. Optimization of the four most effective factors on β-carotene production by Dunaliella salina using response surface methodology. Iranian Journal of Pharmaceutical Research: IJPR. 2019;**18**(3):1566

[11] Zarandi-miandoab L, Hejazi MA. Biodiversity of three Iranian Dunaliella isolates and their potential for biotechnological applications. Journal of Environmental Science and Technology. 2020;**22**(3):69-82

[12] Zarandi-Miandoab L, Hejazi M-A, Bagherieh-Najjar M-B, Chaparzadeh N. Light intensity effects on some molecular and biochemical characteristics of Dunaliella salina. Physiology. 2015;**5**(2):1311-1321

[13] Zarandi Miandoab L, Hejazi MA, Nasiri M. The effect of Cytokinin on growth and physiology of Dunaliella salina. Applied Biology. 2018;**31**(1):121-132

[14] Smith DR, Lee RW, Cushman JC, Magnuson JK, Tran D, Polle JE. The Dunaliella salina organelle genomes: Large sequences, inflated with intronic and intergenic DNA. BMC Plant Biology. 2010;**10**(1):1-14

[15] Zhang Y, Zhang A, Li X, Lu C. The role of chloroplast gene expression in plant responses to environmental stress. International Journal of Molecular Sciences. 2020;**21**(17):6082

[16] Shi C, Wang S, Xia E-H, Jiang J-J, Zeng F-C, Gao L-Z. Full transcription of the chloroplast genome in photosynthetic eukaryotes. Scientific Reports. 2016;**6**(1):1-10

[17] Dobrogojski J, Adamiec M, Luciński R. The chloroplast genome: A review. Acta Physiologiae Plantarum. 2020;**42**(6):1-13

[18] Macadlo LA, Ibrahim IM, Puthiyaveetil S. Sigma factor 1 in chloroplast gene transcription and photosynthetic light acclimation. Journal of Experimental Botany. 2020;**71**(3):1029-1038

[19] Mosa KA, Ismail A, Helmy M. Plant Stress Tolerance: An Integrated Omics Approach. 1st ed. Springer; 24 Jun 2017. ISBN-13: 978-3319593777

[20] Caramori G, Ito K, Adcock IM. Transcription factors in asthma and COPD. IDrugs. Aug 2004;**7**(8):764-70. PMID: 15334310

[21] Liu Y, Li P, Fan L, Wu M. The nuclear transportation routes of membrane-bound transcription factors. Cell Communication and Signaling. 2018;**16**(1):1-9

[22] Latchman DS. Transcription factors: An overview. The International Journal of Biochemistry & Cell Biology. 1997;**29**(12):1305-1312

[23] Roeder RG. The role of general initiation factors in transcription by RNA polymerase II. Trends in Biochemical Sciences. 1996;**21**(9):327-335

[24] Jin J, Zhang H, Kong L, Gao G, Luo J. Plant TFDB 3.0: A portal for the functional and evolutionary study of plant transcription factors. Nucleic Acids Research. 2014;**42**(D1):D1182-D1187

[25] Wani SH. Transcription Factors for Abiotic Stress Tolerance in Plants. London, United Kingdom: Academic Press; 2020

[26] Yoon Y, Seo DH, Shin H, Kim HJ, Kim CM, Jang G. The role of stress-responsive transcription factors in modulating abiotic stress tolerance in plants. Agronomy. 2020;**10**(6):788

Chapter 5

The Solar Saltern of Sfax: Diversity of Hyperhalophilic Microalgae Species as a Promising Naturel Source of Biomolecules

Sana Gammoudi, Neila Annabi-Trabelsi, Mariem Bidhi, Nouha Diwani, Amira Rekik, Hajer Khmekhem, Habib Ayadi, Wassim Guermazi and Jannet Elloumi

Abstract

The saltern of Sfax is a thalasso haline paralic ecosystem were the salinity ranged from 45 to 450 PSU. The microalgae distribution of saltern showed a spatial ecological succession. The specific richness of microalgae decreased with the salinity, accounting 37, 17 and 5 species at three level of salinity from 40 to 80, 80 to 200 and 200 to 450 PSU, respectively. To better understand the behavior of the hyper-halo tolerant microalgae, three autotrophic species *Halamphora* sp. SB1 MK575516 (Diatom), *Phormidium versicolor NCC-466* (Cyanophyceae) and *Dunaliella salina* (Chlorophyceae) were isolated from each level of salinity and they are grown in batch in artificial seawater at laboratory scale. Growth and metabolites synthesized by these microalgae were assessed. Salinity reacts on the physiology of these three species which possess mechanisms of resistance to more or less effective stresses and generally by the synthesis of different biomolecules such as pigments, sugars, proteins and fatty acids.

Keywords: solar saltern, *Halamphora* sp., *P. versicolor*, *D. Salina*, culture, metabolites

1. Introduction

An ecosystem is qualified as extreme when it's physicochemical parameters are most often hostile to life Grégoire, Fardeau, Guasco, Bouanane, Michotey and Bonin [1]. Indeed, any biotope characterized by a very low or very high value of the main parameters that influence their life cycle can be characterized as an extreme environment [2]. These parameters are essentially temperature, salinity, pH, pressure, radiation, desiccation, and oxygenation. Organisms with the ability to live in extreme environments are called "extremophiles". And as a result, several groups have been described taking into account the extreme conditions they can tolerate [1]. These are essentially prokaryotic microorganisms, mostly belonging to the Archaea group. Eukaryotes can also be recorded in extreme environments. They are essentially unicellular algal or fungal organisms [3].

Among these environments, hypersaline ecosystems are very widespread and they can be classified into natural and artificial biotopes. While the natural environments are essentially represented by salt lakes, lagoon and, Sabkhas, the artificial hypersaline environments are represented by saltworks. These latter are transitional ecosystems between the marine and the continental domain [4], consisting of shallow ponds used for the production of halite (NaCl) from seawater which is pumped to the first series of ponds. After an evaporation cause a sufficient increase in salinity, the water is transferred to the next series of ponds, and so on, until brine saturated with NaCl is obtained, from which the halite precipitates in the last series of ponds recognized the crystallization ponds. The salinity in each of the ponds is thus maintained more or less constant over time [5]. This process leads to the selection of the variety of microbial heterotrophs and autotrophs and ciliated protozoa [4]. Species were adapted to different salinity variations [6]. The Sfax solar saltern (Tunisia) is an artificial paralic ecosystem characterized by its floristic and faunal richness [7], as well as by its microalgae richness [8, 9]. This biotope has been the site of several studies since 1998: (i) microalgae [10], (ii) ciliates [11] and (iii) zooplankton [12–14] especially the branchiopod crustacean *Artemia salina* [15, 16]. The cultures of microalgae sampled from the Sfax solar saltern have been the subject of several studies [17–20].

Microalgae are very rarely grouped according to their energy metabolism or even according to their ability to synthesize the necessary metabolites, but rather according to their morphological properties [21]. There are therefore different taxonomic classes of microalgae, the main ones being Rhodophyceae, Chlorophyceae, Bacillariophyceae, Euglenophyceae, Dinophyceae and Cyanobacteria. Microalgae occupy a very important place in nature since they are at the base of a long food chain and contain impressive nutritional proerties [22]. Moreover, they have various fields of exploitation, due to the value-added molecules. The biochemical composition of microalgae proves that they contain high value natural fatty acids (omega-3), which can produce a high value dietary supplement [23]. Furthermore, microalgae contain a high amount of proteins reaching up to 70% of the dry matter for *Spirulina* and also producing mineral elements such as calcium and magnesium [24]. Also, extremophile microalgae have many substances recognized by their bioactive properties such as antiviral, antiproliferative and anticancer properties [25]. These biomolecule possess a powerful antioxidant effect as determined by several authors [26, 27]. Finally, microalgae are largely used in wastewater treatment [28].

In this chapter we will present the biodiversity of the halophilic microalgae of the Sfax solar saltern and the different techniques used for the isolation and valorization of culture or metabolites extracted from three microalgal species.

2. Biodiversity of halophilic microalgae in solar saltern of Sfax

The Sfax solar salternor the Thyna salt works (**Figure 1**) is an artificial system located in the Gulf of Gabes in an arid climate (34° 39′N and 10° 42′E). This system is composed of several interconnected shallow ponds (20 to 70 cm deep) with increasing salinities from the water intake (40 PSU) to the salt Tables (450 PSU) [9]. The saline is separated from the sea by a dam of red silt about 4 m high running along the southern coast of the city of Sfax for about 13 km (**Figure 1**), from the port area to the village of Gargour, occupying an area of 1500 ha [29]. It is one of the most

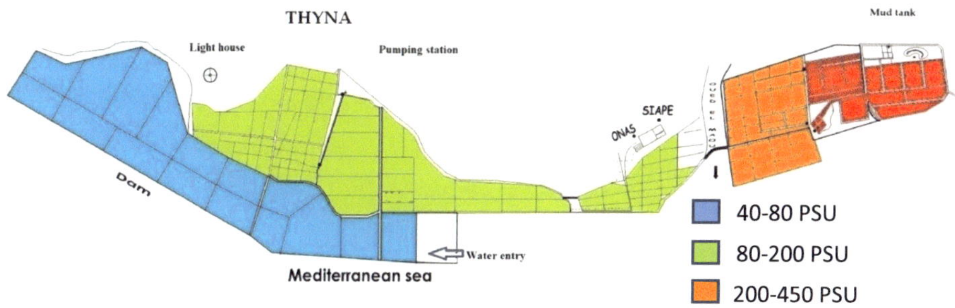

Figure 1.
General map of the Sfax Saltern showing the three levels of increasing salinity.

Salinity (PSU)	Microalgae	Species
40–80	Pennate diatoms	*Achnanthes brevipes*
		Achnanthes sp.
		Cocconeis sp.
		Cylindrotheca closterium
		Diatomella sp.
		Diploneis sp.
		Epithemia sp.
		Lichomphora sp.
		Nitzschia longissima
		Nitzschia ventricosa
		Navicula sp.
		Navicula elegans
		Navicula neoventricosa
		Pinnularia sp.
		Pleurosigma
		Stenopterobia sp.
		Surirella sp.
		Synedra longissima
	Centric diatoms	*Biddulphia* sp.
		Chaetoceros sp.
		Coscinodiscus sp.
		Halamphora sp. SB1 MK 575516*
		Thalassiosira mendiolana
	Dinoflagellates	*Akashiwo sanguinea*
		Gymnodinium sp.
		Mesoporos sp.
		Oxyrrhis marina
		Peridinium afrinacum
		Peridium sp.
		Prorocentrum bipes
		Prorocentrum gracile
		Protoperidinium micans
		Protoperidinium mite
		Protoperidinium pellucidum
		Scrippsiella gregaria
		Scrippsiella trochoïdea
	Euglenophyceae	*Euglena* sp.

Salinity (PSU)	Microalgae	Species
80–200	Pennate diatoms	*Cylindrotheca closterium* *Cymbella* sp. *Epithemia* sp. *Gyrosigma attenuatum* *Nitzschia longissima* *Pleurosigma*
	Dinoflagellates	*Gymnodinium* sp. *Oxyrrhis marina* *Peridium* sp. *Polykrikos* sp. *Protoperidinium micans* *Protoperidinium* sp.
	Cyanobacteria	*Aphanothece* sp. *Phormidium versicolor NCC 466** *Spirulina subsalsa*
	Chlorophyceae	*Chlamydomonas rubrifilum* *Dunaliella salina*
200–450	Cyanobacteria	*Aphanothece* sp. *Phormidium versicolor* *Spirulina subsalsa*
	Chlorophyceae	*Chlamydomonas rubrifilum* *Dunaliella salina**

**Species isolated and cultured in laboratory.*

Table 1.
Classification of microalgae species recorded in Sfax solar saltern according to the salinity gradient [9].

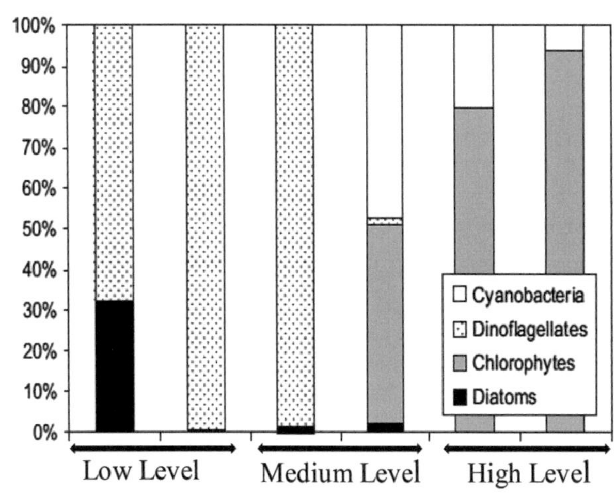

Figure 2.
Mean relative contribution of microalgae groups to total microalgae biomass in six ponds of increasing salinity in the saline of Sfax Tunisia [30].

important salt production areas in Tunisia (300,000 T of salt per year). A total of 45 microalgae taxa were recorded from the Sfax solar saltern and identified belonging to five groups: diatoms, dinoflagellates, Chlorophyceae, Euglenophyceae and

Cyanobacteria. For each group, we clearly observed a marked decrease in the number of taxa with the increase of salinity (**Table 1**, **Figure 2**). Diatoms were dominant in ponds that have salinity ranges from 40 to 80 PSU (67.95% of the microalgae total abundance), whereas the dinoflagellates represented only 22.19% and Euglenophyceae were poorly represented in this pond (1.2%) (**Figure 2**). Dinoflagellates dominated the densities and biomasses of microalgae in the ponds of 80–200 PSU, contributing to 56.7% and 34.4% of the microalgae total abundance, respectively. Chlorophytes largely dominated in the crystallization ponds>200 salinity which accounting for 69.1% of the total microalgae. While Cyanobacteria were relatively abundant in ponds of medium salinity (19.9%) they were rare in hypersaline ponds (0.5%) (**Figure 2**, **Table 1**).

3. Valorization of three algal species

Three different microalgae species were isolated from the three level of salinity of Sfax solar saltern: the diatom *Halamphora* sp. (45–80 PSU) cyanobacterium *Phormidium versicolor* (80–200 PSU) and the chlorophyceae *Dunaliella salina* (200–450 PSU) (**Figures 1** and **3**). These species were cultured in the laboratory. The extraction of the various pigments and biomolecules contained in these species was carried out.

3.1 Isolation and culture conditions

Halamphora sp. was isolated via micromanipulation and serial dilution from a collected water sample. Both *D. salina* and *P. versicolor* were isolated on agar medium. Several antibiotic and antifungal treatments were performed in order to obtain monoclonal and axenic cultures. These three algal species were batch cultured in flasks 500 ml Erlenmeyer flask with artificial sea water of 80 PSU. The cultures were initiated with cell densities of 10^6 cells ml^{-1} for *D. salina* [31], 50,000 cells ml^{-1} for *Halamphora* sp. [19] and an initial concentration of chlorophyll *a* of 0.005 µg ml^{-1} for *P. versicolor* because the numeration was impossible (the filaments intertwine) [9]. *P. versicolor* was cultivated using BG-11 medium as culture medium [32], *D. salina* was grown in Walne's growth medium as modified by Guermazi, Elloumi, Ayadi, Bouain and Aleya [17] and for *Halamphora* sp., the culture was carried out in F/2 Provasoli medium [33]. Cultures were maintained in incubator (FRIOCELL) and incubated in 24-light cycles at (25°C). *Phormidium* and *Halamphora* were reared under light intensity of 130 µmol photons m^{-2} s^{-1}, while Dunalielal was cultured under low irradiance of 27 µmol photons m^{-2} s^{-1}.

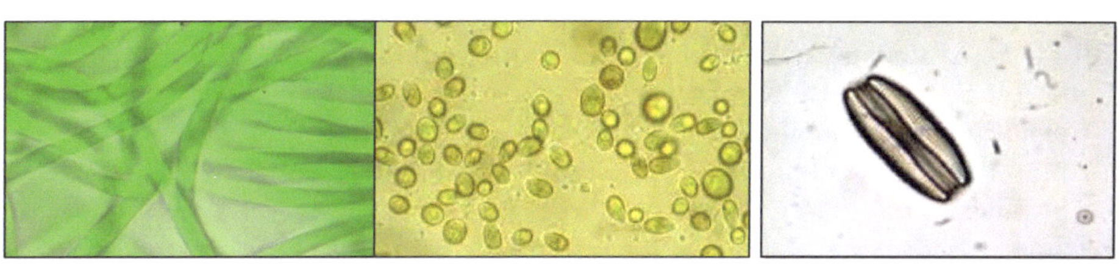

(a) (b) (c)

Figure 3.
Microscopic observation (G ×100) of (a) P. versicolor, (b) D.salina and (c) Halamphora sp. isolated from the Sfax solar salternand cultured in the laboratory.

The biomass of each microalga was separated from the culture media by centrifugation (4500 × g, 10 min), and the pellet was washed with distilled water and centrifuged again at 4500 × g for 10 min (the washing was repeated twice). The pellet was freeze-dried and stored at −70°C.

3.2 Growth tracking

The growth of *D. salina* and *Halamphora* sp. was determined by a daily cell count using a Malassez hemocytometer under a light microscope.

The cyanobacterium *P. versicolor* being filamentous, the chlorophyll *a* was determined to assess the growth of this cyanobacterium according to the equation of Speziale, Schreiner, Giammatteo and Schindler [34]:

$$\text{Chl } a\left(\mu g\, l^{-1}\right) = [11.47\,(OD_{664nm} - OD_{750nm}) - 0.4\,(OD_{664nm} - OD_{750nm})]\,\upsilon / V$$

with.
OD: Optical density.
υ: volume of the acetone extract (ml).
V: volume of the algae suspension (ml).

3.3 Determination of pigments content: chlorophyll *a*, carotenoids and phycocyanin

The dosage of photosynthetic pigments of each microalgal species was carried out after extraction in 90% acetone. The concentration of Chlorophyll *a* was calculated according to the equation of Speziale, Schreiner, Giammatteo and Schindler [34] for the cyanobacterium and conforming to the equations of Jeffrey and Humphrey [35] for Chlorophyceae and the diatom.

Equation of Jeffrey and Humphrey [36] for Chlorophyceae:

$$\text{Chl } a\left(\mu g\, l^{-1}\right) = 11.93OD_{664nm} - 1.93OD_{647nm}$$

Equation of Jeffrey and Humphrey [35] for Diatom

$$\text{Chl } a\left(\mu g\, l^{-1}\right) = 11.47OD_{664nm} - 0.40OD_{630nm}$$

Carotenoid concentrations were calculated according to the equation of Chamovitz, Sandmann and Hirschberg [36] for cyanobacterium and according to Salguero et al. [37] for Chlorophyceae.

Cyanobacteria carotenoid ($\mu g\, l^{-1}$) = ($OD_{461nm} - 0.046OD_{664nm}$) × 4

Chlorophyceae Carotenoid ($\mu g\, l^{-1}$) = 0.0045 ($3000OD_{470nm} - 1.63OD_{750nm}$Chl *a*).

The phycocyanin (C-PC) pigment was isolated from *P. versicolor* using the method developed by Silveira et al. [38] With light modification. The C-PC contents were quantified according the following equation:

$$C-PC\left(mg\, ml^{-1}\right) = \left(OD_{615nm} - 0.474OD_{652nm}\right) / 5.43$$

3.4 Determination of dry matter, proteins, lipids, Total sugars and phenolic compounds

The dry matter of microalgae was determined according to the AOAC standard methods [39]. The protein assay method of Lowry [40] was used by the combination of Folin with Biuret's reagents. The Lipids content was determined gravimetrically after the Soxhlet extraction of dried samples with hexane for 2 hours using Nahita Model 655 (Navarra, Spain). The sugars were estimated by phenol-sulfuric acid method [41] using glucose as a standard. The total phenol content of the *P. versicolor* extract was determined by the method of Singleton and Rossi [42].

3.5 Determination of mineral content of *Halamphora* sp.

The analyses of sodium, potassium, calcium, magnesium, iron, copper, and zinc contents in *Halamphora* sp. were carried out using the inductively coupled plasma optical emission spectrophotometer (ICP-OES) Model 4300 DV, PerkinElmer, Shelton, CT, USA, according to the method of AOAC 1999 [43]. Measurements were done in triplicates.

3.6 Determination of fatty acids profile of *Halamphora* sp. and *D. salina*

For fatty acids analyses, cultures were harvested at the end of the log phase. All lipids were evaporated to dryness with nitrogen and concentrated with hexane. Fatty acids methyl esters (FAMEs) were prepared from the lipid extract by transesterification using a direct transmethylation method according to Lepage and Roy [44]. The FAMEs were then extracted with hexane and determined quantitatively by capillary gas chromatography. We used a Chromopack, CP 9001 gas chromatograph, HPS 5890 series II chromatograph, equipped with a polar 25-m capillary column CP wax 58 (Varian SA, France) (0.32 mm diameter and a layer thickness of 0.52 mm), and a flame ionization detector (FID). We used a splitless injection system with nitrogen as the carrier gas. The oven was programmed to rise from an initial temperature of 180–250°C at rates of

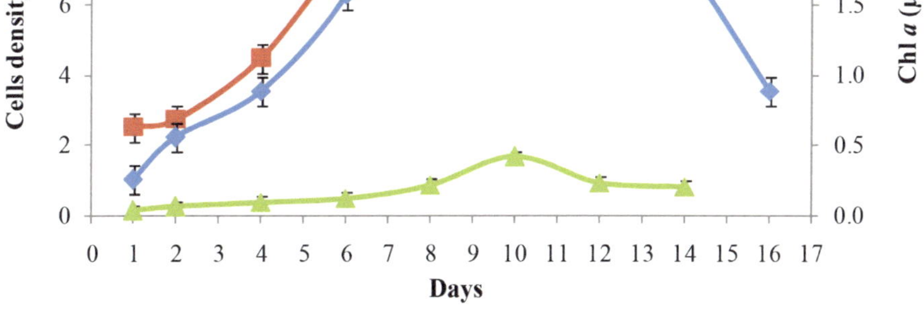

Figure 4.
Growth curves of Halamphora sp., D. salina and P. versicolor cultured batchwise.

10°C min^{-1} (from 180 to 220), 2°C min^{-1} (from 220 to 240), and 5°C min^{-1} (from 240 to 250). Individual FAMEs were identified by comparing retention times with those obtained with laboratory standards and the manufacturer's instructions (Supelco).

3.7 Growth kinetics of three microalgae

The four growth phases—lag, exponential, stationary and decline growth phases—are only observed on growth curves of *Halamphora* sp. (**Figure 4**). While *Halamphora* showed a short lag phase of 2 days, this phase was absent for *Phormidium* and *Dunaliella*. The exponential growth phase was observed for all the microalgae under continuous light but with different slopes. *Phormidium* and *Halamphora* grew faster with similar exponential phase about 5–6 days and reached maximum yield of 2.66 μg. ml^{-1} and 10.22 × 10^6 cells. ml^{-1}, respectively at 8th day. During the exponential phase, the specific growth rate (μ) was about 2.40 and 2.15 day^{-1} for both *Phormidium* and *Halamphora*, respectively. However, it did not exceed 0.7 day^{-1} for *Dunaliella salina*. The growth rate of microalgae is very sensitive to culture conditions, such as irradiance and photoperiod limitation [45]. The density of *Halamphora* sp. is higher than those of *Halamphora acutiuscula* (5.91 × 10^5 cells. ml^{-1}) and *Halamphora coffeaeformis* (6.17 × 10^5 cells. ml^{-1}) which they are cultured in artificial sea water under light-dark (14/10h) cycles at a photon flux density of 300 μmol photons m^{-2} s^{-1} [46].

For *Dunaliella salina*, the maximum cell density was recorded at 10th and did not exceed 1.65 × 10^6 cells. ml^{-1} (**Figure 4**). This value is higher than that reported in solar saltern by Elloumi et al. [47]. Guermazi et al. [31] stated that *D. salina* reached 6 × 10^6 cells. ml^{-1} when reard under 12h/12h light dark regime. Moreover, all microalgae curves are characterized by a short stationary phase (**Figure 4**). It seems that nutrients composition of the culture medium need to be optimized in order to maintain the cells at stationary phase.

3.8 Physicochemical characterization of three microalgal species

Microalgae could be easily grown in a laboratory and used for large-scale cultivation in bioreactors with the ability to control the quality of the cultures by providing purified culture medium that is free of toxic substances. Therefore, microalgae provide a more accessible way to produce qualitative biomolecules of interest [48–50]. Physicochemical characteristics of *D. salina*, *Halamphora* sp. and *P. versicolor* are presented in **Table 2**. The biomass of these microalgae contains moderate amounts of lipids, proteins, carbohydrates and an important percentage of chlorophyll a and carotenoids. The 7% dry matter content of *Halamphora* sp. is close to that found for other strains: 8% for *Halamphora* sp. [51], and for *P. versicolor* content 13% similarly with Singh, Parmar and Madamwar [52], who showed that the dry matter content of *Phormidium ceylanicum* is 10%. However, the lipids and proteins content of *Halamphora* sp. were relatively lower than the values published for other strains of *Halamphora* [53], and it was higher than that of *Amphora coffeaformis* [54]. For *D. salina*, the total lipids increased during growth hereas the amounts of proteins and sugars decreased, while for *P. versicolor*, it recorded a high level of protein (45%). The total sugars content of *Halamphora* sp. was 12.60% DW, which is consistent with that of some microalgae (5–23% DW) [55] and of *P. versicolor* was 21.56%. Moreover, these three algal species were found to be rich in chlorophyll, mainly chlorophyll a, and carotenoids. Continuous illumination favored also the synthesis of these pigments in *D. salina*. In fact, the synthesis of pigments was also stimulated under the

Component	*Halamphora* sp.	*Phormidium versicolor*	*Dunaliella salina*
Dry matter (%Fw)	7 ± 0.45	10 ± 0.66	—
Proteins (%Dw)	27.62 ± 0.33	42 ± 0.78	41.39 ± 6.40
Lipids (%Dw)	11.14 ± 0.19	15.7	27.04 ± 19.74
Total sugars (%Dw)	12.60 ± 0.76	21.56 ± 0.99	13.33 ± 8.06
Aches(% Dw)	37.78 ± 0.43	—	—
Chlorophyll a (%Dw)	4.94 ± 0.27	7.05 ± 0.81	17.02 ± 7.78
Phycocyanin (%Dw)	—	13 ± 0.47	—
Cartenoids(%DW)	1.083 ± 0.05	1.79 ± 0.08	1.22 ± 0.39
Polyphenols(mgGAE g^{-1})	38.27 ± 2.21	408.5 ± 18,18	—
Flavonoides(mgGAE g^{-1})	17.69 ± 0.70	13.67 ± 0.78	—

Data are expressed as mean ± standard deviation of triplicates. FW: fresh weight; DW: dry weight; GAE: gallic acid equivalent; – not realized.

Table 2.
Physicochemical characteristics of three microalgae species.

effect of light and allowing sufficient photosynthetic activity to be maintained for the synthesis of glycerol [56]. Our results also show that the 80% ethanolic extract of Bacillariophyceae and cyanobacterium the highest phenols and flavonoids contents. These high levels may be due to the culture conditions under the high salinity of 80 psu and the extraction conditions. Additionally, a high production of phycocyanin has been proven by the blue microalgae *P. versicolor* content 13%. These results are consisting with the observation by Singh, Parmar and Madamwar [52]. With respect to the ash content of *Halamphora* sp. (37.78% DW) (**Table 2**), it is in line with that found for another *Amphora* strain. Ash content exceeds 50% (55.8 to 67.9%) of the dry weight for some diatoms [54]. *Halamphora* sp. From Sfax solar salter has moderate amounts of sodium, potassium, calcium, and magnesium (**Table 3**). According to Boulay, Abasova, Six, Vass and Kirilovsky [57], the different species of microalgae do not develop the same strategies in order to survive under stressful conditions. We can assume that the species we studied might be a potential candidate for the production of biomolecules for pharmacological purposes.

Mineral	*Halamphora* sp.
Sodium (g Kg^{-1}DW)	1.125 ± 0.2
Potassium (g Kg^{-1}DW)	0.485 ± 0.05
Calcium (g Kg^{-1}DW)	0.584 ± 0.05
Magnesium (g Kg^{-1}DW)	0.747 ± 0.1
Iron (g Kg^{-1}DW)	0.016 ± 0.002
Copper (g Kg^{-1}DW)	0.008 ± 0.001
Zinc (g Kg^{-1}DW)	0.008 ± 0.001

Table 3.
Mineral content of Halamphora sp. [19].

Fatty acids	*Halamphora* sp.	*D. salina*
C14:0	3.623 ± 0.3	2.8 ± 1.2
C15:0	3.418 ± 0.3	—
C16:0	27.427 ± 0.5	21.0 ± 3.5
C17:0	1.664 ± 0.4	—
C18:0	1.974 ± 0.3	9.05 ± 1.0
C20:0	0.734 ± 0.2	—
C24:0	2.468 ± 0.2	—
SFA	**41.308 ± 0.8**	**35.9 ± 0.6**
C14:1	3.386 ± 0.3	9.6 ± 1.1
C16:1	45.089 ± 0.8	2.2 ± 1.2
C17:1	0.521 ± 0.1	—
C18:1	3.658 ± 0.3	14.9 ± 1.2
MUFA	**52.654 ± 1.2**	**30.2 ± 0.8**
C16:2	1.603 ± 0.2	—
C16:3	0.924 ± 0.3	0.1 ± 0.1
C16:4	—	14.0 ± 2.3
C18:2	0.432 ± 0.1	0.9 ± 0.2
C18:3	—	0.8 ± 0.3
C18:4	—	0.6 ± 0.2
C20:4	0.712 ± 0.2	—
C20:5 (EPA)	2.367 ± 0.3	0.1 ± 0.1
C22:6 (DHA)	—	4.3 ± 1.3
PUFA	**6.038 ± 0.5**	**28.66 ± 0.8**

(−): not detected.

Table 4.
Percentage of fatty acids composition of Halamphora sp. and D. salina reared in laboratory [19, 31].

3.9 Fatty acids composition of *Halamphora* sp. and *D. salina*

The fatty acid profile of *Halamphora* sp. and *Dunaliella* sp. was composed of saturated (SFA), monounsaturated (MUFA), and polyunsaturated fatty acids (PUFA) which differed significantly from an alga to another (**Table 4**). The level of SFA recorded in *Halamphora* sp. and *Dunaliella* sp. is high, averaging 41.308 and 35.90% of total fatty acid, respectively. The pattern of SFA show that *Halamphora* is richer in SFA than *Dunaliella*. However, *Dunaliella* and *Halamphora* recorded a high level of palmitic acid (16:0) which accounted 21.0 and 27.42%, respectively. Hence, *Halamphora* sp. could be a suitable producer of SFA, which are easily convertible to biodiesel [58].

Moreover, *Halamphora* exhibited a high amount of palmitoleic acid (C16:1) which reached 45.089%, while that of *Dunaliella* did not exceed 2.2% of total FA. High levels of palmitoleic acid and other bioactive fatty acids were also detected in the fusi form morphotype of the Bacillariophyceae [59]. For *D. salina*, MUFAs were represented by 14:1 (n-5), 16:1 (n-7) and 18:1 (n-9). *D. salina* is an important source of 18:1 (n-9), whichreached14.9 ± 1.2%.

Dunaliella is rich in PUFAs with a percentage of 28.66%, while those of *Halamphora* did not exceed 6% of total fatty acid (**Table 4**). While *Halamphora* produced a noticeable level of EPA (2.367%), *Dunalila* is an important source of DHA reaching 4.3% of total fatty acid. Indeed, it is known that EPA is an important PUFA for health protection from many pathologies, including cardiovascular diseases [60] and cancer [61]. These PUFAs are known to have a number of important nutritional and pharmaceutical applications [62, 63]. They are also known to have beneficial effects on the health of human beings and to play a major role in the prevention of medical disorders in three areas: the heart and the circulation [64], inflammatory conditions and cancers, in particular colon tumorigenesis [65].

4. Conclusion

In conclusion, the saline of Sfax presents a high microalga diversified. Three species *Halamphora* sp., *P. versicolor* and *D. salina* are interesting for biomolecules production such as cartenoid, protein, sugar, fatty acids which they could be valorized in several fields.

Author details

Sana Gammoudi[1*], Neila Annabi-Trabelsi[1], Mariem Bidhi[2], Nouha Diwani[1], Amira Rekik[1], Hajer Khmekhem[1], Habib Ayadi[1], Wassim Guermazi[1] and Jannet Elloumi[1]

1 Laboratory of Marine Biodiversity and Environment, Department of life Sciences, Faculty of Sciences, University of Sfax Tunisia, Tunisia

2 Laboratory of Biochemistry and Enzymatic Engineering of Lipases (LBGEL), National School of Engineers of Sfax (ENIS), Tunisia

*Address all correspondence to: gamoudisana16@gmail.com

References

[1] Grégoire P, Fardeau M, Guasco S, Bouanane A, Michotey V, Bonin P. Les micro-organismes de l'extrême. La Presse Thermale et Climatique. 2009;**146**:49-61

[2] Oarga A. Life in extreme environments, Revista de Biologia e ciencias da. Terrain. 2009;**9**(1):1-10

[3] Cantrell SA, Casillas-Martinez L, Molina M. Characterization of fungi from hypersaline environments of solar salterns using morphological and molecular techniques. Mycological Research. 2006;**110**(8):962-970

[4] Elloumi J, Guermazi W, Ayadi H, Bouain A, Aleya L. Abundance and biomass of prokaryotic and eukaryotic microorganisms coupled with environmental factors in an arid multi-pond solar saltern (Sfax, Tunisia). Journal of the Marine Biological Association of the United Kingdom. 2009;**89**(2):243-253

[5] Oren A. Thoughts on the "missing link" between saltworks biology and solar salt quality. Global NEST Journal. 2010;**12**(4):417-425

[6] Baati H, Amdouni R, Gharsallah N, Sghir A, Ammar E. Isolation and characterization of moderately halophilic bacteria from Tunisian solar saltern. Current Microbiology. 2010;**60**(3):157-161

[7] Chokri MA, Sadoul N, Medhioub K, et al. Analyse comparative de la richesse avifaunistique du salin de Sfax dans le contexte tunisien et méditerranéen. Revue d'Ecologie, Terre et Vie. 2008;**63**(4):351-369

[8] Ayadi H, Abid O, Elloumi J, Bouaïn A, Sime-Ngando T. Structure of the phytoplankton communities in two lagoons of different salinity in the Sfax saltern (Tunisia). Journal of Plankton Research. 2004;**26**(6):669-679

[9] Masmoudi S, Tastard E, Guermazi W, Caruso A, Morant-Manceau A, Ayadi H. Salinity gradient and nutrients as major structuring factors of the phytoplankton communities in salt marshes. Aquatic Ecology. 2014;**49**(1):1-19

[10] Khemakhem H, Elloumi J, Ayadi H, Aleya L, Moussa M. Modelling the phytoplankton dynamics in a nutrient-rich solar saltern pond: Predicting the impact of restoration and climate change. Environmental Science and Pollution Research. 2013;**20**(12):9057-9065

[11] Kchaou N, Elloumi J, Drira Z, Hamza A, Ayadi H, Bouain A, et al. Distribution of ciliates in relation to environmental factors along the coastline of the Gulf of Gabes, Tunisia. Estuarine, Coastal and Shelf Science. 2009;**83**(4):414-424

[12] Toumi N, Ayadi H, Abid O, Carrias J-F, Sime-Ngando T, Boukhris M, et al. Zooplankton distribution in four ponds of different salinity: A seasonal study in the solar salterns of Sfax (Tunisia). Hydrobiologia. 2005;**534**(1):1-9

[13] Thabet R, Leignel V, Ayadi H, Tastard E. Interannual and seasonal effects of environmental factors on the zooplankton distribution in the solar saltern of Sfax (South-Western Mediterranean Sea). Continental Shelf Research. 2018;**165**:1-11

[14] Ladhar C, Tastard E, Casse N, Denis F, Ayadi H. Strong and stable environmental structuring of the zooplankton communities in interconnected salt ponds. Hydrobiologia. 2015;**743**(1):1-13

[15] Guermazi W, Elloumi J, Ayadi H, Bouain A, Aleya L. Coupling changes in fatty acid and protein composition of Artemia salina with environmental factors in the Sfax solar saltern (Tunisia). Aquatic Living Resources. 2008;**21**(1):63-73

[16] Guermazi W, Ayadi H, Aleya L. Correspondence of the seasonal patterns of the brine shrimp, Artemia salina (LEACH, 1819)(Anostraca) with several environmental factors in an arid solar saltern (Sfax, southern Tunisia). Crustaceana. 2009;**82**(3):327-348

[17] Guermazi W, Elloumi J, Ayadi H, Bouain A, Aleya L. Rearing of Fabrea salina Henneguy (Ciliophora, Heterotrichida) with three unicellular feeds. Comptes Rendus Biologies. 2008;**331**(1):56-63

[18] Gammoudi S, Athmouni K, Nasri A, Diwani N, Grati I, Belhaj D, et al. Optimization, isolation, characterization and hepatoprotective effect of a novel pigment-protein complex (phycocyanin) producing microalga: Phormidium versicolor NCC-466 using response surface methodology. International Journal of Biological Macromolecules. 2019;**137**:647-656

[19] Boukhris S, Athmouni K, Hamza-Mnif I, Siala-Elleuch R, Ayadi H, Nasri M, et al. The potential of a brown microalga cultivated in high salt medium for the production of high-value compounds. BioMed Research Internationa. 2017;**2017**:1-10

[20] Belhaj D, Frikha D, Athmouni K, Jerbi B, Ahmed MB, Bouallagui Z, et al. Box-Behnken design for extraction optimization of crude polysaccharides from Tunisian Phormidium versicolor cyanobacteria (NCC 466): Partial characterization, in vitro antioxidant and antimicrobial activities. International Journal of Biological Macromolecules. 2017;**105**:1501-1510

[21] Fogg GE, Westlake D. The importance of extracellular products of algae in freshwater: With 3 figures and 2 tables in the text. Internationale Vereinigung für theoretische und angewandte Limnologie: Verhandlungen. 1953;**12**(1):219-232

[22] Santini A, Tenore GC, Novellino E. Nutraceuticals: A paradigm of proactive medicine. European Journal of Pharmaceutical Sciences. 2017;**96**:53-61

[23] Harun R, Singh M, Forde GM, Danquah MK. Bioprocess engineering of microalgae to produce a variety of consumer products. Renewable and Sustainable Energy Reviews. 2010;**14**(3):1037-1047

[24] Marfaing H, Lerat Y. Les algues ont-elles une place en nutrition? Phytothérapie. 2007;**5**(1):2-5

[25] Zeid A, Sigamani S. Pharmaceutically valuable bioactive compounds of algae. 2016;**9**(6):43-47

[26] Shariati M, Hadi MR. Microalgal biotechnology and bioenergy in Dunaliella. In: Progress in Molecular and Environmental Bioengineering-from Analysis and Modeling to Technology Applications. IntechOpen; 2011

[27] Li H-B, Cheng K-W, Wong C-C, Fan K-W, Chen F, Jiang Y. Evaluation of antioxidant capacity and total phenolic content of different fractions of selected microalgae. Food Chemistry. 2007;**102**(3):771-776

[28] Suali E, Sarbatly R. Conversion of microalgae to biofuel. Renewable and Sustainable Energy Reviews. 2012;**16**(6):4316-4342

[29] Kobbi-Rebai R, Annabi-Trabelsi N, Khemakhem H, Ayadi H, Aleya L.

Impacts of restoration of an uncontrolled phosphogypsum dumpsite on the seasonal distribution of abiotic variables, phytoplankton, copepods, and ciliates in a man-made solar saltern. Environmental Monitoring and Assessment. 2013;**185**(3):2139-2155

[30] Elloumi J, Guermazi W, Ayadi H, Bouaïn A, Aleya L. Detection of water and sediments pollution of an arid saltern (Sfax, Tunisia) by coupling the distribution of microorganisms with hydrocarbons. Water Air and Soil Pollution. 2008;**187**(1):157-171

[31] Guermazi W, Masmoudi S, Boukhris S, Ayadi H, Morant-Manceau A. Under low irradiation, the light regime modifies growth and metabolite production in various species of microalgae. Journal of Applied Phycology. 2014;**26**(6):2283-2293

[32] Stanier R, Kunisawa R, Mandel M, Cohen-Bazire G. Purification and properties of unicellular blue-green algae (order Chroococcales). Bacteriological Reviews. 1971;**35**(2):171-205

[33] Guillard RR. Culture of phytoplankton for feeding marine invertebrates. In: Culture of Marine Invertebrate Animals. Boston, MA: Springer; 1975. pp. 29-60

[34] Speziale BJ, Schreiner SP, Giammatteo PA, Schindler JE. Comparison of N, N-dimethylformamide, dimethyl sulfoxide, and acetone for extraction of phytoplankton chlorophyll. Canadian Journal of Fisheries and Aquatic Sciences. 1984;**41**(10):1519-1522

[35] Jeffrey ST, Humphrey G. New spectrophotometric equations for determining chlorophylls a, b, c1 and c2 in higher plants, algae and natural phytoplankton. Biochemie und Physiologie der Pflanzen. 1975;**167**(2):191-194

[36] Chamovitz D, Sandmann G, Hirschberg J. Molecular and biochemical characterization of herbicide-resistant mutants of cyanobacteria reveals that phytoene desaturation is a rate-limiting step in carotenoid biosynthesis. Journal of Biological Chemistry. 1993;**268**(23):17348-17353

[37] Salguero A, de la Morena B, Vigara J, Vega JM, Vilchez C, León R. Carotenoids as protective response against oxidative damage in Dunaliella bardawil. Biomolecular Engineering. 2003;**20**(4-6):249-253

[38] Silveira ST, Burkert JDM, Costa JAV, Burkert CAV, Kalil SJ. Optimization of phycocyanin extraction from Spirulina platensis using factorial design. Bioresource Technology. 2007;**98**(8):1629-1634

[39] Horwitz W, Albert R. The Horwitz ratio (HorRat): A useful index of method performance with respect to precision. Journal of AOAC International. 2006;**89**(4):1095-1109

[40] Classics Lowry O, Rosebrough N, Farr A, Randall R. Protein measurement with the Folin phenol reagent. The Journal of Biological Chemistry. 1951;**193**(1):265-275

[41] Dubois M, Gilles KA, Hamilton JK, Rebers PT, Smith F. Colorimetric method for determination of sugars and related substances. Analytical Chemistry. 1956;**28**(3):350-356

[42] Singleton VL, Rossi JA. Colorimetry of total phenolics with phosphomolybdic-phosphotungstic acid reagents. American Journal of Enology and Viticulture. 1965;**16**(3):144-158

[43] Cunniff P, Washington D. Official methods of analysis of aoac international. Journal of AOAC International. 1997;**80**(6):127A

[44] Lepage G, Roy CC. Improved recovery of fatty acid through direct transesterification without prior extraction or purification. Journal of Lipid Research. 1984;**25**(12):1391-1396

[45] Ho SH, Huang SW, Chen CY, Hasunuma T, Kondo A, Chang JS. Characterization and optimization of carbohydrate production from an indigenous microalga *Chlorella vulgaris* FSP-E. Bioresour Technol. 2013;**135**:157-165

[46] Nguyen-Deroche TLN, Caruso A, Le TT, Bui TV, Schoefs B, Tremblin G, et al. Zinc affects differently growth, photosynthesis, antioxidant enzyme activities and phytochelatin synthase expression of four marine diatoms. The Scientific World Journal. 2012;**2012**:1-15

[47] Elloumi J, Carrias JF, Ayadi H, Sime-Ngando T, Bouaïn A, Communities structure of the planktonic halophiles in the solar saltern of Sfax. Tunisia, Estuarine: Coastal and Shelf Science; 2009;**81**(1):19-26

[48] Corrêa PS, Morais Júnior WG, Martins AA, Caetano NS, Mata TM. Microalgae biomolecules: Extraction, separation and purification methods. PRO. 2021;**9**(1):10

[49] Widowati I, Zainuri M, Kusumaningrum HP, Susilowati R, Hardivillier Y, Leignel V, et al. Antioxidant activity of three microalgae Dunaliella salina, Tetraselmis chuii and Isochrysis galbana clone Tahiti. In: IOP Conference Series: Earth and Environmental Science. Vancouver, BC, Canada: IOP Publishing; 2017. p. 012067

[50] Pouvreau J-B, Taran FDR, Rosa P, Pin S, Fleurence Jl, Pondaven P. Antioxidant and free radical scavenging properties of marennine, a blue-green polyphenolic pigment from the diatom Haslea ostrearia (Gaillon/Bory) Simonsen responsible for the natural greening of cultured oysters. Journal of Agricultural and Food Chemistry. 2008;**56**(15):6278-6286

[51] Chtourou H, Dahmen I, Jebali A, Karray F, Hassairi I, Abdelkafi S, et al. Characterization of Amphora sp., a newly isolated diatom wild strain, potentially usable for biodiesel production. Bioprocess and Biosystems Engineering. 2015;**38**(7):1381-1392

[52] Singh NK, Parmar A, Madamwar D. Optimization of medium components for increased production of C-phycocyanin from Phormidium ceylanicum and its purification by single step process. Bioresource Technology. 2009;**100**(4):1663-1669

[53] Khatoon H, Banerjee S, Yusoff F, Shariff M. Evaluation of indigenous marine periphytic Amphora, Navicula and Cymbella grown on substrate as feed supplement in Penaeus monodon postlarval hatchery system. Aquaculture Nutrition. 2009;**15**(2):186-193

[54] Lee S-H, Karawita R, Affan A, Lee J-B, Lee K-W, Lee B-J, et al. Potential of benthic diatoms Achnanthes longipes, Amphora coffeaeformisand Navicula sp.(Bacillariophyceae) as antioxidant sources. Algae. 2009;**24**(1):47-55

[55] Brown MR. The amino-acid and sugar composition of 16 species of microalgae used in mariculture. Journal of Experimental Marine Biology and Ecology. 1991;**145**(1):79-99

[56] Liska AJ, Shevchenko A, Pick U, Katz A. Enhanced photosynthesis and redox energy production contribute to salinity tolerance in Dunaliella as revealed by homology-based proteomics. Plant Physiology. 2004;**136**(1):2806-2817

[57] Boulay C, Abasova L, Six C, Vass I, Kirilovsky D. Occurrence and function of the orange carotenoid protein in photoprotective mechanisms in various cyanobacteria. Biochimica et Biophysica Acta (BBA)-Bioenergetics. 2008;**1777**(10):1344-1354

[58] Gao C, Zhai Y, Ding Y, Wu Q. Application of sweet sorghum for biodiesel production by heterotrophic microalga Chlorella protothecoides. Applied Energy. 2010;**87**(3):756-761

[59] Desbois AP, Walton M, Smith VJ. Differential antibacterial activities of fusiform and oval morphotypes of Phaeodactylum tricornutum (Bacillariophyceae). Journal of the Marine Biological Association of the United Kingdom. 2010;**90**(4):769-774

[60] Iglesia I, Huybrechts I, Gonzalez-Gross M, Mouratidou T, Santabárbara J, Chajès V, et al. Folate and vitamin B12 concentrations are associated with plasma DHA and EPA fatty acids in European adolescents: The healthy lifestyle in Europe by nutrition in adolescence (HELENA) study. British Journal of Nutrition. 2017;**117**(1):124-133

[61] Zheng W, Wang X, Cao W, Yang B, Mu Y, Dong Y, et al. E-configuration structures of EPA and DHA derived from Euphausia superba and their significant inhibitive effects on growth of human cancer cell lines in vitro. Prostaglandins, Leukotrienes and Essential Fatty Acids. 2017;**117**:47-53

[62] Apt KE, Behrens PW. Commercial developments in microalgal biotechnology. Journal of Phycology. 1999;**35**(2):215-226

[63] Ward OP, Singh A. Omega-3/6 fatty acids: Alternative sources of production. Process Biochemistry. 2005;**40**(12):3627-3652

[64] Mozaffarian D, Wu JH. Omega-3 fatty acids and cardiovascular disease: Effects on risk factors, molecular pathways, and clinical events. Journal of the American College of Cardiology. 2011;**58**(20):2047-2067

[65] Bourre J-M. Dietary omega-3 fatty acids for women. Biomedicine & Pharmacotherapy. 2007;**61**(2-3):105-112

Chapter 6

Algae Toxins and Their Treatment

Ahmed Aidan Al-Hussieny

Abstract

Algae are distributed worldwide in the sea, in freshwater and in wet situations on land. Most are microscopic algae, but some of them are so large, also some marine seaweeds that can exceed 50 m in length. The algae have chlorophyll and can make their own food through the steps of photosynthesis. Recently they are classified in the kingdom of protested, which include a variety of unicellular and some basic multinuclear and multicellular eukaryotic organisms that have cells. Algal poisoning is an intense, often lethal condition caused by high concentrations of toxic blue-green algae (more commonly known as cyanobacteria—literally blue-green bacteria) in drinking water as well as in water used for recreation, agriculture and aquaculture. The study cur in the productive dangerous from the algae toxin that productive from cyanobacteria in aquatic environment. The important contamination for water source identification and non-identification and identify on algae that responsible on productive of toxin in water that represented by *Cylindrospermum*, *Aphanizomenon Anabaena*, *Microcystis*, *Lyngbya*, *Oscillatoria*, *phormidium*, and suitable environment for algae to productive toxin. Such as temperature, pH, nutrient, salinity, density identify on the toxin concentration in water that content organisms that productive toxin between (1–100 mg/l). With the use of different methods of treating algal toxins such as (potassium permanganate, activated carbon, oxidation, chlorine and ozone), and the best treatment was the use of potassium permanganate at a concentration (2 mg/l), which gave the best treatment while preserving the ecosystem.

Keywords: microalgae, toxin, treatment, harmful algal bloom, cyanobacterial toxins, potassium permanganate

1. Introduction

The enrichment of lakes and reservoirs with nutrients leads to an increase in the growth of algae, especially cyanobacteria, forming floating masses on the surface, causing a decrease in the concentration of dissolved oxygen and death in fish, and the death of livestock and other animals as a result of ingesting algae toxins. Filamentous cyanobacteria and green algae (Chlorophytes) cause clogging in filters of water treatment systems or problems in industrial systems when such water is used. The dinoflagellates are another group of phytoplankton that can secrete toxic substances. One of the by-products of algal blooms are high concentrations of organic carbon [1]. Increased phosphorous concentration and low P:N ratio are major factors for such a condition, and several studies indicate that toxins from cyanobacteria pose a health risk. According to the World Health Organization (WHO), the maximum acceptable concentration of the toxic substance (Microcystin-LR) in tank water that may be used for drinking is

IntechOpen

(0.5–1.0 μg/cubic decimeter), as exposure to an increase of this substance causes liver cancer. Human exposure to this type of poison is possible because it is difficult to carry out a complete treatment of cyanobacterial toxins in drinking water plants. Cyanobacteria also cause the death of animals when they ingest these toxins and also lead to a lack of oxygen and the death of fish [2]. Algae are distributed worldwide in the sea, in freshwater and in wet situations on land. Most are microscopic algae, but some of them are so large, also some marine seaweeds that can exceed 50 m in length. The algae have chlorophyll and can make their own food through the steps of photosynthesis. Recently they are classified in the kingdom of protiste, which include a variety of unicellular and some basic multinuclear and multicellular eukaryotic organisms that have cells. Algal poisoning is an intense, often lethal condition caused by high concentrations of toxic blue-green algae (more commonly known as cyanobacteria—literally blue-green bacteria) in drinking water as well as in water used for recreation, agriculture and aquaculture. Severe illness of livestock and Fatalities, birds, pets, fish and wildlife from high growths of cyanobacteria water blooms occur almost in all of the countries in the world. Severe deadly poisonings have also been notarized in people. Poisoning usually comes during warm seasons when the water blossom are more acute and of longer duration. Almost poisonings come among animals drinking cyanobacteria infested freshwater, but aquatic animals, mostly mariculture fish and prawn, are also affected. The toxins of cyanobacteria comprise six special chemical classes collectively called cyanotoxins [3]. Toxic algae, micro-algal blooms, phytoplankton blooms, red tides, or harmful algae, are all terms for normally occurring phenomena. Around 300 species of micro algae are notify at times to form mass appearance, so called blooms. About one fourth of these species are recognized to produce toxins. The scientific society points out to these events with a generic term, 'Harmful Algal Bloom' (HAB), understanding that, because a wide range of organisms are implicated and some species have toxic impacts at low cell intensity, not all HABs are 'algal' and not all occur as 'blooms' [4]. Many of the organisms in charge for red tides are closely distributed and, in recent years, the organisms appear to be markedly spreading. Natural events such as hurricanes can spread over organisms, and it is doubtful that some organisms may be moved long distances in ship ballast waters. Another factor that may motivate algal proliferation in both freshwater and marine systems is augmentation nutrient loading. Certain algae occur more usually in some areas than others and it is useful to know which ones are problems in particular locations. Good sources of information about algal blooms are the State public health department or the State division of marine resources or marine fisheries [5].

2. Pollutant removal

2.1 Sources of water pollution

Water, especially surface water, is exposed to the dangers of pollution, as the water source is considered polluted when it directly or indirectly changes its composition or condition as a result of human action, that is, if it becomes less suitable for some or all uses. These include sewage, household waste, hospital wastewater, and rainwater, where these pollutants are loaded with large quantities of various organic and inorganic materials and many types of microorganisms that cause many diseases, as well as urination, defecation, and throwing dead animals into the water, especially in rural areas, as well as picnic places represented by excreta. and food waste, where the type of sewage network systems plays a major role in the aggravation of these pollutants, as there are two types of sewage network systems in Baghdad, which are the separate system and the combined system. Also, the term (red tide) is considered to refer to types of Plankton (phytoplankton) that are spread in high density in any water body (may reach more than ten million cells per liter) and are known as harmful algae. Of the (4400 species) of Plankton plant there are only from (50–60 poisonous species). One of the most important spread of the phenomenon of red tide is a defect in environmental factors, including the difference in the ratio of phosphorous to nitrogen through sewage pollution. The higher the phosphorous rate in the water mass, the higher the rate of red tide or harmful algae appearing with the availability of other environmental conditions. The effects of the red tide are the destruction of fish from economic fish farms if this phenomenon spreads in the farm and costs thousands or even millions of dollars annually. The eyes, nose and mouth are irritated during the presence of algae toxins in the tourist beaches. The toxins increase the toxicity rate of marine food, which is transmitted to humans through marine meals (**Figure 1**).

2.2 Definite sources pollution

Industrial and domestic flows result in high concentrations of pollutants that find their way into natural waters, and they are among the primary factors that lead to the deterioration of water quality. As it is known, one of the reasons for the occurrence of nutritional enrichment is the influx of these pollutants containing high concentrations of nutrients estimated at more than (four times) than what is present in natural waters. Therefore, reducing or limiting these sources of pollution is the first successful and important step for water quality management. It is easy to control the known sources of pollution using treatment methods, the most important and most effective of which is the establishment of wet areas (Wetlands) (**Figure 2**).

2.3 Non-definite sources pollution

Sources that cause problems that are difficult to control and unknown sources of pollution prepare more nutrients resulting from the modification that humans make

Figure 1.
Fish damage through the spread of toxin-producing algae.

Figure 2.
Nutritional enrichment in the waters of the Iraqi marshes.

in nature such as deforestation, agriculture, industrial and urban development. These sources supply fresh water with low concentrations of nutrients, but it is difficult to control because they are transferred to water bodies from the vast surrounding lands. The preparation of these sources increases in the heavy rainy season, which leads to erosion in the surface of the soil and then filtering of these nutrients from the soil to the aquatic organisms. Man directly and indirectly affects the occurrence of the phenomenon of food enrichment from several aspects, in addition to the industrial, household and agricultural waste it raises. There are many sources of pollutants affecting the Tigris River, including human pollutants, industrial pollutants, agricultural pollutants, and pollutants resulting from groundwater, and thus affected the physical, chemical and biological characteristics, which can be broadly classified into:

A. **Pollutants affecting the physical properties**: Some toxic pollutants lead to a change in the color, turbidity, taste or temperature of the water.

B. **Pollutants affecting chemical properties**: They represent organic and inorganic substances that make water toxic and dangerous and affect public health and aquatic life through their impact on the pH value or the water content of salts and minerals.

C. **Pollutants affecting biological properties:** Pollutants that affect aquatic organisms in general, including microorganisms that have an impact on human health (bacteria and viruses). These pollutants also play a major role in the phenomenon of eutrophication of some algae and aquatic plants.

2.4 Cyanobacterial toxins

Cyanobacteria create a variety of toxic secondary metabolites that are detrimental to a variety of other organisms. Scientists feel that these pollutants are currently posing a serious threat to society's health in numerous parts of the world. The toxins produced by cyanobacteria and the species that produce them are listed in **Table 1**. Toxins produced by cyanobacteria are divided into two categories: cytotoxins and biotoxins [7]. The varieties of algal toxins produced by various blue-green algae, as well as their impacts, are listed in **Table 1**.

2.4.1 Cytotoxins

This type of toxin is produced by some marine cyanobacteria species, and while it has no harmful effects on animals, it is poisonous to cells generated in cell cultures and inhibits the growth of a wide spectrum of microorganisms. Toxins such as these are examples (Tolytoxin, Tubericidin, Scytophycins and Actiphycins). Indolocarbazoles, tautazoles, microbilinisonitriles, and paracyclophares are more examples).

2.4.2 Biotoxins

These compounds are made by cyanobacteria, which can have a variety of negative health consequences on humans and animals, and are frequently lethal. Toxins are categorized into three categories: neurotoxins, hepatotoxins, and endotoxins [7].

2.4.3 Neurotoxins

Alkaloids are poisonous in a short amount of time and are usually lethal, as they cause paralysis of the surrounding skeletal muscles, followed by paralysis of the respiratory muscles, resulting in incapacity to breathe and death. Several forms of these toxins have been identified (e.g., *Oscillatoria* and *Trichodesmium*).

i. **Anatoxin**: It is produced by the species (Anabaena flos-aquae) with a molecular weight of (765 Da), and it is also produced by species of the genus (*Oscillatoria*) as well.

Toxin types	Cyanobacteria genera	Effect of toxins
Hepatoxins Microcystin Nodularin	*Anabaena, Anabaenopsis, Aphanocapsa, Hapalosiphon, Microcystis, Nostoc, Oscillatoria, Nodularia*	These toxins directly affect the zooplankton community, especially those that prefer cyanobacteria species as an important food source for them
Neurotoxins Anatoxin-a Homoanatoxin-a Anatoxin-a (s) Saxitoxin	*Anabaena, Aphanizomenon, Cylindrospermum, Microcystis, Oscillatoria, Phormidium. Anabaena, Oscillatoria, Anabaena, Aphanizomenon, Cylindrospermopsis Lyngbya*	Infection with these types of toxins leads to the inability to breathe and then death through paralysis of the respiratory muscles
Cytotoxin Cylindrospermopsin	*Aphanizomenon, Cylindrospermopsis, Umezakia*	Its toxic effect on the liver, but it was found that it affects the kidneys as well, causing the destruction of the tissues that it attacks
Lipopolysaccharides (LPS)	Many species of Cyanobacteria	These substances have a toxic effect that causes ill health in humans, and it was also found that they are fatal to mice when injected into the peritoneal membranes

Table 1.
Cyanotoxins with public health significance from acute exposures [6].

ii. **Homoantoxin**: It is also produced by the type (*Oscillatoria rubescence*), and it is less toxic than the first type [8].

iii. **Anatoxin-a(s)**: It is produced by species of the genus (*Anabaena*), which is about ten times more toxic than the first and differs from it in its chemical composition and has a molecular weight of 252 Da.

iv. **Paralytic Shellfish Poisons (PSPs)**: The complicated set of 18 toxins that paralyze crustaceans is grouped into three classes (saxitoxin, gongyautoxins, and C-toxins), and these toxins are usually produced by species (*Aphanizomenon flos-aquae* and *Anabaena circinalis*). These toxins are thought to have a quick neurological effect by interrupting nerve communication by closing sodium channels, but they have no effect on potassium permeability.

2.5 Hepatotoxins

Toxins of this sort cause a variety of types and strains that belong to the genera (*Microcystis*, *Anabaena*, *Nodularia*, *Cylindrospermopsis*, *Oscillatoria* and *Nostoc*). These toxins are the most frequent among cyanobacterial toxins, and they have a strong toxic impact. However, they take longer to kill than neurotoxins, and death can take anywhere from 5 min to a few days, depending on numerous factors such as the animal's weight, the type of poison, and the dose. These toxins are divided into three categories [9].

2.5.1 Microcystins

They're monocyclic seven-chain peptides with an unique amino acid named Adda connecting the side chains. Because the peptide ring comprises five amino acids that are used in the synthesis of all forms of microcystins produced by the species (*Microcystis aeruginosa*), it was named after it. Other species from the genera (*Oscillatoria*, *Nodularia*, *Anabaena*, *Nostoc*) and others are used to make it. So far, more than a million microcystins have been identified (60 species). Microcystins MC-LR, MC-RR, and MC-YR are the most frequent and poisonous kinds of microcystins. Microcystins have a molecular weight of 909–1044 Da, depending on the species. Microcystins are known for their long-term resilience to high temperatures, however it has been discovered that they can tolerate boiling without denaturation. It withstands pH changes and dissolves easily in water, ethanol, methanol, and acetone, and cells require energy to consume the poison [10].

2.5.2 Nodularin

It's a pentacyclic monocyclic peptide that looks a lot like MC–LR but is smaller. The peptide ring has a molecular weight of 824 Da and comprises amino acids similar to those found in MC–LR. Only one variety has been identified as being produced by the species (*Nodularia spumigena*), and this proliferation has a poisonous impact identical to that of MC-LR.

2.5.3 Cylindrospermopsin

It is one of the toxins produced by the type (*Cylindrospermopsis mceberskii*), and it is the only alkaloid compound among the hepatotoxicants, as it shares this characteristic with neurotoxins, and its toxic effect is not limited to the liver, as it has been discovered that it also affects the kidneys, causing tissue destruction [11].

2.6 Endotoxins

It refers to the lipo polysaccharide (LPS) that forms the cell wall of pan cyanobacteria, and it has been discovered that these chemicals are hazardous to humans. When injected into the peritoneal membranes at a dose of (1–1.2 mg/kg after 48 h), it was likewise proven to be deadly to rats [12].

3. Toxic effect of microcystins

Many researches have confirmed that the compounds produced by various varieties of cyanobacteria are harmful to many animals and humans, as they have been discovered to cause the death of many creatures in many parts of the world, including cattle, horses, dogs, birds, fish, and crocodiles. These toxins have a direct impact on society Zooplankton, particularly those that prefer cyanobacteria species as a major food source, such as the genus Daphnia, where it was discovered that low concentrations of these toxins reduce the ability of these organisms to reproduce new generations, as well as their members' growth rates. Toxic levels above a certain threshold cause death [13]. The use of water contaminated with microcystins in the dialysis process in a hospital's hemodialysis unit, which resulted in damage Acute hepatitis, confirms the seriousness of these toxins, as the worst accident recorded so far occurred in 1996, with the victim (60 patients) due to the use of water contaminated with microcystins in the dialysis process. Due to the usage of recycled water for drinking, Western culture is one of the societies most exposed to these poisons, as proven by two occurrences in Australia. Due to pollution of drinking water with these chemicals, the first victim (139 children) and a number of adults became infected, resulting in severe liver, hematuria, renal failure, and death. The high concentration of toxin in the drinking water was caused by the chemical treatment of water containing blooming algae (*Microcystis aeruginosa*) with copper sulfate $CuSO_4$, which is commonly employed to kill algae cells. High body temperature, skin rash, enteritis, general weakness, lack of appetite, pallor of mucous membranes, vomiting, diarrhea, liver poisoning, and death within hours or days, depending on the amount of dose given and the weight of the animal, are common signs of microcystin poisoning. The most common cause of death is an abrupt bleeding within the liver. Microcystins, according to the majority of researchers in this subject, have a stimulating influence on the growth of malignant tumors when consumed low concentrations and for long periods [14].

4. Hazards resulting from algal toxins

Environmental parameters such as temperature, illumination, pH, salinity, macro- and micronutrients are among the most important regulators of toxins production from blue-green algae, according to environmental studies in continuous cultures of algae. Microcytins and Anatoxin-a are maintained to a significant extent inside the cell. Because the degree of (Microcytins) harm to the environment and neighboring creatures increases as the logarithmic length of growth increases, the presence of (Microcytins) cannot be overlooked. According to certain research, the quantities of toxins in water containing creatures that emit toxins range from (1–100 g/l) and can be higher, and as a result, (microcytins) are among the contents that pose a health risk and have an effect if the water has been ingested from change the detoxification of cyanobacteria and their cells [15].

5. Algae producing toxins

Algae producing toxins:

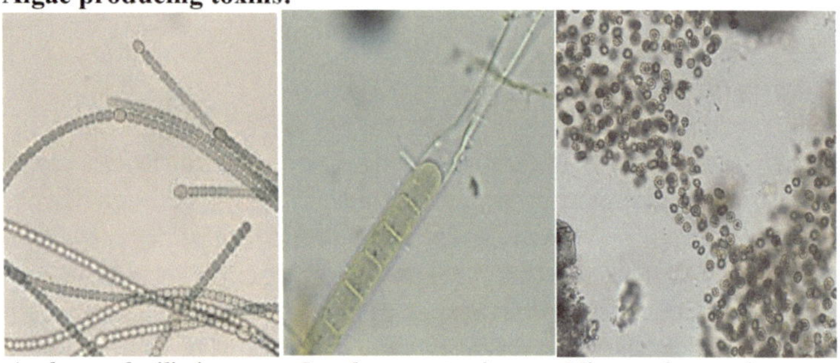

Anabaena fertilissima. *Lyngbya martensiana* *Microcystis aeruginosa*

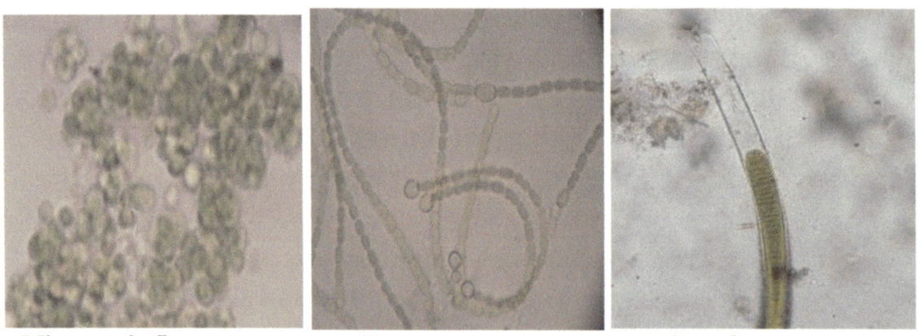

Microcystis flos-aquae *Nostoc carneum* *Lyngbya connectens*

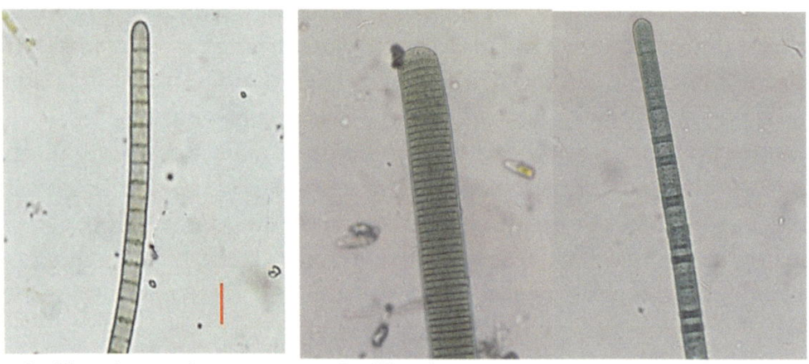

Oscillatoria curviceps *Oscillatoria sancta* *Oscillatoria tenuis*

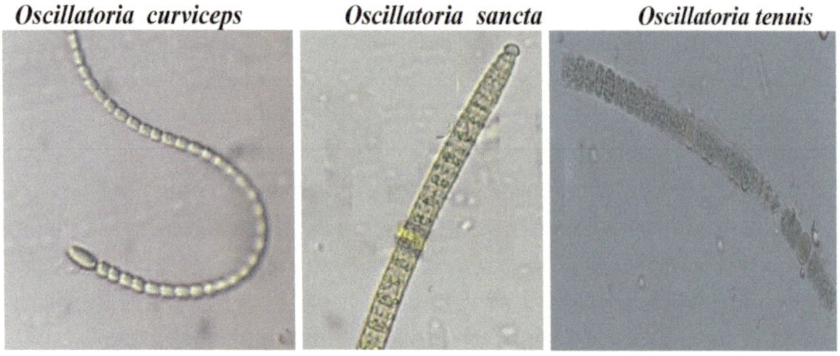

Anabaena oscillorides *Phormidium favosum* *Nostoc punctiforme*

6. Methods for treating algal toxins

6.1 Treatment of algal toxins

6.1.1 potassium permanganate

After reducing the quantity of algae that produces food enrichment (excessive growth), a potassium permanganate mixture with a concentration of 30 mg/l textured alum as coagulate was used. The concentrations of chlorophyll algal toxins per transaction were examined and compared to the standard at the end of the 72-h experiment mediated by the GC/MS device, in addition to reducing the productivity of the initial algae by reducing the concentrations of chlorophyll algal toxins per transaction. The results revealed the presence of algal toxins belonging to the groups Neurotoxins, Hepatotoxins, Pyriproxyfen, Emodin, Brevetoxins-10 (A), and Cytotoxins in the standard treatment, with a note detoxification algal 100% concentrations of 8 and 16 mg/l, respectively, textured potassium permanganate in comparison to the standard, which contained a lot of chemical compounds to algal toxins (**Figure 3**).

A result of the cessation of photovoltaic installation process stops the outer wall systems (systems enzymatic) to withdraw nutrients that enter into the composition of the algal toxin combination the non-arrival of light to stop light receptors [16], and the concentrations of 2 and 4 mg/l for the same article have some toxic compounds converted into non-toxic compounds and **Figures 4–8** describe them. Were treated toxins algal belonging to the group Neurotoxins a Besnfein Anatoxin-a, Homoanatoxin-a and the various toxins which is alkaline compounds Alkaloids with effect very quickly called and can be fatal in most cases where the cause muscle surrounding paralysis followed by a respiratory muscle paralysis, which leads to an inability to breathe then death.

These toxins, as seen in Hepatotoxins group Class Microcystin-LA, are the most common among toxins Cyanobacteria and have a severe impact, but death takes longer than nerve toxins, ranging from 5 min to a few days depending on several factors such as the weight of the animal, the type of poison, and the dosage, among others. Other forms of toxins, such as Pyriproxyfen, Emodin, and Brevetoxins-10, were shown to be effective in removing algal toxins that had emerged within the

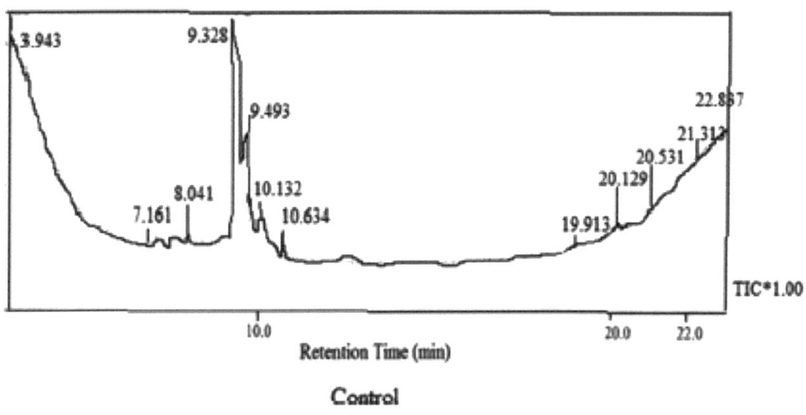

Figure 3.
GC-MASS of algal toxins within the treated standard.

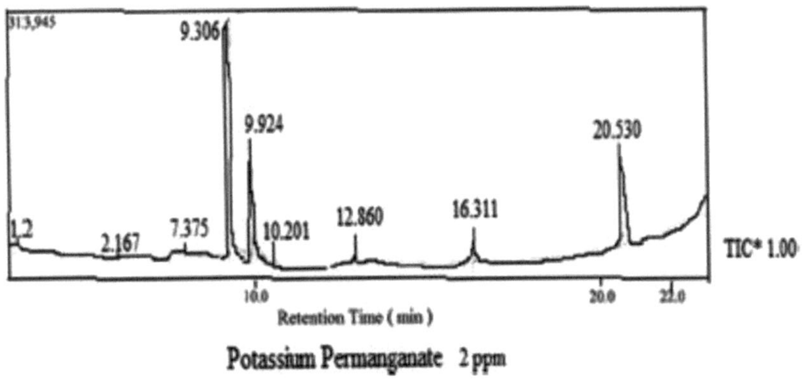

Figure 4.
GC-MASS of algal toxins within concentration of 2 mg/l.

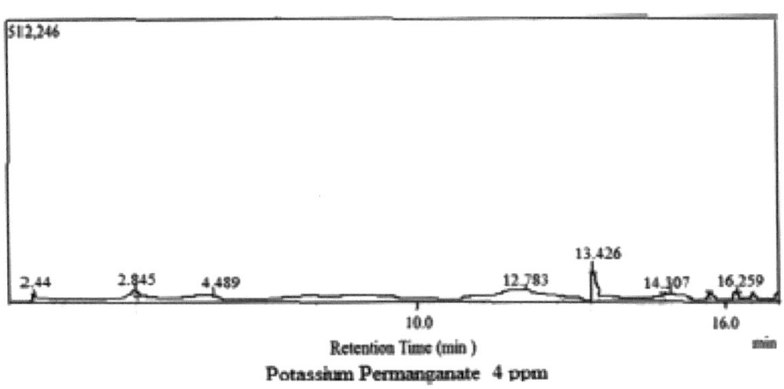

Figure 5.
GC-MASS of algal toxins within concentration of 4 mg/l.

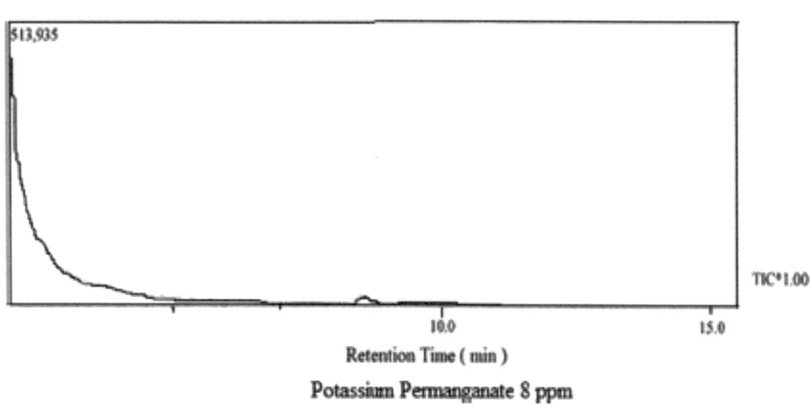

Figure 6.
GC-MASS of algal toxins within concentration of 8 mg/l.

DOI: http://dx.doi.org/10.5772/intechopen.102909

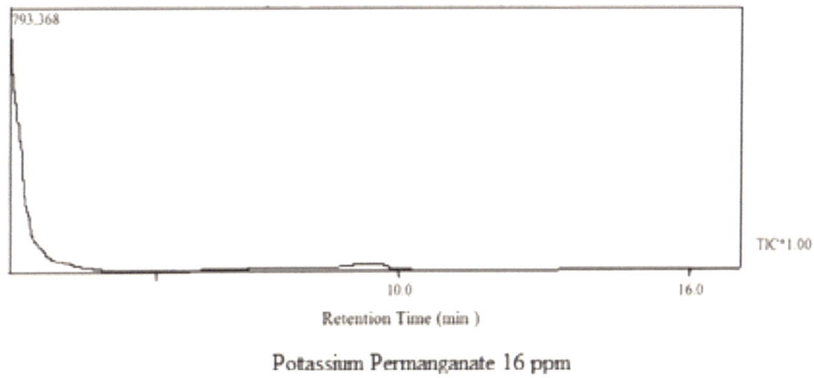

Figure 7.
GC-MASS of algal toxins within concentration of 16 mg/l.

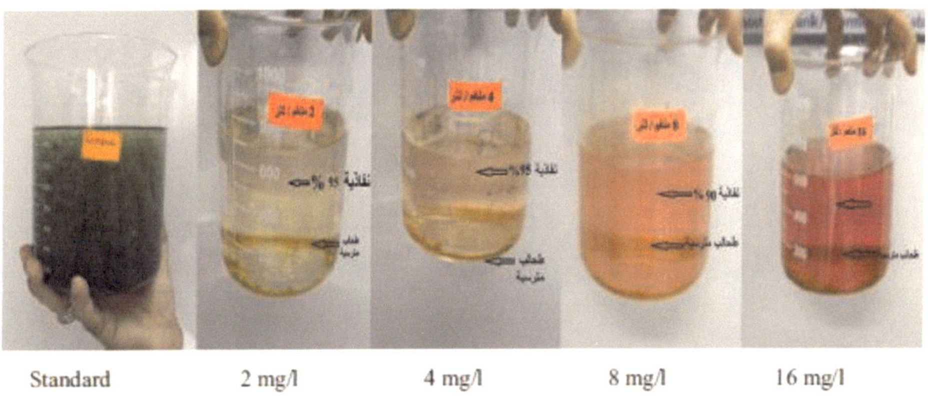

Figure 8.
Concentrations of potassium permanganate and its impact on algae society compared to standard.

conventional treatment of 8 and 16 mg/l potassium permanganate, as shown in **Table 2**.

The same material with concentrations of 2 and 4 mg/l has been converted into non-toxic and hazardous chemicals, as indicated in **Table 2**. The current study also suggests that potassium permanganate at concentrations of 8 and 16 mg/l, respectively, could be used to treat algal cells by stopping photosynthesis and disabling all vital events without tearing the outer wall of the moss, and then deposition blocs blooms to the pelvic floor and a rise in turbidity levels in the water column, as opposed to the standard, which shows a rise in biomass value and low turbidity **Figure 8**. Furthermore, all potassium permanganate concentrations of 2, 4, 8, and 16 mg/l had no effect on the algae's outer wall. This is because to the potassium permanganate mixture's precise concentration of alum, which caused the algae's exterior wall to not tear, preserving the outer toxic blooming. It inhibits photosynthesis, resulting in a decrease in the primary productivity of chlorophyll-producing algae, solution Alum is necessary for the production of potassium permanganate and aids in the sintering, coagulation, and sedimentation processes [17]. Furthermore, potassium permanganate affects some algal toxins but not others, as it affects the toxins anatoxin-a, cylindrospermopsin, and microcystin and analyses have valued the final removal,

Algal toxins	Synthetic version of the toxin	Standard	Concentrations of potassium permanganate (mg/l)			
			2	4	8	16
1-Neurotoxins Anatoxin-a	$C_2H_3C_{12}NO$	*	-	-	-	-
	$C_9H_{13}NO_2$	*	-	-	-	-
	$C_{18}H_{27}NO_3$	*	-	-	-	-
	$C_{11}H_{12}N_2O_6$	*	-	-	-	-
	$C_{11}H_{17}N_3O$	*	-	-	-	-
	$C_{10}H_{17}N_3O$	*	-	-	-	-
	$C_9H_{15}Br_2NO$	*	-	-	-	-
Homoanatoxin-a	$CH_4N_2O_2$	*	-	-	-	-
	$C_{11}H_{17}NO_2$	*	-	-	-	-
	$C_{13}H_9BrN_2O_3$	*	-	-	-	-
2-Hepatotoxins Microcystin- LA	$C_3H_7NO_4S$	*	-	-	-	-
3-Pyriproxyfen	$C_{20}H_{29}NO_3$	*	-	-	-	-
4-Emodin	$C_{15}H_{10}O_5$	*	-	-	-	-
5-Brevetoxins-10(A)	$C_4H_8O_2$	*	-	-	-	-
6- Cytotoxins	$C_2H_2Cl_3NO$	*	-	-	-	-

-, absence of toxins.
**The presence of toxins.*

Table 2.
Treatment of algal toxins of different concentration of potassium permanganate with chemical formulations statement mediated by GC-MASS.

while it does not affect saxitoxin toxin, despite the fact that it is produced by algae greens blue—this is what the organization confirms [18].

6.1.2 Activated carbon

The use of activated carbon for the removal of a wide range of organic compounds, including numerous algal toxins, is a well-accepted treatment approach. Powdered activated carbon (PAC) and granular activated carbon (GAC) can both be used as a physical process to adsorb toxins in source water, while GAC can also be used as a biological process to degrade toxins by allowing bacteria to grow on GAC media (instead of sand or anthracite) in rapid gravity filters. The nature of the source water, particularly the sort of toxins and competing natural organic matter (NOM) constituents present, has a significant impact on the efficacy of the adsorption or biological process. Operators should consider improving reactivation and replacement frequency based on the seasonal prevalence of blue-green algae if GAC is currently being used. Because algal toxin occurrences occur on a regular or seasonal basis, the use of PAC can be beneficial because it may be introduced sporadically to the traditional treatment procedure to react to the presence of toxins in a relatively cost-effective manner.

PAC can be added before coagulation and removed in the settling tanks, or it can be added after coagulation and removed by filtering. When employing PAC, keep in mind that it must be removed by a downstream operation and discarded, as PAC is rarely reused or regenerated. When utilizing PAC, detention times must be considered to ensure that enough time is provided for adequate adsorption removal. Prior to coagulation, PAC basins are occasionally employed, but care must be taken to ensure that the PAC adsorption rate correctly accounts for any NOM compound competition for adsorption sites. For details on selection and dosing, contact your PAC supplier.

6.1.3 Oxidation

Chlorination (gaseous elemental chlorine, liquid sodium hypochlorite, or calcium hypochlorite), chloramines, chlorine dioxide, potassium permanganate, and ozone are all examples of oxidation in this section. UV with hydrogen peroxide is also demonstrated. Because most oxidants will lyse the blue-green algae cells present and release their toxins, peroxidation (the administration of an oxidant at any point in the treatment process prior to filtering) is not suggested. To keep the cell structure intact and the toxins contained, blue-green algae cells should be removed during the coagulation process before adding an oxidant if at all possible (i.e., intracellular). Water systems should consider using a weaker oxidant such as potassium permanganate if pre-oxidation is required for acceptable turbidity and/or organic carbon removal.

6.1.4 Chlorine

Chlorine reacts with microcystins, cylindrospermopsin, and saxitoxins to a lesser amount. Anatoxin-a does not seem to react well with chlorine. In addition, saxitoxin inactivation works best at higher pH levels, whereas microcystin inactivation works best at lower pH levels. The pH of the water and the presence of NOM affect the reactivity of chlorine with contaminants. Depending on the water quality circumstances, the contact time (CT) values necessary for the elimination of microcystins with free chlorine may be many times higher than those required for the surface water treatment rule. Chloramine and chlorine dioxide in commonly used levels have not been shown to be effective against any of the four poisons. Chloramines are effective against microcystins at very high doses and over long periods.

6.1.5 Ozone

Microcystins, anatoxin-a, and cylindrospermopsin react more quickly with ozone than with other common oxidants. Saxitoxin is the one that is least affected by ozone. Only 20% of the saxitoxins present would be destroyed under equivalent settings when microcystins would be fully removed. Although hydrogen peroxide alone is ineffective in removing pollutants, ozone combined with hydrogen peroxide is significantly more powerful (**Table 3**).

	Anatoxin-a	Cylindrospermopsin	Microcystin	Saxitoxin
Chlorine	Not effective	Effective	Effective	Somewhat effective
Chloramine	Not effective	Not effective	Not effective	Inadequate information
Chlorine dioxide	Not effective	Not effective	Not effective	Inadequate information
Potassium permanganate	effective	Date ranges from not effective to possibly effective	Effective	Not effective
Ozone	Effective	Effective	Effective	Not effective
UV/advanced oxidation	Effective	Effective	Not effective	Inadequate information

Table 3.
General effectiveness of blue-green algal toxin inactivation with specific oxidants [19].

Author details

Ahmed Aidan Al-Hussieny
Ministry of Science and Technology, Baghdad, Iraq

*Address all correspondence to: ahmed.edan85@gmail.com

References

[1] Tarczyn'ska M, Romanowska-Duda Z, Jurczak T, Zalewski M. Toxic cyanobacterial blooms in a drinking water reservoircauses, consequences and management strategy. Water Science and Technology: Water Supply. 2001;**1** (2):237-246

[2] Reynolds CS. Ecology and control of cyanobacteria (Blue – green algae). In: Codd GA, Robert C, editors. Public Health Aspects of Cyanobacteria (Blue – green algae), Proceedings of a Seminar in London, 14 June 1991. PHLS Microbiol. Digest. 1991;**8**(3):87-90

[3] Carmichael WW. Overview of Algae Poisoning. Merck Manuals. 2013. Available from: mvm/toxicology/algal_poisoning/overview_of_algal_poisoning.html

[4] UNESCO, IOC, Intergovernmental Oceanographic Commission of UNESCO. Harmful Algal Bloom Programme. Oostende, Belgium: Project Office for IODE; 2015

[5] Creekmore LH. Field Manual of Wildlife Diseases: General Field Procedures and Diseases of Birds. USGS Science for a Changing World. Funded by the U.S. Fish and Wildlife Service, Division of Federal Aid, Administrative Grant No. AP95-017

[6] Stewart I, Webb PM, Schluter PJ, Shaw GR. Recreational and occupational field exposure to freshwater cyanobacteria—A review of Anedotal and case reports, epidemiologic assessment. Environment Health: A Global Access Science Source. 2006;**5**:6

[7] Rapala J, Lahti K, Sivonen K, Niemela SI. Biodegradability and adsorption on lake sediments of cyanobacterial hepatotoxins and anatoxin-a. Letters in Applied Microbiology. 1994;**19**:423-428

[8] Rapala J, Sivonen K. Assessment of environmental conditions that favour hepatotoxic and neurotoxic *Anabaena* spp. strains in cultured under light-limitation at different temperatures. Microbial Ecology. 1998;**36**:181-192

[9] Carmichael WW. Cyanobacteria secondary metabolites the cyanotoxins. Journal of Applied Bacteriology. 1992;**72**: 445-459

[10] Watanabe MF, Watanabe M, Kato T, Harada K, Suzuki M. Composition of cyclic peptide toxins among strains of *Microcystis aeruginosa* (Blue – green algae, Cyanobacteria). Botanical Magazine (Tokyo). 1991;**104**:49-57

[11] Mazur H, Plinski M. Stability of cyanotoxins, microcystin – LR, microcystin – RR and nodularin in Sea water and BG – medium of different salinity. Oceanologia. 2001;**43**(3):329-339

[12] Codd GA, Bell SG, Brooks WP. Cyanobacterial toxin in water. Water Science Technology. 1989;**21**(3):1-13

[13] Hietala J. Life history responses of *Daphnia* to toxic Cyanobacteria [Ph.D. thesis]. Finland: University of Turku; 1996

[14] Pouria S, Audrade A, Barosa J, Cavalcanti RL, Barreto VTS, Preiser W, et al. Fatal microcystin in toxication in haemodialysis unit in Caruau, Brazil. The Lancet. 1998;**352**:21-26

[15] Humpage AR, Flaconer I. Microcystin – LR and liver tumor promotion: Effect on cytokinesis, ploidy, and apoptosis in cultured hepatocytes. Environment Toxicology. 1999;**14**:61-75

[16] Ohio, Environmental Protection Agency. White Paper on Algal Toxin Treatment, American Water Works Association Ohio Section Technology Committee. 2011

[17] Al-Hussieny AA, Sameer E, Fahim R, Zainab Hani J. Reduction in chlorophyll productivity in different algae species by potassium permanganate and alum concentration matters. Ministry of Science and Technology. 2014;**5**(2):62-71

[18] Section Technology Committee, Draft White Paper on Cyano Toxin Treatment, Environmental Protection Agency, American Water Works Association Ohio Section Technology Committee. 2015. pp. 1-15

[19] Ohio Section Technology Committee. White Paper on Algal Toxin Treatment, Environmental Protection Agency. 2011. pp. 1-6

Chapter 7

Potential of Native Microalgae from the Peruvian Amazon on the Removal of Pollutants

Marianela Cobos, Segundo L. Estela, Carlos G. Castro, Miguel A. Grandez, Alvaro B. Tresierra, Corayma L. Cabezudo, Santiago Galindo, Sheyla L. Pérez, Angélica V. Rios, Jhon A. Vargas, Roger Ruiz, Pedro M. Adrianzén, Jorge L. Marapara and Juan C. Castro

Abstract

Environmental pollution is a severe and common problem in all the countries worldwide. Various physicochemical technologies and organisms (e.g., plants, microorganisms, etc.) are used to address these environmental issues, but low-cost, practical, efficient, and effective approaches have not been available yet. Microalgae offer an attractive, novel, and little-explored bioremediation alternative because these photosynthetic organisms can eliminate pathogenic microorganisms and remove heavy metals and toxic organic compounds through processes still under study. Our research team has conducted some experiments to determine the bioremediation potential of native microalgae on some pollutant sources (i.e., leachate and wastewater) and its ability to remove hazardous chemical compounds. Therefore, in this chapter, we provide the results of our research and updated information about this exciting topic. Experiments were conducted under controlled culture conditions using several native microalgae species, variable time periods, different pollutant sources, and hazardous chemicals such as ethidium bromide. The results indicated that native microalgae can remove pollutants (i.e., phosphorus, ammonia, etc.) of wastewater, leachate, and some hazardous chemical compounds such as ethidium bromide. In conclusion, native microalgae have an excellent potential for removing several pollutants and, consequently, could be used to develop bioremediation technologies based on native microalgae from the Peruvian Amazon.

Keywords: bioremediation, native microalgae, leachate, pollutants, wastewater

1. Introduction

Microalgae have aroused the scientific community's interest by their biotechnological potential and increased commercial demand because these microorganisms

 IntechOpen

are an excellent source of a wide range of chemicals with biomedical interest (e.g., carotenoids, essential fatty acids, polyphenols, polysaccharides, etc.) [1–3]. In addition, they are helpful for bioremediation applications in wastewater treatment and other decontamination applications [4, 5]. Some advantages of this biological system are that bioremediation reinforces biogeochemical processes, toxic chemicals are degraded and not simply physically separated from the environment, and the process requires less energy than other technologies and uses less manual supervision. Furthermore, the bioaccumulation of heavy metals by microalgae cells may represent a feasible method for the treatment of leachates and wastewater containing bioavailable heavy metals [6–10].

Additionally, microalgae could be cultivated in wastewater lagoons with small nutrient requirements for their maintenance and development. This component usually constitutes the final step to completing the decontamination process in many wastewater treatment systems [11–13]. Therefore, massive cultivation of microalgae using wastewater as a source of nutrients is a cost-effective approach due to the simplicity of the technology allowing both pollutants (i.e., biological and chemical) remotion and the obtention of a valuable microalgae biomass rich in proteins, lipids, pigments, bioactive chemicals, etc. [14–17].

In this context, this chapter aims to provide updated information based on the results of investigations conducted by our research team using some strains from the freshwater microalgae collection culture native from the Peruvian Amazon.

2. Use of native microalgae for pollutants removal

2.1 Leachate treatment from an open-air garbage dump

Solid waste production in Iquitos city and other cities worldwide has been increased in direct relation to the demographic explosion. Commonly, cities such as Iquitos and other main cities of the Peruvian Amazon have an inefficient garbage collection system, and their main streets and popular markets are often full of garbage (**Figure 1**). In addition, these cities do not have a proper garbage disposal approach, and landfill sites are missing; consequently, the solid wastes are directly deposited in open-air garbage dumps (**Figure 1**).

In these open-air garbage dumps, the solid wastes can be dispersed and degraded by abiotic and biotic factors, producing a gamma of solid, gaseous, and liquid products; the latter is known as leachate. This wastewater flows out from a landfill or an open-air garbage dump sites due to precipitation, ground-water intrusion, moisture content of waste, and rate of evaporation [18]. The volume and pollutant composition of this leachate wastewater fluctuate over time; therefore, in the early acid phase, there exists a high concentration of the four groups of pollutants (dissolved organic matter, heavy metals, inorganic macrocomponents, and xenobiotic organic compounds); finally, in the long methanogenic phase, the leachate liquid has a lower concentration of the four groups of pollutants and is characterized by its very low concentration of heavy metals and biochemical oxygen demand/chemical oxygen demand (BOD/COD) ratio [19–21]. In addition, leachate liquid has a great diversity and composition of bacterial and archaeal populations of the members *Alphaproteobacteria*, *Betaproteobacteria*, *Gammaproteobacteria*, and *Epsilonproteobacteria*, among others [22–24]. For these reasons, leachate liquid should be appropriately disposed and treated to keep away ecotoxicological and environmental damage [25].

Figure 1.
Solid waste accumulation in the main streets and popular markets of Iquitos city (A, B, and C) and its final disposal in an open-air garbage dump (D).

According to these necessities, our research team evaluated the potential use of native microalgae of the Peruvian Amazon for leachate treatment generated in an open-air garbage dump. To do these experiments, leachate liquid samples (3 L) were collected from leachate pools generated from an open-air garbage dump of Nauta city, Loreto, Peru. After, leachate liquid samples were subsequently filtered through 0.45- and 0.25-μm filter membranes to remove particulate matter and microorganisms.

The experiments were conducted for 5 days with three native microalgae strains (*Ankistrodesmus* sp., *Chlorella* sp., and *Scenedesmus* sp.) of the freshwater microalgae collection culture native from the Peruvian Amazon. Each experiment included a control group (microalgae strain cultured with Chu-10 medium) and two treatments. The first treatment contained 100% leachate and the second 50% leachate and 50% Chu-10 culture medium. Assays were conducted by triplicate in 500-mL Erlenmeyer flasks and started using in each one a 200-mL final culture volume, 3×10^8 microalgae cells, a light intensity of 100 $\mu E \cdot m^2 \cdot s^{-1}$, a photoperiod regime of 12–12 h (light-dark), ambient temperature at 25°C ± 2°C, and constant homogenization at 200 rpm.

We evaluated the microalgae capabilities for chemical pollutant removal in leachates by quantifying these pollutants in the culture medium at the beginning and on the 5th day of the experiments, using standardized methods with the multiparameter LaMotte 3633-04 Fresh Water Aquaculture Test Kit. In addition, phosphate was quantified using a spectrophotometric method [26].

The results showed that the three microalgae strains were able to eliminate chemical pollutants in leachate (**Table 1**). Ammonium was efficiently removed from 90%

Pollutant compound/ chemical parameter	Microalgae strain	Concentration at the beginning of the experiments (mg/L)	Concentration on the 5th day of the experiments (mg/L)	Percentage decrease
Ammonium	*Ankistrodesmus* sp.	0.20 ± 0.01	0.00 ± 0.000	100 ± 0.00
	Chlorella sp.	0.20 ± 0.03	0.02 ± 0.001	90 ± 0.95
	Scenedesmus sp.	0.20 ± 0.01	0.01 ± 0.000	95 ± 0.22
Nitrite	*Ankistrodesmus* sp.	0.50 ± 0.05	0.05 ± 0.002	90 ± 0.72
	Chlorella sp.	0.50 ± 0.01	0.04 ± 0.001	92 ± 0.16
	Scenedesmus sp.	0.50 ± 0.06	0.03 ± 0.002	94 ± 0.33
Chloride	*Ankistrodesmus* sp.	24 ± 1.00	2 ± 0.10	91.7 ± 0.68
	Chlorella sp.	24 ± 1.15	2 ± 0.21	91.7 ± 0.49
	Scenedesmus sp.	24 ± 0.58	2 ± 0.10	91.7 ± 0.33
Phosphate	*Ankistrodesmus* sp.	100 ± 5.03	10 ± 0.26	90 ± 0.39
	Chlorella sp.	100 ± 3.61	10 ± 0.17	90 ± 0.21
	Scenedesmus sp.	100 ± 1.00	10 ± 0.20	90 ± 0.30
Carbon dioxide	*Ankistrodesmus* sp.	37 ± 1.00	0.0 ± 0.0	100 ± 0.0
	Chlorella sp.	37 ± 1.73	0.0 ± 0.0	100 ± 0.0
	Scenedesmus sp.	37 ± 0.58	0.0 ± 0.0	100 ± 0.0
Calcium and magnesium salts (hardness)	*Ankistrodesmus* sp.	160 ± 1.53	28 ± 0.50	82.5 ± 0.24
	Chlorella sp.	160 ± 0.58	28 ± 0.92	82.5 ± 0.59
	Scenedesmus sp.	160 ± 1.53	48 ± 1.00	70.0 ± 0.90
Carbonate and bicarbonate salts (alkalinity)	*Ankistrodesmus* sp.	180 ± 0.58	76 ± 1.00	57.8 ± 0.44
	Chlorella sp.	180 ± 1.00	96 ± 0.50	46.7 ± 0.29
	Scenedesmus sp.	180 ± 1.15	96 ± 1.00	46.7 ± 0.31
pH	*Ankistrodesmus* sp.	9 ± 0.50	9 ± 0.50	0.0 ± 0.0
	Chlorella sp.	9 ± 0.51	9 ± 0.50	0.0 ± 0.0
	Scenedesmus sp.	9 ± 0.50	9 ± 0.00	0.0 ± 0.0

Table 1.
Decrease in pollutant compound concentration and some chemical parameter values in landfill leachate cultures (100% leachate) of three native microalgae strains from the Peruvian Amazon.

(*Chlorella* sp.) to 100% (*Ankistrodesmus* sp.). These microalgae strains displayed similar pollutant elimination capabilities for nitrite, chloride, phosphate, carbon dioxide, and other chemical pollutants. CO_2, carbonate, and bicarbonate decrease can be related to its consumption by the microalgae cells in the photosynthetic process.

2.2 Wastewater treatment

The generation of great volumes of wastewater in the main cities of the Peruvian Amazon is increasing notably in the past 20 years. This environmental issue is associated with the intense migration of people from rural to urban areas with the hope to get better opportunities to improve their life qualities. This unplanned migration is generating unorganized human settlements in the marginal areas of the big cities, which lack basic services, such as electric fluid, potable water, and sewage system (**Figure 2**). In addition, none of these cities have wastewater treatment plants; then, wastewater is directly disposed into the main rivers of the Amazon basin, causing significant pollution of the aquatic ecosystems and affecting the aquatic flora, fauna, microbiota, and, of course, the human settlements located along the main rivers.

In this context, with a view to alleviate this pollution problem, we need to investigate eco-friendly, efficient, and low-cost options to treat wastewater. In this sense, we did experiments to determine whether native microalgae are useful to decontaminate wastewater generated in Iquitos city because there are several successful experiences around the world using these microorganisms [4, 5, 11, 27].

Therefore, to do the experiments, wastewater samples (5 L) were collected from the two main wastewater drainage systems of Iquitos city (Moronacocha and Huequito), Loreto, Peru. Furthermore, particulate matter and microorganisms were removed from the wastewater samples using the same previously described filtration approach (item 2.1) and were sterilized by autoclaving at 121°C for 30 min.

The experiments were conducted for 15 days with two native microalgae strains (*Ankistrodesmus* sp. and *Chlorella* sp.) of the freshwater microalgae collection culture native from the Peruvian Amazon. Each experiment included a control group (microalgae strain cultured with Chu-10 medium) and three treatments. The first treatment contained 100% wastewater from the Moronacocha wastewater drainage system and the second one contained 100% wastewater from the Huequito wastewater drainage system. Assays were conducted by triplicate in 250-mL Erlenmeyer flasks and started using in each one a 100-mL final culture volume, 4×10^{10} microalgae cells, a light

Figure 2.
Typical open-air sewage systems in Iquitos city and other main cities of the Peruvian Amazon.

intensity of 150 μE·m^2·s^{-1}, a photoperiod regime of 12–12 h (light-dark), an ambient temperature at 27°C ± 2°C, a relative humidity at 83%, and constant aeration with an air pump system.

We evaluated the microalgae capabilities for chemical pollutants removal in wastewater by quantifying these pollutants in the culture medium at the beginning and on the 15th day of the experiments, using standardized methods with the multiparameter LaMotte 3633-04 Fresh Water Aquaculture Test Kit. In addition, phosphate was quantified using a spectrophotometric method [26].

The results showed that the two microalgae strains were capable to remove chemical pollutants from the two wastewater samples (**Tables 2** and **3**). However, there are marker differences; for example, ammonium was efficiently removed from wastewater of the Huequito wastewater drainage system (from 97.2% to 100%); in contrast, this pollutant was poorly removed from wastewater of the Moronacocha wastewater drainage system (only 20% with both microalgae strains).

2.3 Ammonium removal using an immobilized microalgae

Ornamental fish export is an important economic activity in Iquitos city, they provide benefits to several families dedicated to this area. A frequent problem during the process of ornamental fish transportation is high mortality rate, which could be attributable to decrease in water quality during transportation. These changes are due to the accumulation of toxic and metabolites of the fish catabolic process, such as ammonium [28], which, in turn, alkalinizes the pH and decreases the dissolved oxygen concentration in the aqueous medium [29]. Oxygen deficiency, toxin

Pollutant compound/ chemical parameter	Microalgae strain	Concentration at the beginning of the experiments (mg/L)	Concentration on the 15th day of the experiments (mg/L)	Percentage decrease
Ammonium	*Ankistrodesmus* sp.	50 ± 1.00	40 ± 1.00	20 ± 0.40
	Chlorella sp.	50 ± 1.50	40 ± 2.00	20 ± 3.47
Chloride	*Ankistrodesmus* sp.	24 ± 0.76	2 ± 0.10	91.7 ± 0.57
	Chlorella sp.	24 ± 0.76	2 ± 0.26	91.7 ± 1.29
Phosphate	*Ankistrodesmus* sp.	2 ± 0.10	1.10 ± 0.20	45.0 ± 0.83
	Chlorella sp.	2 ± 015	1.05 ± 0.07	47.5 ± 0.72
Carbon dioxide	*Ankistrodesmus* sp.	40 ± 1.53	8.5 ± 0.57	78.8 ± 2.08
	Chlorella sp.	40 ± 1.53	5.0 ± 0.55	87.5 ± 1.33
Calcium and magnesium salts (hardness)	*Ankistrodesmus* sp.	65 ± 0.70	60 ± 1.73	7.7 ± 1.86
	Chlorella sp.	65 ± 1.76	52 ± 2.29	20.0 ± 4.63

Table 2.
Decrease in pollutant compound concentration in wastewater cultures obtained from the Moronacocha wastewater drainage system using two native microalgae strains from the Peruvian Amazon.

Pollutant compound/ chemical parameter	Microalgae strain	Concentration at the beginning of the experiments (mg/L)	Concentration on the 15th day of the experiments (mg/L)	Percentage decrease
Ammonium	*Ankistrodesmus* sp.	36 ± 1.00	0.0 ± 0.00	100 ± 0.00
	Chlorella sp.	36 ± 1.73	1.0 ± 0.10	97.2 ± 0.39
Chloride	*Ankistrodesmus* sp.	38 ± 1.00	36 ± 2.00	5.3 ± 0.06
	Chlorella sp.	38 ± 2.00	28 ± 1.05	26.3 ± 3.28
Phosphate	*Ankistrodesmus* sp.	1.9 ± 0.10	1.5 ± 0.10	21.1 ± 4.94
	Chlorella sp.	1.9 ± 0.20	1.2 ± 0.10	36.8 ± 1.25
Carbon dioxide	*Ankistrodesmus* sp.	50 ± 3.46	14 ± 1.59	72 ± 1.30
	Chlorella sp.	50 ± 1.73	4 ± 0.15	92 ± 0.54
Calcium and magnesium salts (hardness)	*Ankistrodesmus* sp.	60 ± 1.73	38 ± 1.01	36.7 ± 3.28
	Chlorella sp.	60 ± 2.00	32 ± 2.00	46.7 ± 1.56

Table 3.
Decrease in pollutant compound concentration in wastewater cultures obtained from the Huequito wastewater drainage system using two native microalgae strains from the Peruvian Amazon.

accumulation, and an increase in total ammonium concentration in the water are believed to be the main cause of fish mortality during transportation [30].

To help solve this problem, our research team evaluated the hypothesis that by using immobilized microalgae, the ammonium concentration decreased significantly. To test the formulated hypothesis, the experiments were conducted by triplicate for 2 weeks with one native microalgae strain (*Chlorella* sp.) of the freshwater microalgae collection culture native from the Peruvian Amazon. To do the experiment, first, *Chlorella* sp. (**Figure 3**) was cultured in increasing volumes of BG-11 medium (100, 250, and 500 mL) to obtain sufficient microalgal biomass to begin the experiments. Once sufficient microalgal biomass was generated, the microalgae cells were harvested by centrifugation. The culture conditions were a light intensity of 100 $\mu E \cdot m^2 \cdot s^{-1}$, a photoperiod regime of 12–12 h (light-dark), an ambient temperature at 27°C ± 2°C, a relative humidity at 83%, and constant aeration with an air pump system, after microalgae cells were immobilized (**Figure 3**) according to Zamani et al. [31]. Finally, 2×10^3 alginate beads with trapped microalgae cells were transferred to polypropylene boxes of 40 × 50 × 40 cm (W × H × L) with 5 L of distilled water and ammonium chloride (NH_4Cl) at 800 μM. These immobilized cells were cultured for 2 weeks under the conditions described earlier, monitoring in the culture supernatant each 24 h the ammonium levels according to Solórzano [32, 33].

The results showed that immobilized *Chlorella* sp. can efficiently remove the toxic ammonium from the aqueous medium. Thus, 33.07% and 76.10% of ammonium were removed from the culture system on the 7th and 14th days of culture, respectively (**Figure 4**). These results are similar to previously reported studies that showed that microalgae of the genus *Chlamydomonas*, *Chlorella*, *Scenedesmus*, *Picochlorum,* and others can efficiently remove ammonium ions from several kinds of wastewater [34–39].

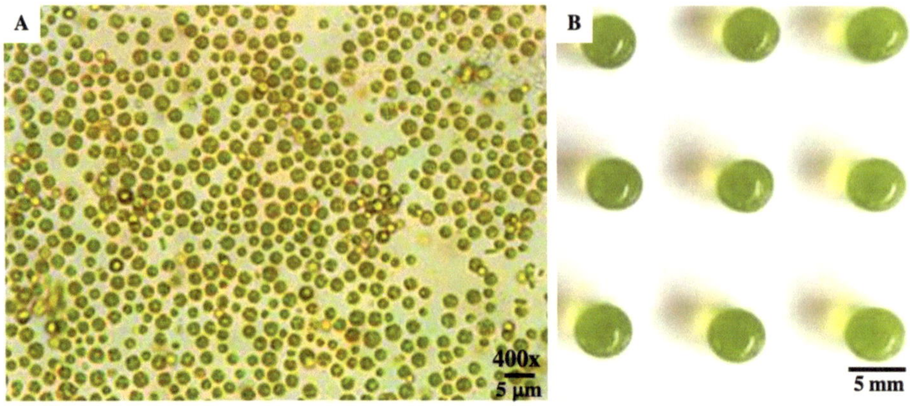

Figure 3.
Microphotography of Chlorella sp. cells (A) and immobilized microalgae cells in alginate beads (B).

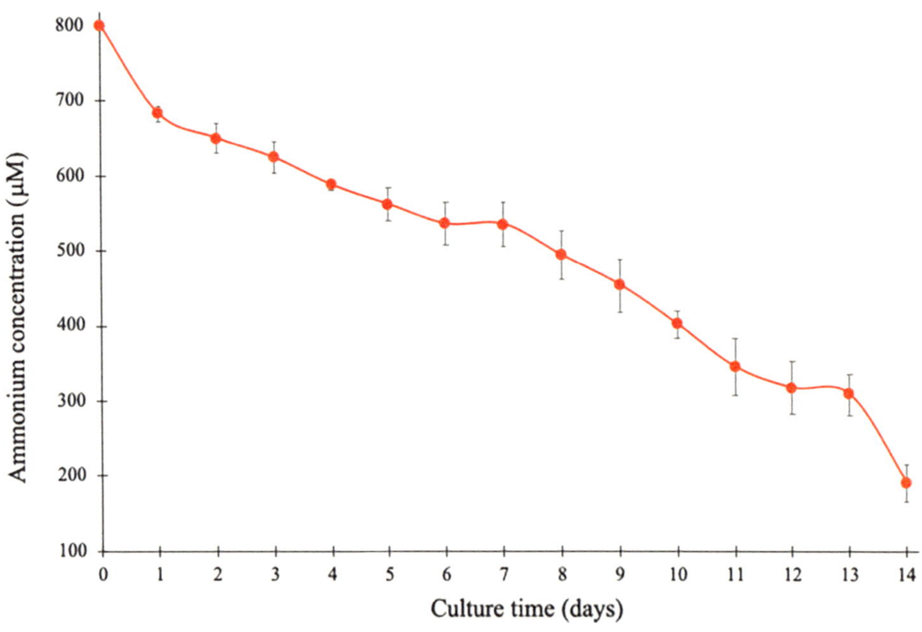

Figure 4.
Ammonium removal from the aqueous medium by Chlorella sp. immobilized in alginate beads.

Ammonium ions enter microalgae cells through ammonium transporters/ammonia permeases (AMTPs) embedded into the plasmatic membrane. These membrane-spanning proteins be made of 11 highly conserved transmembrane domains that fold into a channel across ammonia or ammonium translocates [40, 41]. According to X-ray crystallographic studies of some prokaryotic partners of these protein transporters, these are characterized as a compact trimer with 11 transmembrane helices per monomer and a narrow, mainly hydrophobic, channel for substrate conduction, located at the center of each monomer of the trimeric molecule. In addition, at the

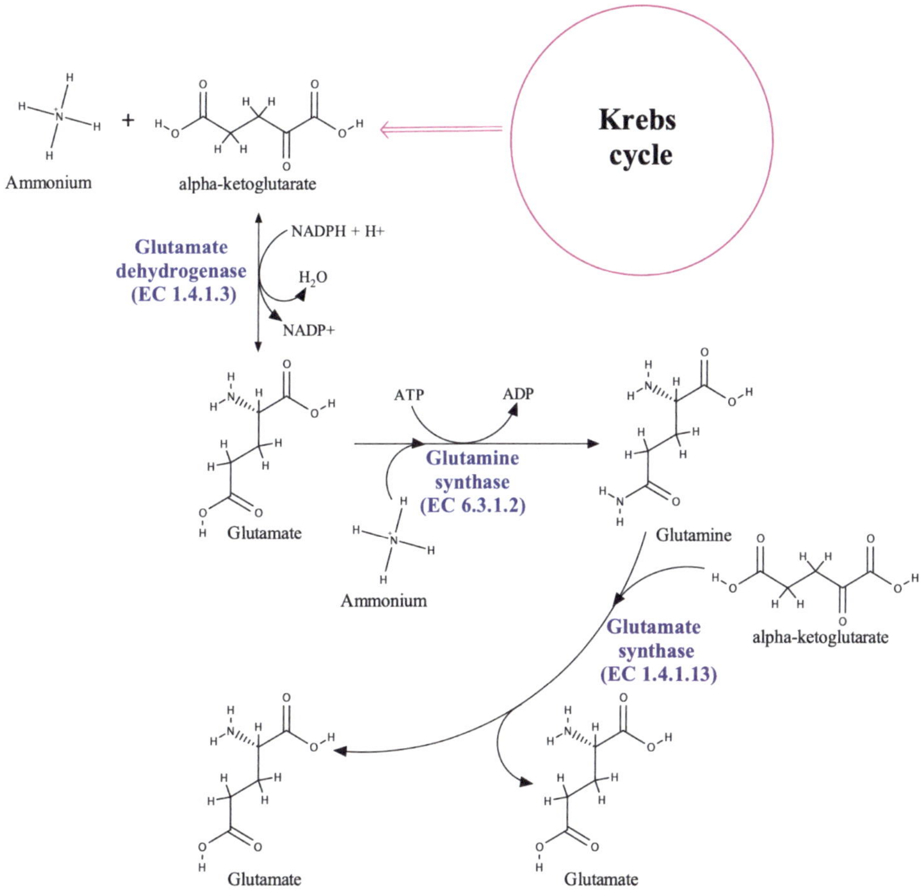

Figure 5.
Three key enzymes of microalgae involved in the incorporation of ammonium into amino acids and proteins.

periplasmic side of the transporter protein, a binding site for NH_4^+ is observed [42, 43]. In the particular case of *Chlamydomonas reinhardtii*, this microalga possesses the largest family of ammonium transporters consisting of eight members, which have complexly and finely regulation mechanisms at transcriptional and post-translational levels (**Figure 5**) [44–46].

According to Ahmad and Hellebust [47], the microalga *Chlorella autotrophica* can use two mechanisms to incorporate inorganic nitrogen sources into amino acids and proteins, which are related to the levels of the enzymes glutamate dehydrogenase (GDH) and glutamine synthetase (GS). Thus, GS levels are high in microalgae cells grown in nitrate and under nitrogen-starved conditions. However, in cells growing on ammonium, the GDH catalytic activity is increased. Both the enzymes require ammonium as a secondary substrate (**Figure 6**). Frequently, plant and microalgae cells prefer ammonium (NH_4^+) since it has the lowest metabolic energy cost than other inorganic nitrogen forms [36] because it can be directly ligated to amino acids by the action of three enzymes: glutamate dehydrogenase, glutamine synthetase, and glutamate synthase.

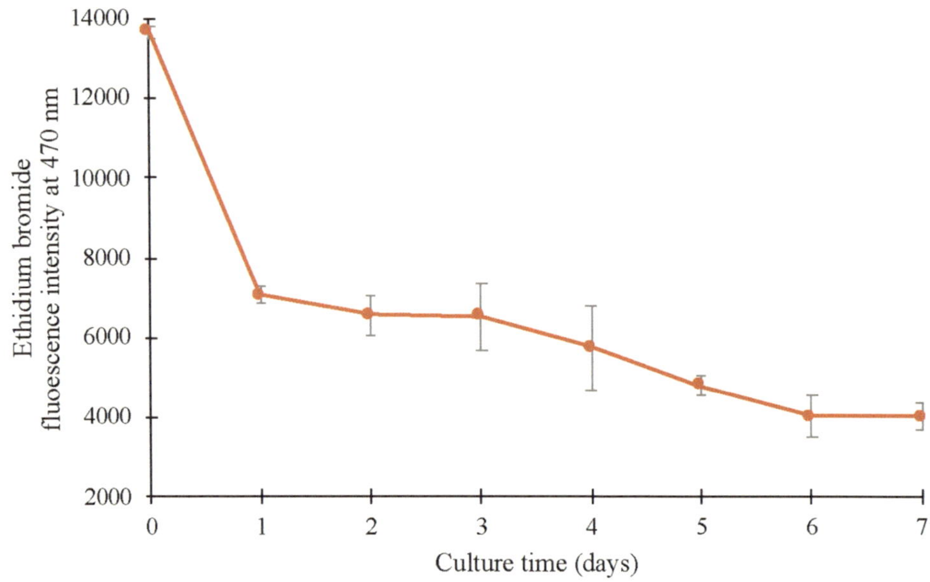

Figure 6.
Ethidium bromide removal kinetics by a microalgae consortium from the Peruvian Amazon.

2.4 Ethidium bromide removal using microalgae

The release of untreated effluent from research laboratories in our country and worldwide into water bodies is a major threat to the environment and human health. Commonly, effluent from laboratories and other research facilities is rich in toxic organic compounds, such as dyes used in the nucleic acid analysis, especially ethidium bromide, which is considered a serious biohazard due to its mutagenic, carcinogenic, teratogenic, and very toxic potentials when inhaled, ingested, or absorbed through the skin, and can irritate the eyes, mouth, and upper respiratory tract [48, 49]. To overcome these pollution problems, ethidium bromide and other toxic compounds could be partially or completely degraded to nontoxic forms before disposal. Consequently, some research laboratories worldwide are testing the biodegradation of ethidium bromide using plants and various kinds of microorganisms, including bacteria and microalgae [50–53], to develop, in the next future, modern, cost-effective, and eco-friendly bioremediation approaches.

In this context, our research team has evaluated the ability of a microalgae consortium for the removal of ethidium bromide from aqueous medium. For this experiment, three previously cultured native microalgae strains *Ankistrodesmus* sp., *Chlorella* sp., and *Scenedesmus* sp. were proportionally mixed (10^6 microalgae cells per milliliter of each strain) and transferred by triplicate into 250-mL Erlenmeyer flasks until a 100-mL final culture volume of BG-11 medium containing ethidium bromide at 1 mg/mL. In the experiments, a control group containing the same quantity of heat-inactivated microalgae cells and ethidium bromide at equal concentrations was included. Then, the assays were conducted for 7 days with a light intensity of 150 $\mu E \cdot m^2 \cdot s^{-1}$, a photoperiod regime of 12–12 h (light-dark), an ambient temperature at 27°C ± 2°C, a relative humidity at 83%, and constant aeration with an air pump system. Ethidium bromide concentrations were monitored every day measuring the

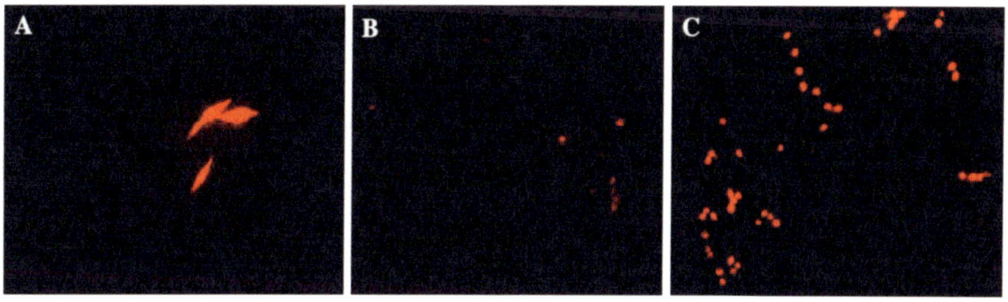

Figure 7.
Fluorescence microphotography of three native microalgae cells exposed to ethidium bromide. Ankistrodesmus sp. (A), Chlorella sp. (B), and Scenedesmus sp. (C).

intensity of fluorescence emission at 470 nm with a Qubit™ 4 Fluorometer (Thermo Fisher Scientific, Inc., Waltham, MA, USA).

The results showed that on the 7th day of starting the experiments, it was evidenced that the microalgal consortium was able to decrease the ethidium bromide concentration (directly related to fluorescence intensity) in the culture supernatant until 70.5% (**Figure 6**). These results corroborate the previous report of Cavalcante de Almeida et al. [52]. These authors also evaluated the capability of the microalgae *Chlorella vulgaris*, *Desmodesmus subspicatus,* and *Raphidocelis subcapitata* separately and in a consortium for ethidium bromide removal from an aqueous medium [52]. Their results strongly suggest the great potential of these microalgae species for phycoremediation application for ethidium bromide removal, which was directly dependent on the microalgae biomass.

Probably, the phycoremediation process used for microalgae to remove ethidium bromide is similar to several detoxification strategies used against aromatic organic pollutants (e.g., polycyclic aromatic hydrocarbons, phenolic compounds, dyes, etc.), including biosorption, bioaccumulation, biotransformation, and biodegradation [54, 55]. The first one is a metabolically independent process, which is a physicochemical phenomenon, supported by a gamma of mechanisms comprising absorption, adsorption, surface complexation, ion exchange, and precipitation [56]. The second one consists in the selective transportation by the monovalent cation uptake transport system [57] and other unidentified transporters, followed for its accumulation into some organelles such as nucleus, mitochondria, and chloroplast, which can be intercalated with DNA molecules (**Figure 7**). Finally, biotransformation and biodegradation are dependent on the metabolic capabilities of the microalgae cells, which are determined for their genomic background that codes a repertory of required enzymes [54]. To date, however, none of the metabolic pathways for ethidium bromide biodegradation has been described.

3. Conclusions

Native microalgae isolated from the Peruvian Amazon have a potential biotechnological application in the remotion of diverse chemical pollutants. These microorganisms showed abilities to remove pollutants contained into leachate generated in an open-air garbage dump and from two wastewaters from Iquitos city. In addition, an immobilized version of the microalgae *Chlorella* sp. was capable to remove ammonium efficiently. Finally, a microalgae consortium composed of three microalgae from the genus *Ankistrodesmus* sp., *Chlorella* sp., and *Scenedesmus* sp. was competent to remove the toxic compound ethidium bromide. Together, these experimental pieces of evidence indicate that native microalgae have an excellent potential for removing several pollutants and, consequently, could be used to develop bioremediation technologies based on native microalgae from the Peruvian Amazon.

Acknowledgements

This research was supported by the Peruvian funding agency Consejo Nacional de Ciencia, Tecnología e Innovación Tecnológica (CONCYTEC) through the Programa Nacional de Investigación Científica y Estudios Avanzados (PROCIENCIA), Funding Award Contract No. 018-2018-FONDECYT/BM-Improvement of Research Infrastructure (Scientific Equipment), the Scientific University of Peru and the Specialized Biotechnology Unit of the Natural Resources Research Center, National University of the Peruvian Amazon.

Conflict of interest

The authors declare no conflict of interest.

Author details

Marianela Cobos[1,2,3]*, Segundo L. Estela[3], Carlos G. Castro[3], Miguel A. Grandez[3], Alvaro B. Tresierra[3], Corayma L. Cabezudo[3], Santiago Galindo[3], Sheyla L. Pérez[3], Angélica V. Rios[3], Jhon A. Vargas[2,4], Roger Ruiz[5], Pedro M. Adrianzén[1,2], Jorge L. Marapara[1,2] and Juan C. Castro[1,2]*

1 Faculty of Biological Sciences, Academic Department of Biomedical Sciences and Biotechnology, National University of the Peruvian Amazon (UNAP), Iquitos, Peru

2 Specialized Unit of Biotechnology, Research Center of Natural Resources of the Amazon (CIRNA), National University of the Peruvian Amazon (UNAP), Iquitos, Peru

3 Laboratory of Biotechnology and Bioenergetics, Scientific University of Peru (UCP), Iquitos, Peru

4 São Carlos Institute of Physics, University of São Paulo, São Carlos, SP, Brazil

5 Faculty of Food Industries, National University of the Peruvian Amazon (UNAP), Iquitos, Peru

*Address all correspondence to: marianela.cobos@unapiquitos.edu.pe and juan.castro@unapiquitos.edu.pe

References

[1] Cobos M, Paredes JD, Maddox JD, Vargas-Arana G, Flores L, Aguilar CP, et al. Isolation and characterization of native microalgae from the Peruvian Amazon with potential for biodiesel production. Energies. 2017;**10**(2):224

[2] Castro JC, Cobos M. Chapter 15 - Biochemical profiling, transcriptomic analysis, and biotechnological potential of native microalgae from the Peruvian Amazon. In: Ahmad A, Banat F, Taher H, editors. Algal Biotechnology. Amsterdam, Netherlands: Elsevier; 2022. pp. 305-321

[3] Cobos M, Pérez S, Braga J, Vargas-Arana G, Flores L, Paredes JD, et al. Nutritional evaluation and human health-promoting potential of compounds biosynthesized by native microalgae from the Peruvian Amazon. World Journal of Microbiology and Biotechnology. 2020;**36**(8):121

[4] Ahmed SF, Mofijur M, Parisa TA, Islam N, Kusumo F, Inayat A, et al. Progress and challenges of contaminate removal from wastewater using microalgae biomass. Chemosphere. 2022;**286**:131656

[5] Aketo T, Hoshikawa Y, Nojima D, Yabu Y, Maeda Y, Yoshino T, et al. Selection and characterization of microalgae with potential for nutrient removal from municipal wastewater and simultaneous lipid production. Journal of Bioscience and Bioengineering. 2020;**129**(5):565-572

[6] Iqbal J, Javed A, Baig MA. Heavy metals removal from dumpsite leachate by algae and cyanobacteria. Bioremediation Journal. 2022;**26**(1):31-40

[7] Blanco-Vieites M, Suárez-Montes D, Delgado F, Álvarez-Gil M, Battez AH, Rodríguez E. Removal of heavy metals and hydrocarbons by microalgae from wastewater in the steel industry. Algal Research. 2022;**64**:102700

[8] Arunakumara KKIU, Zhang X. Heavy metal bioaccumulation and toxicity with special reference to microalgae. Journal of Ocean University of China. 2008;**7**(1):60-64

[9] Perales-Vela HV, Peña-Castro JM, Cañizares-Villanueva RO. Heavy metal detoxification in eukaryotic microalgae. Chemosphere. 2006;**64**(1):1-10

[10] Leong YK, Chang JS. Bioremediation of heavy metals using microalgae: Recent advances and mechanisms. Bioresource Technology. 2020;**303**:122886

[11] You X, Yang L, Zhou X, Zhang Y. Sustainability and carbon neutrality trends for microalgae-based wastewater treatment: A review. Environmental Research. 2022;**209**:112860

[12] Ji B. Towards environment-sustainable wastewater treatment and reclamation by the non-aerated microalgal-bacterial granular sludge process: Recent advances and future directions. Science of The Total Environment. 2022;**806**:150707

[13] Singh V, Mishra V. Evaluation of the effects of input variables on the growth of two microalgae classes during wastewater treatment. Water Research. 2022;**213**:118165

[14] Malik S, Shahid A, Betenbaugh MJ, Liu CG, Mehmood MA. A novel wastewater-derived cascading algal biorefinery route for complete valorization of the biomass to biodiesel and value-added bioproducts. Energy

Conversion and Management. 2022;**256**:115360

[15] Chew KW, Yap JY, Show PL, Suan NH, Juan JC, Ling TC, et al. Microalgae biorefinery: High value products perspectives. Bioresource Technology. 2017;**229**:53-62

[16] Sivaramakrishnan R, Suresh S, Kanwal S, Ramadoss G, Ramprakash B, Incharoensakdi A. Microalgal biorefinery concepts' developments for biofuel and bioproducts: Current Perspective and Bottlenecks. International Journal of Molecular Sciences. 2022;**23**(5):2623

[17] Ummalyma SB, Sirohi R, Udayan A, Yadav P, Raj A, Sim SJ, et al. Sustainable microalgal biomass production in food industry wastewater for low-cost biorefinery products: A review. Phytochemistry Reviews. 2022:1-23

[18] Nawaz T, Rahman A, Pan S, Dixon K, Petri B, Selvaratnam T. A review of landfill leachate treatment by microalgae: Current status and future directions. PRO. 2020;**8**(4):384

[19] Kjeldsen P, Barlaz MA, Rooker AP, Baun A, Ledin A, Christensen TH. Present and long-term composition of MSW Landfill Leachate: A review. Critical Reviews in Environmental Science and Technology. 2002;**32**(4): 297-336

[20] Chu LM, Cheung KC, Wong MH. Variations in the chemical properties of landfill leachate. Environmental Management. 1994;**18**(1):105-117

[21] Salem Z, Hamouri K, Djemaa R, Allia K. Evaluation of landfill leachate pollution and treatment. Desalination. 2008;**220**(1):108-114

[22] Deng C, Zhao R, Qiu Z, Li B, Zhang T, Guo F, et al. Genome-centric metagenomics provides new insights into the microbial community and metabolic potential of landfill leachate microbiota. Science of The Total Environment. 2022;**816**:151635

[23] Remmas N, Roukouni C, Ntougias S. Bacterial community structure and prevalence of Pusillimonas-like bacteria in aged landfill leachate. Environmental Science and Pollution Research. 2017;**24**(7):6757-6769

[24] Passarini MRZ, Moreira JVF, Gomez JAM, Bonugli-Santos RC. DNA metabarcoding of the leachate microbiota from sanitary landfill: Potential for bioremediation process. Archives of Microbiology. 2021;**203**(8):4847-4858

[25] Teng C, Zhou K, Peng C, Chen W. Characterization and treatment of landfill leachate: A review. Water Research. 2021;**203**:117525

[26] Motomizu S, Wakimoto T, Tôei K. Spectrophotometric determination of phosphate in river waters with molybdate and malachite green. The Analyst. 1983;**108**(1284):361-367

[27] Do CVT, Pham MHT, Pham TYT, Dinh CT, Bui TUT, Tran TD, et al. Microalgae and bioremediation of domestic wastewater. Current Opinion in Green and Sustainable Chemistry. 2022;**34**:100595

[28] Randall DJ, Tsui TKN. Ammonia toxicity in fish. Marine Pollution Bulletin. 2002;**45**(1-12):17-23

[29] Thurston RV, Russo RC, Vinogradov GA. Ammonia toxicity to fishes. Effect of pH on the toxicity of the unionized ammonia species. Environmental Science & Technology. 1981;**15**(7):837-840

[30] Amend DF, Croy TR, Goven BA, Johnson KA, McCarthy DH.

Transportation of fish in closed systems: Methods to control ammonia, carbon dioxide, pH, and bacterial growth. Transactions of the American Fisheries Society. 1982;**111**(5):603-611

[31] Zamani N, Noshadi M, Amin S, Niazi A, Ghasemi Y. Effect of alginate structure and microalgae immobilization method on orthophosphate removal from wastewater. Journal of Applied Phycology. 2012;**24**(4):649-656

[32] Solórzano L. Determination of ammonia in natural waters by the phenolhypochlorite method. Limnology and Oceanography. 1969;**14**(5):799-801

[33] Ivančič I, Degobbis D. An optimal manual procedure for ammonia analysis in natural waters by the indophenol blue method. Water Research. 1984;**18**(9):1143-1147

[34] Su Y. Revisiting carbon, nitrogen, and phosphorus metabolisms in microalgae for wastewater treatment. Science of The Total Environment. 2021;**762**:144590

[35] Goswami RK, Agrawal K, Verma P. Phycoremediation of nitrogen and phosphate from wastewater using *Picochlorum* sp.: A tenable approach. Journal of Basic Microbiology. 2022;**62**(3-4):279-295

[36] Salbitani G, Carfagna S. Ammonium Utilization in Microalgae: A Sustainable Method for Wastewater Treatment. Sustainability. 2021;**13**(2):956

[37] de Bashan LE, Trejo A, Huss VA, Hernandez JP, Bashan Y. Chlorella sorokiniana UTEX 2805, a heat and intense, sunlight-tolerant microalga with potential for removing ammonium from wastewater. Bioresource Technology. 2008;**99**(11):4980-4989

[38] Zhou Y, He Y, Xiao X, Liang Z, Dai J, Wang M, et al. A novel and efficient strategy mediated with calcium carbonate-rich sources to remove ammonium sulfate from rare earth wastewater by heterotrophic Chlorella species. Bioresource Technology. 2022;**343**:125994

[39] Wang Q, Yu Z, Wei D. High-yield production of biomass, protein and pigments by mixotrophic Chlorella pyrenoidosa through the bioconversion of high ammonium in wastewater. Bioresource Technology. 2020;**313**:123499

[40] McDonald TR, Dietrich FS, Lutzoni F. Multiple horizontal gene transfers of ammonium transporters/ammonia permeases from prokaryotes to eukaryotes: Toward a new functional and evolutionary classification. Molecular Biology and Evolution. 2012;**29**(1):51-60

[41] Kumar A, Bera S. Revisiting nitrogen utilization in algae: A review on the process of regulation and assimilation. Bioresource Technology Reports. 2020;**12**:100584

[42] Conroy MJ, Durand A, Lupo D, Li XD, Bullough PA, Winkler FK, et al. The crystal structure of the Escherichia coli AmtB-GlnK complex reveals how GlnK regulates the ammonia channel. Proceedings of the National Academy of Sciences of the United States of America. 2007;**104**(4):1213-1218

[43] Andrade SLA, Dickmanns A, Ficner R, Einsle O. Crystal structure of the archaeal ammonium transporter Amt-1 from Archaeoglobus fulgidus. Proceedings of the National Academy of Sciences. 2005;**102**(42):14994-14999

[44] González-Ballester D, Camargo A, Fernández E. Ammonium transporter genes in Chlamydomonas: The nitrate-specific regulatory gene Nit2 is involved

in Amt1;1 expression. Plant Molecular Biology. 2004;**56**(6):863-878

[45] Ermilova EV, Zalutskaya ZM, Nikitin MM, Lapina TV, Fernández E. Regulationbylightofammoniumtransport systems in Chlamydomonas reinhardtii. Plant, Cell & Environment. 2010;**33**(6):1049-1056

[46] Sanz-Luque E, Ocaña-Calahorro F, Llamas A, Galvan A, Fernandez E. Nitric oxide controls nitrate and ammonium assimilation in Chlamydomonas reinhardtii. Journal of Experimental Botany. 2013;**64**(11):3373-3383

[47] Ahmad I, Hellebust JA. Nitrogen metabolism of the marine microalga Chlorella autotrophica. Plant Physiology. 1984;**76**(3):658-663

[48] Lee WK, Ciarimboli G, Koepsell H, Thevenod F. Transport of mutagenic vital dye ethidium bromide by the human organic cation transporters hOCT2 and hOCT3. The FASEB Journal. 2008;**22**(S1):1217

[49] Saeidnia S, Abdollahi M. Are other fluorescent tags used instead of ethidium bromide safer? DARU Journal of Pharmaceutical Science. 2013;**21**(1):71

[50] Debroy A, Yadav M, Dhawan R, Dey S, George N. DNA dyes: Toxicity, remediation strategies and alternatives. Folia Microbiologica. 2022:1-13

[51] Salomão AL, de Almeida HC, Pereira JL, Teixeira LC, Marques M. Biotechnology applied for removal of organic compounds by phycoremediation. Linnaeus Eco-Technology. 2018:109

[52] de Almeida HC, de Salomão AL, Lambert J, Teixeira LC, Marques M. Phycoremediation potential of microalgae species for ethidium bromide removal from aqueous media. International Journal of Phytoremediation. 2020;**22**(8):1-7

[53] Gandhi VP, Kesari KK, Kumar A. The identification of ethidium bromide-degrading bacteria from laboratory gel electrophoresis waste. Biotech. 2022;**11**(1):4

[54] Touliabah HES, El-Sheekh MM, Ismail MM, El-Kassas H. A review of microalgae- and cyanobacteria-based biodegradation of organic pollutants. Molecules. 2022;**27**(3):1141

[55] Mondal M, Halder G, Oinam G, Indrama T, Tiwari ON. Chapter 17 - Bioremediation of organic and inorganic pollutants using microalgae. En: Gupta VK, Pandey A, New and Future Developments in Microbial Biotechnology and Bioengineering Amsterdam: Elsevier; 2019. pp. 223-235

[56] Fomina M, Gadd GM. Biosorption: Current perspectives on concept, definition and application. Bioresource Technology. 2014;**160**:3-14

[57] Peña A, Ramírez G. Interaction of ethidium bromide with the transport system for monovalent cations in yeast. Journal of Membrane Biology. 1975;**22**(1):369-384

Chapter 8

Mixotrophyc Culture of *Dunaliella salina* in Cuban Fishing Wastewaters

Gerardo Suárez Álvarez and Teresita de Jesús Romero López

Abstract

This report presented all the referring information about the organic culture of microalgae *Dunaliella salina*, Teodorescu, 1905, produced in organic effluents of Cuban Fishing Industry, and in a synthetic organic medium (MIP-1) and in control (Johnson medium). Here were studied the possibilities possessed by its algae as high-added value, in food industry and pharmaceutical products. A comparative study is offered on 69 bullfights (57 at laboratory scale and 8 in pilot plant), with different combinations and offered an economic valuation on the production of the algal bulk in form of paste in a pilot plant and to industrial scale as well as some possibilities for its use. This report suggest that the culture medium MIP-1 resulted advantageous, which is capable of achieving algae concentrations about 500,000 to 870,000 cell/mL during 15 days of be inoculated, represented a yield between 1,7 and 2,2 g/L on the dry weight base. According to the result data from the culture of this microalgae, to produce 1 kg of wet bulk was calculated a cost that fluctuates between \$6.40 and 6.50, where the culture means in the case of the MIP-1 alone represents \$0.09/m^3 of culture medium.

Keywords: Dunaliella, technology, fishing effluent, culture medium, ß-carotene, proteins, bromate ecological analysis, algal growth

1. Introduction

This research is in a pilot's plant phase and with the same one it is sought to interpret the behavior of the growth, carotenogenesis and protein contain of the microalga *Dunaliella salina* cultivated in synthetic medium with addition of a source of inorganic or organic carbon. The growth was measured with the purpose of evaluating its development under different culture conditions, since this can facilitate the decrease of the production costs in a climbed bigger. With the same principle the carotenogenosis was evaluated, to look for that fertilizer or which are those that increase the concentration of this pigment so important for the human health for its high nutritious value. If one wants to achieve a quick growth and development of this algae is an indispensable requirement the employment of near areas to salines, fundamentally those that are not subjected to the commercial exploitation of the salt, to have bigger area readiness, what represented a bigger net production. The culture medium that intends with this work is since of novel character, until the moment it has been planted in the existent bibliography that this microalgae species is not able to grow in a nutritive medium that its base of carbon is organic, reason because it is the first time that is cultivated in this way.

Among their perspectives uses, the culture in areas near natural saline, is the best option for commercial exploitation, also as high rate ponds. The lack of vitamins in the human feeding is a problem of world character, for what is indispensable to increase the sources of these, to achieve the man's bigger survival. A way of achieving this end is with the culture of *D. salina*, which is able to accumulate a great quantity of Beta-carotene, a biological precursor of the vitamin "A", which transforms it in this when are being ingested by the man and it is not toxic to the human organism like in the case of the ingest of synthetic vitamin "A" pills; for what that biological practice is of fundamental importance.

In the current world, the microalgae cultivation is of great interest, due to the uses but diverse that one makes of these, since they can be used in the production of food animal and human, as aide of high value, as chemical and biochemical products, as fertilizer and in the purification of polluted waters, among others [1, 2].

The microalgae *D. salina* that belong to the class Chlorophyceae and to the order Dunaliellales is of great importance for the man, since it is the fundamental natural source of the ß-carotene or provitamin "A". The ß-carotene is recognized by its high one to be able to as antioxidant, what is the same thing an anti-carcinogenic of having proven effectiveness that can be used also as coloring [3], in the alimentary industry (mayonnaise's, pastry, bakery, soups, juices, jells, etc.).

At the present time several countries are devoted to the commercial exploitation of the *D. salina*, among them the main ones are Australia and Israel, both although they use different technologies, they have been able to obtain a sustained production of this microalga.

Australia, the main producer of natural ß-carotene, not uses salt lakes dedicated to the exploitation of the salt, with very little energy expense but with use of big extensions of lands of which they prepare with easiness in this country. Israel uses high-speed lagoons with land saving but with more energy expense. Other countries have also attempted new cultivation technologies to increase the production of this microalga with the smallest possible cost.

In Cuba we have been carried out several studies to find the solutions, as culture medium, but economic and offer a high yield, for this in the Fisheries Industry Ministry (MIP) the work has been guided in two ways: one with a medium of where the source of carbon is organic and it comes from the residual waters of the fishing industry and another with a inorganic carbon that tries to substitute the conventional nutrients for its use in the salines. This in turn can use a source of inorganic or organic carbon, in dependence of the requirements of the cultivation.

2. Materials and methods

To know the effect of different synthetic culture medium, on the growth and development of *D. salina*, were carried out 57 testes, according to the established experimental design as it is shown in the **Table 1**. As control the modified Johnson medium (MJ), was used [4], that is it usually used for the commercial culture of this microalgae. The employee inoculates belongs to the strain HFI-1; obtained in the MIP by sexual crossing among a strain from the Algal Culture Collection (ACC) of the "Centro Nacional de Investigaciones Científicas" (CENIC) in Cuba, whose origin is Chilean and another (MUR-8) from ACC at Algal Biotechnology Laboratory (ABL) of Murdoch University, Western Australia. The seeding, was carried out adding a seed of 300,000 cel/mL in erlenmeyers flask of 200 mL with 100 mL of salininzated medium with addition of sodium chloride (NaCl), until obtaining a concentration of 20%. The culture mother's cells were in the phase of exponential growth and Betacarotene production.

The incident illumination received on all the experimental flasks, was continuous, during all day, with fluorescent light tubes of 40 Watts and the rotation of the flasks

No.	Design	Key
1	H_2O + KNO_3 0.30 g/l	41
2	H_2O + KNO_3 0.30 g/l	9
3	H_2O + KNO_3 0.30 g/l + Az 0.1 g/l	11
4	H_2O + KNO_3 0.30 g/l + NH_4NO_3 0.05 g/l	42
5	H_2O + KNO_3 0.30 g/l + KH_2PO_4 0.035 g/l	43
6	H_2O + KNO_3 0.30 g/l + KH_2PO_4 0.035 g/l + 0.05 g/l NH_4NO_3	44
7	H_2O + KNO_3 0.50 g/l	10
8	H_2O + KNO_3 0.50 g/l	35
9	H_2O + KNO_3 0.50 g/l + Sugar 0.1 g/l	12
10	H_2O + KNO_3 0.50 g/l + Sugar 0.1 g/l	36
11	H_2O + KNO_3 0.50 g/l + PO_4-S 0.05 g/l	33
12	H_2O + KNO_3 0.50 g/l + PO_4-S 0.05 g/l	13
13	H_2O + KNO_3 0.50 g/l + PO_4-S 0.05 g/l + 0.1 g/l Sugar	14
14	H_2O + KNO_3 0.50 g/l + PO_4-S 0.05 g/l + 0.1 g/l Sugar	34
15	H_2O + KNO_3 0.50 g/l + Urea 0.1 g/l + 0.05 g/l PO_4-S + 0.1 g/l Sugar	17
16	H_2O + KNO_3 0.50 g/l + Urea 0.1 g/l + 0.05 g/l PO_4-S	16
17	H_2O + KNO_3 0.50 g/l + Urea 0.1 g/l + 0.05 g/l PO_4-S + 0.1 g/ Sugar	18
18	H_2O + KNO_3 0.30 g/l + Urea 0.1 g/l + 0.03 g/l PO_4-S	39
19	H_2O + $NaNO_3$ 0.30 g/l + Urea 0.1 g/l + 0.03 g/l PO_4-S + 0.1 g/l Sugar	40
20	H_2O + $NaNO_3$ 0.50 g/l	19
21	H_2O + $NaNO_3$ 0.50 g/l + Sugar 0.1 g/l	20
22	H_2O + $NaNO_3$ 0.50 g/l + Sugar 0.1 g/l	21
23	H_2O + $NaNO_3$ 0.50 g/l + Sugar 0.1 g/l	22
24	H_2O + $NaNO_3$ 0.50 g/l + Sugar 0.1 g/l + $CaCl_2$(MJ) + KH_2PO_4	26
25	H_2O + $NaNO_3$ 0.50 g/l + Sugar 0.1 g/l + KCl(MJ) + KH_2PO_4	25
26	H_2O + $NaNO_3$ 0.50 g/l + Sugar 0.1 g/l + KH_2PO_4 (MJ)	28
27	H_2O + $NaNO_3$ 0.50 g/l + Sugar 0.1 g/l + $MgCl_2$(MJ) + KH_2PO_4	23
28	H_2O + $NaNO_3$ 0.50 g/l + Sugar 0.1 g/l + $MgSO_4$(MJ) + KH_2PO_4	24

No.	Design	Key
29	H_2O + $NaNO_3$ A11 + Sugar 0.1 g/l + $NaHCO_3$(MJ) + KH_2PO_4	27
30	H_2O + Urea 0.02 g/l + Sugar 0.1 g/l	48
31	H_2O + Urea 0.02 g/l + PO_4-S 0.05 g/l + 0.1 g/l Sugar	52
32	H_2O + Urea 0.05 g/l + Sugar 0.1 g/l	47
33	H_2O + Urea 0.05 g/l + PO_4-S 0.05 g/l + 0.1 g/l Sugar	51
34	H_2O + Urea 0.10 g/l + Azúcar 0.1 g/l	46
35	H_2O + Urea 0.10 g/l + PO_4-S 0.05 g/l	15
36	H_2O + Urea 0.10 g/l + PO_4-S 0.05 g/l	37
37	H_2O + Urea 0.10 g/l + PO_4-S 0.05 g/l + 0.1 g/l Sugar	38
39	H_2O + Urea 0.10 g/l + PO_4-S 0.05 g/l + 0.1 g/l Miel de Purga	53
40	H_2O + Urea 0.20 g/l + Sugar 0.1 g/l	45
41	H_2O + Urea 0.20 g/l + PO_4-S 0.05 g/l	6
42	H_2O + Urea 0.20 g/l + PO_4-S 0.05 g/l	8
43	H_2O + Urea 0.20 g/l + PO_4-S 0.05 g/l + 0.1 g/l Sugar	7
44	H_2O + Urea 0.20 g/l + PO_4-S 0.05 g/l + 0.1 g/l Sugar	49
45	MJ	54
46	MJ	55
47	MJ	56
48	MJ	57
49	MJ	1
50	MJ	2
51	MJ	3
52	MJ	29
53	MJ + Sugar 0.1 g/l	4
54	MJ + Sugar 0.1 g/l	5
55	MJ + Sugar 0.1 g/l	30
56	MJ + KNO3 0.5 g/l	31
57	MJ + KNO3 A7 + 0.1 g/l Sugar	32

Table 1.
All test carry out.

facilitated that it's received a mean value of 9. 85 Klux. The luminous intensity was measured with a battery lux meter with three work ranges.

All the experimental series to laboratory scale got ready with distilled water, as diluent of the reagents that conform the synthetic medium. According to the applied reaction, was used commercial sugar cane like source of organic carbon; also for a specific culture molasses was used and for another, sodium bicarbonate ($NaHCO_3$) to carry out a comparison among the different experiments, diverse culture medium were used with the unlike reagents that compose the Johnson medium in an individual way.

Of the obtained results, with the purpose of checking to a bigger scale the best results, was carried out an experiment to pilot plant scale with a lagoon that contained

1000 liters of culture, whose dilution water was fresh water with addition of commercial salt until reaching a concentration of 20% (Weight/Volume).

Cell counts were determined daily by a Neubauer cell hemocytometer with 1 mm^2 of useful area. The pigments ß-carotene and chlorophyll a and b, was extracted with acetone and were determined according to the standard method of the ABL [5], in a cellular button after centrifuging during 8 minutes to 4500 r.p.m. A volume of 5 mL of blended cultivation with 5 mL of distilled water (H_2O), for this way to diminish the salinity of the sample and to avoid that the cells that lack rigid cellular membrane break, being ignored this way the spill of the liquid, plasmatic to the water of the means. The calculations were carried out according to:

*Total carotene = ABS452 * 3.86 (Vol. extract/Vol. it shows)*

*Chlorophyll to = 11.93 * ABS664 - 1.93 * ABS647*

*Chlorophyll b = 20.36 * ABS647 - 5.50 * ABS664*

The primary data of the different series of growth in number and the ß-carotene concentrations did not resist a test of normality, and when relating the logarithm of the variances of each data series against the logarithms of the stockings of the cel/mL counts, a pending "b" of 2.47 was obtained since all the values were normalized according to expression log10 (x + 1). The slope for the regression of the ß-carotene concentrations was 2.33; because the same transformation was applied to normalize the data. Total proteins were determinate following the method of Lowry et al. [6]. The dry weight was analyzed by gravimeter. The validity of the results was analyzed applying an analysis of variance (ANOVA) of double classification with Duncan and Tuckey tests using a significance level of 95% and were solved by means of the packages of programs StatWin 8.0 and Excel 2016.

3. Results and discussion

In all the experimental series to laboratory scale, the temperature of the cultivation stayed between 25 and 30°C. The rotation of the different experimental flasks favored in the luminous intensity of the same ones (80000–140,000 lux). The growth curves of 57 experimental series, with a salinity of 20%, are shown in the **Figure 1**; where clearly it is understood that there is a separation of the same ones in 2 groups, in those which the growths improve they evidenced in those that contemplated the addition in common of urea. When carrying out an analysis of hierarchical classification with the strategy of the complete binding with the Euclidean distances, must be proven, the separation in two groups, one that understands to the cases 37; 38; 39; 40; 50; 51 and 52; all with inclusion of urea and other big with three subdivisions that they offered significant statistical differences among them according to an analysis of variance of double classification. The F calculated for the 57 treatments was of 7.29 and the F of chart of 1.34 with 95% of probability. For the effect of the time of reaction the calculated F was of 15.10 against an F of chart of 1.53; for what thinks about that significant statistical differences exist among them.

Of these groups, 12 cases were chosen (MJ; 10; 12; 19; 20; 33; 39; 40; 48; 49; 50 and 51) that represent the groups where the rates of more growth were presented as one observes in the **Figure 2**. For the condition MJ, a curve average of 5 repetitions

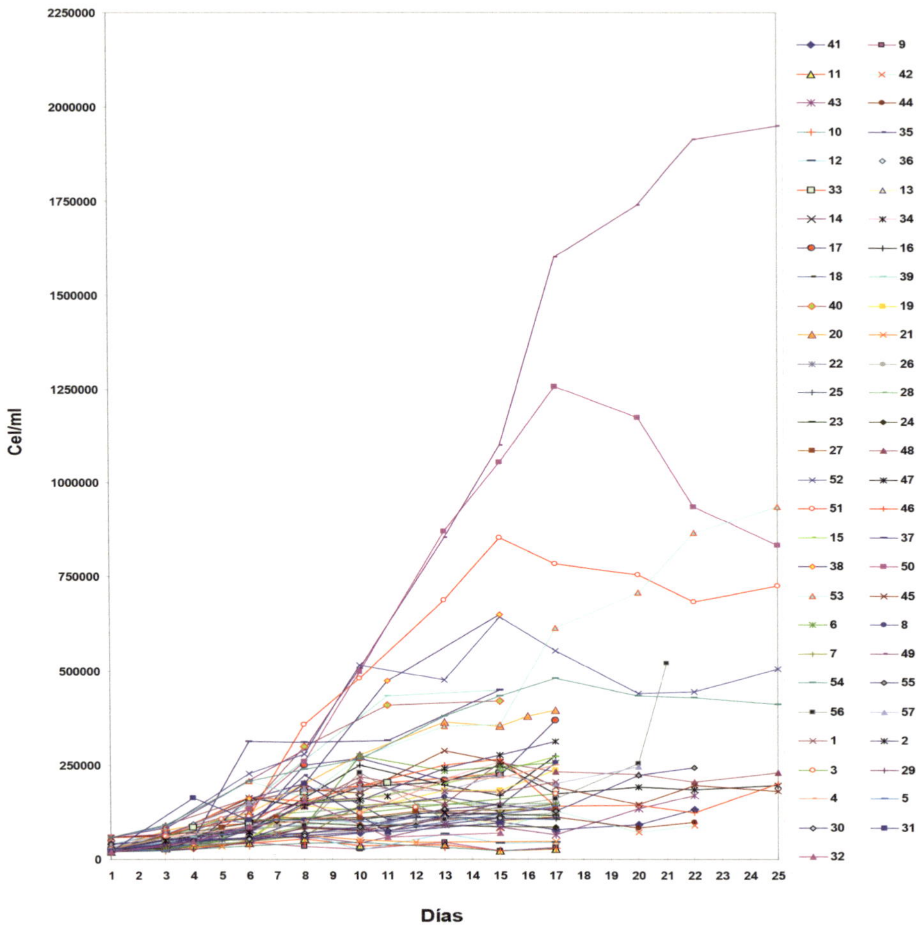

Figure 1.
The growth curves of 57 experimental series, salinity of 20%.

was used where an ANOVA offered a smaller Fc 1.08, at 2.26 of chart to indicate that they did not have statistical differences at a level of 95% among them. The conditions 10; 12 and 33 represent to the experiments where the fundamental nutriment was the potassium nitrate (KNO_3) and they embrace the highest values (33); half and under (12 and 10). The conditions 19; 20; 39 and 40 represent the experiment that the fundamental nutriment have possessed sodium nitrate ($NaNO_3$) and finally the 48; 4; 50 and 51, those the urea was the fundamental nutritional component. With these 12 groups to be carried out an analysis of hierarchical classification that it offered the cluster that is presented in the **Figure 3**, and 2 subgroups was obtained, one that represents the use of the urea and the other one subdivided in three, has in an end to the group that uses the Johnson medium and the $NaNO_3$ to reason of 0.5 g/L, to which unites the condition 33 that is constituted by potassium nitrate (KNO_3) and superphosphate (PO_4-S). Among these, there are others represented by the other nutritional medium as it was presented in the **Figure 2**. A factor that apparently has favored the acceleration of the speed of growth in number, is the addition of sugar like source of organic carbon. The osmoregulation of this microalga species depends fundamentally on the glycerol production that accumulates inside the own cell and

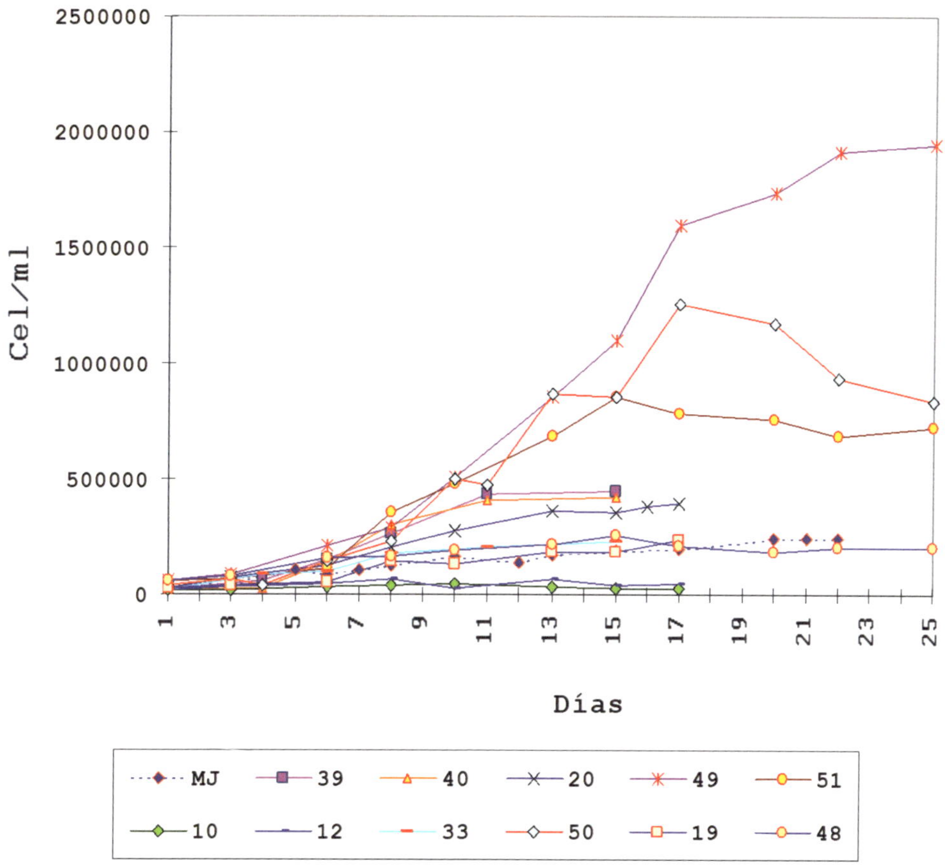

Figure 2.
Experimental groups where the high rates of growth were presented.

it allows him to survive drastic changes of salinity and bigger salinity the ß-carotene production it is bigger that in turn allows him to support bigger intensity of luminous radiation, but Ben-Amotz and Avron [7], had pointed out that the polysaccharides facilitate the glycerol synthesis for that, apparently the addition of sugar facilitates the whole process of the cycles of life and ß-carotene production.

In any case, any company dedicated to the commercial exploitation of the *D. salina* in the world uses the addition of carbon in its organic form, because traditionally they have thought about according to the revision given by Borowitzka and Borowitzka [8] and Ben-Amotz and Avron [7, 9] that this species is unable to use this source of carbon, being emphasized that the same, alone can use it in its inorganic form, for what this form constitutes something new for this species; being significant in all the experiments carried out in the MIP, those series that used the Johnson medium always reached never bigger values to 500,000 cel/mL and with the addition of sugar were arrived until almost 2,000,000 cel/mL, what represents a great advantage for obtaining of biomass protein.

With the obtained results at the laboratory scale, were carried out two tests at pilot plant level in a high-speed lagoon, with paddles and with a 1000 L of culture medium (**Figure 4**) where used medium was fresh water salinized up to 20% with NaCl to which was added 0.5 g/L of $NaNO_3$ and $NaHCO_3$ to the concentration of the Johnson

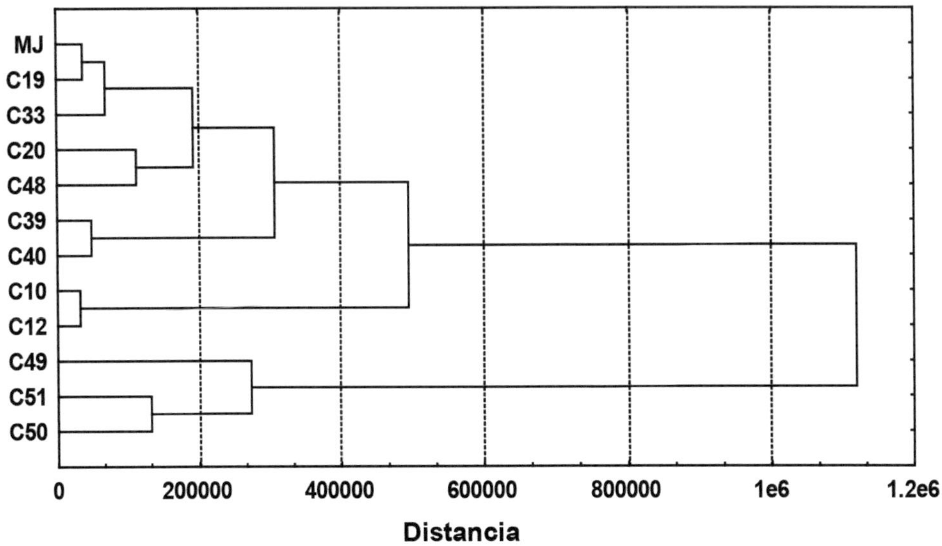

Figure 3.
Analysis of hierarchical classification that it offered the cluster more effective.

medium in a case (0.043 g/L) and 0.1 g/L of sugar. In the other case urea was used to reason of 0.2 g/L; 0.02 g/L of superphosphate and 0.1 g/L of sugar. These two variants were adopted because they represent to the highest results obtained regarding the growth in number of the cells of *D. salina*. In the lagoon with $NaNO_3$ the speed of growth from a beginning was bigger, to reach values from 800,000 to 900,000 cel/mL among the day's 8 and 14; but in the lagoon with urea and superphosphate, although the speed of growth was not presented so quick the cellular concentrations reached the 900,000 cel/mL equally, to the 14 days to arrive until near values to 1,200,000 cel/mL, to the 20 days, for what anyone of these culture can be used for the commercial exploitation of this microalga; since generally the companies that market it harvest to the 20 days of initiate the cultivation and with near concentrations to the 500,000 cel/mL.

The illumination in the experiments outdoors, in the lagoons to plant pilot's scale, had fluctuated between 120,000 and 180,000 lux, with picks at 12 in the day; indicating that the received illumination was adapted for the growth of this microalga species, according to the data that report Ben-Amotz et al. [10] and Borowitzka and Borowitzka [4].

Regarding the pH, it is understood that the same one during the first hours of the day this in the surroundings of 8.15 to be increased up to 8.3 as it lapses the day and this is product of the consumption of CO_2 for the algae during the hours of light [11]. These values are also among those understood among the good ones for the development of this species [8].

In the case of ß-carotene production, concentrations were reached of up to 11 μg ß-carotene/mL, of them those that understood addition of KNO_3 arrived up to 6.5 μg ß-carotene/mL, those that had alone addition of $NaNO_3$ arrived up to 3.5 μg ß-carotene/mL, those that used Johnson medium, 8 μg ß-carotene/mL was reached, but those that used urea arrived to concentrations of 11 μg ß-carotene/mL, those represent the best option for the cultivation to commercial scale.

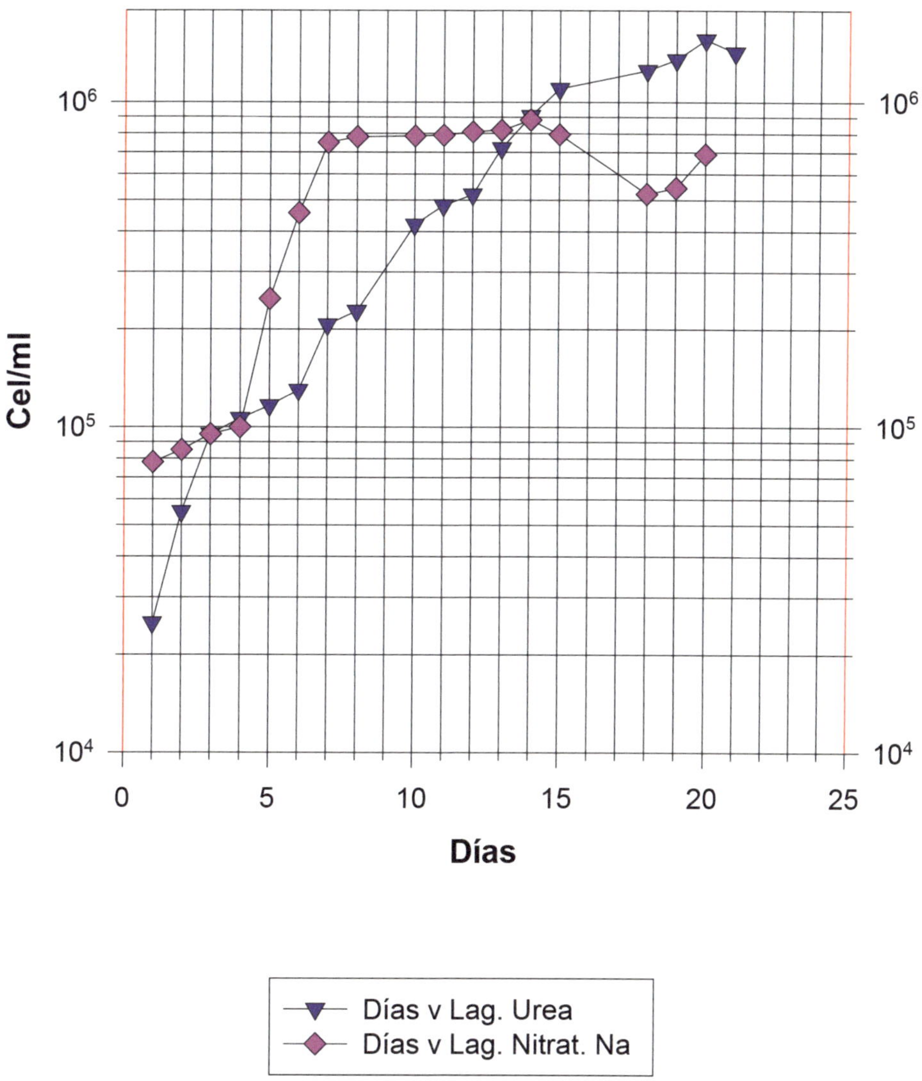

Figure 4.
Two tests at pilot plant level in a high-speed lagoon, with paddles.

For the 12 groups obtained by means of the analysis of hierarchical classification, regarding the growth in number, that offered significant statistical differences with 95% of probability according an ANOVA of double classification, for the different types of used medium ($Fc = 2.52 > Ft = 1.99$) and for the days of cultivation ($Fc = 10.35 > Ft = 2.25$); the evolution of the production of ß-carotene was analyzed (**Figure 5**), of which is understood that the biggest concentration (10–12 of μg Bc/mL) of this pigment, it happened in the cases 49; 50 and 51 all with use of the urea and the superphosphate like nutritious medium, with sugar as organic carbon. Results that they did not offer significant statistical differences with the cultivation carried out with the Johnson medium, reason because it is feasible to use the urea to commercial scale. For the rest of the cases the levels of ß-carotene did not surpass the 3 μg ß-carotene/mL, being among them those that used $NaNO_3$ that so good results offered for the production of biomass.

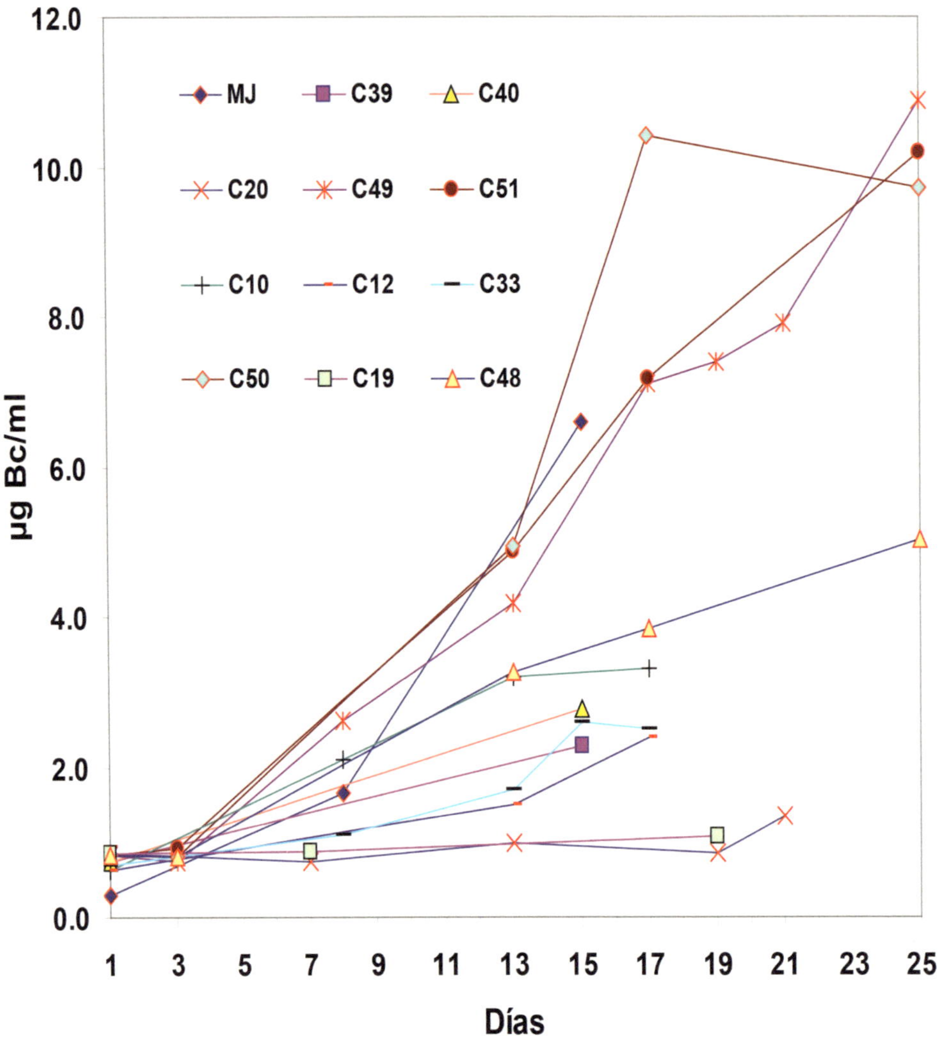

Figure 5.
Evolution of the production of ß-carotene, with the better results.

Analyzing the production of carotene for the results of the cultures employees for the lagoons to pilot scale (**Figure 6**), the biggest evolution was reported for the lagoon with addition of urea (0.2 g/L); superphosphate (0.05 g/L) and sugar (0.1 g/L).

The content of proteins in this microalga strain fluctuated between 50 and 88%, being obtained the highest values with those that were in cultivation that it was added $NaNO_3$. Concentration is it presents it like an alternative in the animal or human feeding.

The dry weight of the *D. salina* corresponded to values between 10 and 40 pg./cel, like to yields between 1.7 and 2.2 g of cells per liter of culture, which are inside the highest among those reported in the literature like commercial scale. The highest coincide with the culture those was added $NaNO_3$ or urea and sugar.

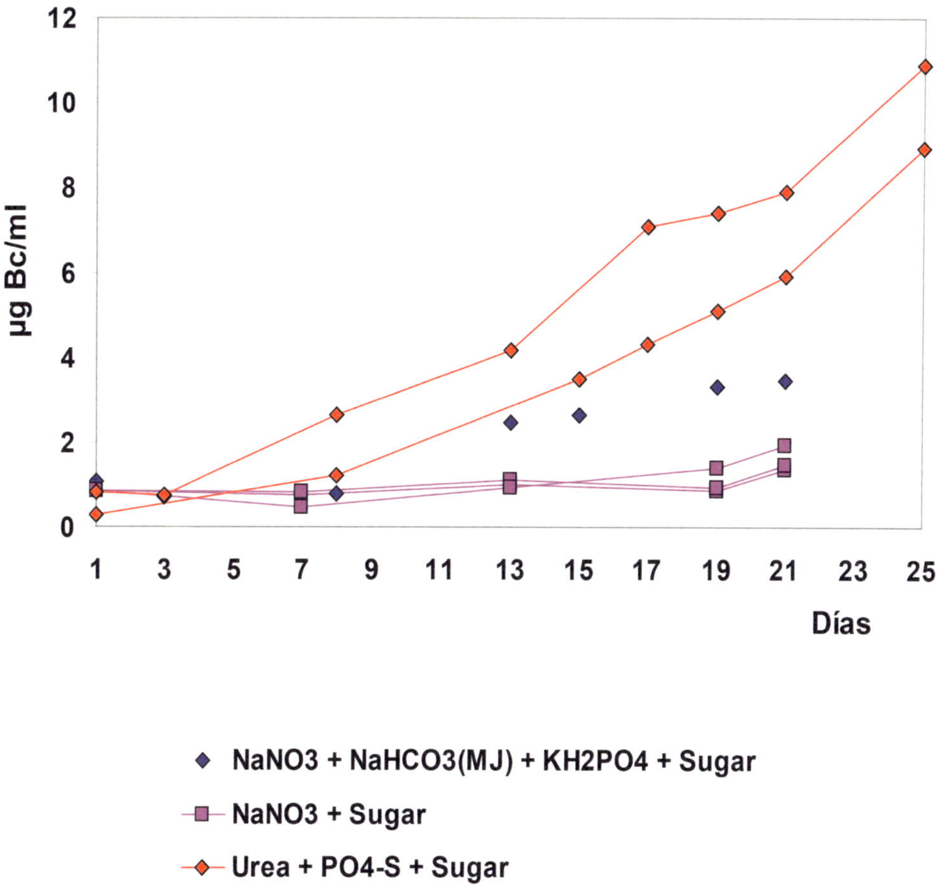

Figure 6.
Carotene for the lagoons to pilot scale, the biggest evolution was reported for the lagoon with addition of urea (0.2 g/l); superphosphate (0.05 g/l) and sugar (0.1 g/l).

4. Economic evaluation

The evaluation of the economic feasibility of the process of obtaining of the microalgas *D. salina* was carried out on the base of the experience gathered until that moment with the one made for the Chlorella sp. [12, 13], and keeping in mind the pilot plant capacity installed at "Industrial de alimentos" (INDAL).

In this calculations was considerate:

a—a volume of cultivation of 3 m^3.

b—a yield of 0.8 kg of dry alga for m^3 of culture medium.

c—a productivity of humid alga that represents a biomass from 1.6 to 2.1 kg/day.

d—a retention time of 15–20 days for each harvesting, representing a crop cycle every two weeks.

e—an available time of 300 days troops, foreseeing 60 days of bottom of technological requirement and other causes.

The methodology for these calculations were carries out using the pattern of economic evaluation created by the ABL of the University of Murdoch, Western Australia, being calculated the cost of a kg of alga for the pilot plant installed at INDAL industry, as well as for an industrial plant. In the **Table 2** a summarized information of the calculation about the unitary cost is presented. The cost obtained to produce 1 kg of alga, as it has been calculated with the pattern in question it was among $6.4 and $4.0 USD.

5. Cost of the means of cultivation

The fundamental culture medium for the commercial cultivation of the *D. salina* was the modified Johnson (Borowitzka, [4]); the Ben-Amotz 1 [7]; the Ben-Amotz 2 [9] and those of the MIP with urea and the sodium that are presented in this work. The reagents to prepare the different culture that can be used by *D. salina* are presented in the **Table 3** and their respective unitary costs in the **Table 4**.

Inputs	**Pilot plant**	**Industrial plant**
Number of lagoons	2.0	2.0
Total Area (m^2)	19.4	720.0
Total Volume (m^3)	1.9	216.0
Outputs	**Pilot plant**	**Industrial plant**
Annual productivity (ton)	1.4	52.0
Total Power centrifugation (kW/d)	1.8	82.76
Losses for evaporation (m^3/año)	11.7	1798.8
Hours for lagoon crop (h)	1.5	13.7
Centrifugal required	1.0	3.0
Volume harvested by day (m^3/día)	3.7	205.20
Volume harvested by year (m^3/año)	1170	65304.4
Biomass harvested by day (Kg/day)	1.2	68.2
Total biomass harvested (Kg/year)	389	21718.5
Cost for nutritious ($)	0	0.03
Construction	**Pilot plant**	**Industrial plant**
Land preparation ($/ha)	50.0	5000.0
Cultivation system ($/ha)	838.0	50000.0
Crop system ($/ha)	2102.0	459.8
Contingency ($/ha)	30.2	29.6
Total capital ($)	3050.0	621936.0
Relative cost	**Pilot plant**	**Industrial plant**
Work (%)	0.01	2.7

Inputs	**Pilot plant**	**Industrial plant**
Energy (%)	5.0	13.7
Maintenance (%)	6.2	70.3
Over heads (%)	7.00	5.8
Capital (%)	0.03	7.34
Operation	**Pilot plant**	**Industrial plant**
Energy cost ($/kWh)	0.02	0.07
Maintenance (% of capital)	5.0	10.0
Average cost yearly	**Pilot plant**	**Industrial plant**
Algae cost Kg ($)	6.4	4.0

Table 2.
Summarized information of the calculation about the unitary.

Reagents	**1**	**2**	**3**	**4**	**5**
	g/l	**mM**	**mM**	**g/l**	**g/l**
$MgCl_2.6H_2O$	1.5	—	—	—	—
$MgSO_4.7HH_2O$	0.5	5	5	—	—
KCl	0.2	—	—	—	—
$CaCl_2.H_2O$	0.2	0.2	0.3	—	—
KNO_3	1.0	0.75	0.5	—	—
$NaHCO_3$	0.043	50	50	—	—
KH_2PO_4	0.035	0.2	0.2	—	—
$NaNO_3$	—	—	—	—	—
$CuCl_2$	—	0.001	—	—	—
Traces	10 ml/l	—	—	—	—
Sol. $FeCl_2$	10 ml/l	—	—	—	—
NaCl	200	200	200	200	200
Sugar	—	—	—	0.1	0.1
Urea	—	—	—	0.2	—
PO_4-S	—	—	—	0.05	—

Table 3.
Reagents to prepare the different culture medium.

According to the unitary costs that are presented in the **Table 4**, for the cultivation means used for the microalgae *D. salina* the most economical is since in fact the one that offered better results in the production of ß-carotene, without considering the addition of alone NaCl it represents $0.09 USD, for each m^3 of cultivation. The cost of the salt (NaCl) that ascends $5.94 USD, for m^3 of cultivation represent an initial expense, because when recycling 95% of the medium after the harvesting of microalgae, alone it would be necessary to add 5% of this cost ($0.30/$m^3$) to the represented value, but it must de adding to all the culture medium, because for the comparison, all they use it.

Reagents	$/ton	
Urea	260.0	
PO4-S	239.0	
KNO3	667.5	
NaNO3	456.0	
MgCl2	4400.0	
KH2PO4	700.0	
MgSO4	599.8	
KCl	26200.0	
CaCl2	480.0	
NaHCO3	302.2	
NaCl	33.0	
Sugar	225.0	
Culture Medium	**$/m3**	**NaCl $/m3**
Johnson	5.93	5.94
Ben-Amotz "1"	1.93	5.94
Ben-Amotz "2"	1.91	5.94
MIP (urea)	0.09	5.94
MIP (sodium)	0.22	5.94

Table 4.
The most economical reagents.

6. Conclusions

1. The addition of a nitrate source like $NaNO_3$ or Urea, it increased the speed of growth of this microalgae species with relationship to other cultures employees for *D. salina*.

2. With the substitution of the nutrients proposed in this work it is possible to diminish the cycle of life of this species since about 15 days, less than the 20; that it is when is commercially harvested.

3. The concentration of ß-carotene obtained by means of the medium of proposed culture is at the same level of the yields that were obtained in the world with this species.

4. The protein levels for this microalgae on the base of dry weight are reported between 50 and 88%, what allows their use like source of animal or human feeding.

5. Yields were obtained among 1.6 and 2.1 kg/m^3 of culture with the medium that used urea, superphosphate or sodium nitrate, in substitution of the Johnson medium.

6. The additions of sugar as source of organic carbon accelerate the growth and carotenogenesis in the cultivation of this species.

7. *D. salina* can grow and to be developed heterotrophic, besides autotrophic like are recognized by all the authors.

8. According to the used evaluation pattern, the production of 1 kg of alga, is considered as $6.50 USD and the culture medium proposed for the commercial production of *D. salina* without considering the addition of alone NaCl represent $0.09 USD for each m^3 of cultivation.

7. Recommendations

It is recommended to use a synthetic culture medium constituted by Urea, superphosphate and sugar cane, like Cuban MIP-1 medium for the commercial exploitation of microalga *D. salina*, as alternative for their cultivation to great scale, in areas near to natural salines that it does not explode commercially and it would allow their constant production for the ß-carotene extraction or use like biomass protein for the human feeding or animal.

Due to the carotene concentrations that can be obtained of this strain, their cultivation is recommended to great scale, for its later extraction and commercialization.

We intends to use sugar cane, like source of carbon to accelerate the growth and the carotenogenesis of this species and to diminish their cycle of life to less than 20 days, that which will benefit their commercial exploitation.

Acknowledgements

We thank to all the workers of the fishing industry INDAL, especially for the help that they have lent us for the realization of this work, and especially to Prof. M. Borowitzka for his important comments.

Author details

Gerardo Suárez Álvarez[1]* and Teresita de Jesús Romero López[2]

1 Centro de Investigaciones Pesqueras, Ministerio de la Industria Alimentaria, Barlovento. Sta. Fe. Playa, La Habana, Cuba

2 Centro de Investigaciones Hidráulicas (CIH), Universidad Tecnológica de La Habana, La Habana, Cuba

*Address all correspondence to: gerardoeloy650@gmail.com

References

[1] Hernández-Pérez A, Labbé J. Microalgas, cultivo y beneficios. Microalgae, culture and benefits. Revista de Biología Marina y Oceanografía. 2014;**49**(2):157-173. DOI: 10.4067/S0718-19572014000200001

[2] Oscanoa A, Cervantes M, Fabre P. Manual Para la producción de biomasa microalgal en condiciones de invernadero. Handbook for the production of microalgal biomass under greenhouse conditions. Inf Inst Mar Perú. 2020;**47**(3):332-356

[3] Borowitzka MA. Products from algae. In: Phang SM, Yuan KL, Borowitzka MA, Whitton BA, editors. Algal Biotechnology in the Asia-Pacific Regions. 1994. Kuala Lumpur, Malasya: University of Malaya; ISBN-983-803-026-8. pp. 5-15

[4] Borowitzka MA, Borowitzka LJ. *Dualilla*: 27-58. In: Borowitzca MA, Borowitzca LJ, editors. Micro Algal Biotecnology. UK: Cambridge University Press; 1988. p. 477. ISBN: 052-132-349-5

[5] Vonshak A, Borowitzka M. Laboratory manual. In: Research Seminar and Workshop on Mass Cultures of Microalgae. Algal Biotechnology Laboratory. W. Australia: Murdoch University; 1991. p. 220

[6] Lowry OH, Rosebrough NJ, Farr AL, Randall RJ. Protein measurement with the Folin phenol reagent. Journal of Biological Chemistry. 1951;**193**:265-275

[7] Ben-Amotz A, Avron M. The biotechnology of mass culturing *Dunaliella* for products of commercial interest. Algal and Cyanobacterial Biotechnology. 1989;**7**(2):90-114

[8] Borowitzka MA, Borowitzka LJ. Micro Algal Biotecnology. Cambridge Univ. Press; 1988. p. 477

[9] Ben-Amotz A, Avron M. The biotechnology of cultivating the halotolerant alga *Dunaliella*. Trends in Biotechnology. 1990;**8**(5):121-126

[10] Ben-Amotz A, Katz A, Avron M. Accumulation of β-carotene in halotolerant algae: Purification and characterization of β-carotene-rich globules from *Dunaliella bardawill* (Chlorophyceae). Journal of Phycology. 1982;**18**:529-537

[11] Gloyna E. Estanques de estabilización de aguas residuales. Organización Mundial de la Salud. Ginebra. Serie de Monografias. 1973;**60**:192

[12] Romero LT. Tecnología de cultivo de *Chlorella vulgaris* en los efluentes líquidos de la industria pesquera y subproductos derivados. Anais do congresso Latinoamericano de Ficología. 1998;**I**:475-495

[13] Borowitzka MA, Borowitzka LJ. Effects of salinity increase on carotenoid accumulation in the green algae *Dunaliella salina*. Jounal of Applied Phycology. 1990;**2**:111-119

Chapter 9

Water Cleaning by Means of Microalgae in the Channels of Xochimilco, Mexico

Saúl Almanza Encarnación, María Guadalupe Figueroa Torres, María Jesús Ferrara Guerrero, Aída del Rosario Malpica Sánchez and José Roberto Angeles Vázquez

Abstract

The lake of Xochimilco, Mexico presents a high degree of chemical contamination, despite belonging to a Protected Natural Area and RAMSAR site, due to this, it was decided to evaluate the quality of its waters and propose solutions to its contamination. The objective of this work was to know the relationship that exists between the microalgae species associated with a dump that affects the site, coming from a Wastewater Treatment Plant and the physicochemical factors coupled with the development of bioassays. Water sampling was carried out in the site and adjacent areas to know the microalgae species and their cleaning role, in addition to laboratory bioassays to verify the results. There are 88 species of microalgae and their influence on water cleaning. In laboratory, it was confirmed that some algae species were purifiers of contamination by nutrients.

Keywords: residual water treatment, microalgae, physicochemical factors, Xochimilco, Mexico

1. Introduction

Nowadays the majority of water bodies are with some degree of affectation, in Mexico City one of the main aquatic ecosystems are Xochimilco channels located in the political delegation of the same name, this zone have different appointments including that of Natural and Cultural Heritage of Humanity in 1987 by UNESCO, it was recognized as Protected Natural Area in 1992, also, is considered an ecological conservation area, it is part of the RAMSAR wetlands of global importance, and is cataloged as a GIAHS site [1].

Previously, Lake Xochimilco received water from various rivers and springs, which was mainly used for crop irrigation; besides being the habitat of native and migratory birds [2]; nevertheless, nowadays it is threatened by several problems, mainly by overexploitation of water and contamination by use of agrochemicals [3, 4].

Also, pollutant sources have been observed such as clandestine discharges of sewage from human settlements, and the deposit of solid waste [5]. Since the growth of the urban spot, there have been changes in the productive activities causing diverse problems in the area such as the excessive growth of introduced aquatic vegetation like water lilies. Water control structures have been affected in the area and it has been observed a decrease in the extension of the main and secondary channels. This has caused that in places with little or no movement of water, there is an increase in eutrophication levels and flooding during the rainy season [6].

Due to the above, the aquatic communities have been affected, one of these communities are the microalgae which, by having short life cycles, present changes in its structure and dynamics in short period of time. Also, microalgae have a great importance for Xochimilco channels, despite their small size and being unnoticed by most people, because they provide important environmental services, such as: CO_2 capture, oxygen liberation from photosynthesis, natural cleaning of the water of the channels (deseutrophication), in addition to reduce the concentration of heavy metals, which are harmful to the ecosystem [7–11].

On the other hand, microalgae serve as food for other species present in water bodies, such as copepods, crustaceans, small fish, and some amphibians in juvenile stages, helping to the conservation of biodiversity. Also, can be used as biological indicators to monitor water quality [4].

Considering the above, the aim of this investigation was to know the role of microalgae in the purification of water from a Wastewater Treatment Plant, under field and laboratory conditions.

2. Material and methods

2.1 Study area

The lake system of Xochimilco is located south in Mexico City, surrounded by a mountainous area formed by the hills Xochitepec, Cantil and the volcanoes Teoca, Zompole and Teutli [12]. It is in the geographic coordinates 19° 00′ and 19° 20’ North Latitude; 99° 00′ and 99° 16’ West Latitude, with an approximate surface of 2657 ha, at an altitude between 2240 and 2500 m [13–15].

The climate is sub-humid temperate, with rains in summer and an average annual temperature varying between 8 and 18°C. The average rainfall is of 620 mm/year, the most abundant rains occur between the months of June and September and the minimum from December to February [16].

Among the most important channels are Cuemanco, Canal Nacional, Chalco, del Bordo, Apatlaco, San Sebastián, Apampilco, Texhuilo and Japón. Also, the main lagoons are Tlilac, del Toro, Huetzalin, Apampilco, Texhuilo and the Lake of conservation of flora and fauna of San Gregorio Atlapulco [17].

2.2 Field work

Sampling was carried out at the dump of water from Cerro de la Estrella wastewater treatment plant, located in the old channel of Cuemanco. The samples of microalgae were collected directly from the outlet of the dump water pipe, just at the drop and at the distances of 10, 20, 40 and 60 m (**Figure 1**).

Figure 1.
Study zone and sampling points map, based on Google earth, 2020.

For microalgae study two types of samples were taken, for the quantitative and qualitative analysis. For the first, samples were taken with the aim of a Van Dorn bottle, placed in 500 mL containers with lugol at 1% solution. For qualitative samples it was used a trawl net with a mesh opening of 54 μm, this samples were placed in amber jars of 30 mL and it was added formalin at 4%. In each sampling point it was recorded the pH, temperature, conductivity, depth, and turbidity. Also, were taken water samples of 100 mL to determine nutrients concentrations of NO_2-, NO_3-, NH_4+, and PO_4^{3-} in laboratory.

On the other hand, water samples with live organisms were stored for the isolation of three species for use in wastewater purification bioassays from the Cerro de la Estrella treatment plant.

2.3 Laboratory work

The sample review of microalgae was carried out in Phycology and Phytopharmacology laboratory from UAM Xochimilco, using a Zeiss optical microscope model Axiostar. Aliquots of 0.1 mL were taken and reviewed with the scanning technique [18], which consist on locate a starting point and make the revision in the form of "transects", from each sample the necessary aliquots were revised until no new organism was observed.

For the isolation of microalgae different techniques were used, including capillary pipetting, seeding in agar plates, and reseeding in liquid medium.

The nutrients (NO_2-, NO_3-, NH_4+, PO_4^{3-}, TP, and TN) were evaluated using a multiparametric photometer HI 83200 [19]. Initial parameters were valuated, which were the different forms of inorganic nitrogen (NO_2-, NO_3-, and NH_4+), phosphate as orthophosphate and heavy metals (Pb, Ni and Cu), using a multiparametric photometer HI 83200 and a spectrophotometer HACH 3900.

Bioassays of removal of nutrients and heavy metals were conducted, for which the isolated microalgae were used in the samples of treated wastewater from the Cerro de la Estrella treatment plant. To bioassays, manual agitation twice a day were given to avoid sedimentation of microalgae and water nutrients.

To provide the necessary amount of light, white light bulbs were used and controlled at intervals of 12 hours light and 12 hours dark with a timer clock.

2.4 Data analysis

In order to know the differences in the values of nutriment concentrations and parameters taken *in situ* in the sampling points located at different distances, were obtained a coefficient of variation, standard deviation, and arithmetic mean, using the program Excel 2013, in addition to a linear correlation analysis to know the interaction with the physicochemical parameters registered in field.

On the other hand, counts were made for 10 days and plotted to know the growth curve of each isolated microalgae.

Regarding bioassays, with the obtained values population growth graphs were made for each microalga; also, it was made a comparison between physicochemical parameters measured in each bioassay. In addition, this data was analyzed to observe the change in nutrient and heavy metal concentrations in each bioassay and evaluated the purification capacity of each microalga.

3. Results

3.1 Microalgae

A total of 88 species belonging to five Divisions were found, being Bacillariophyta the most diverse Division with 35 species, followed by Chlorophyta Division with 27 species, while Euglenophyta, Cyanoprokaryota and Pyrrophyta Divisions had 16, 6, and 4 species respectively (**Figure 2**).

The sampling point with the highest number of species was the one located at 20 meters away from the dump, presenting 56 species, while the sampling point with the lowest quantity of species was the one located directly under the dump with only 16 species observed (**Figure 3**).

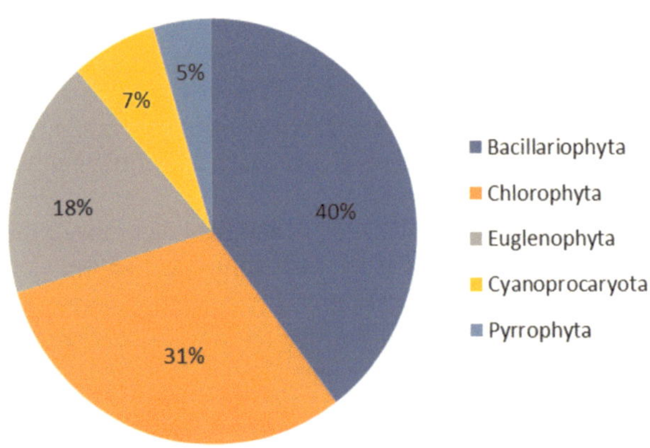

Figure 2.
Percentage of algae species by division found in the "Cerro de la Estrella" treatment plant dump.

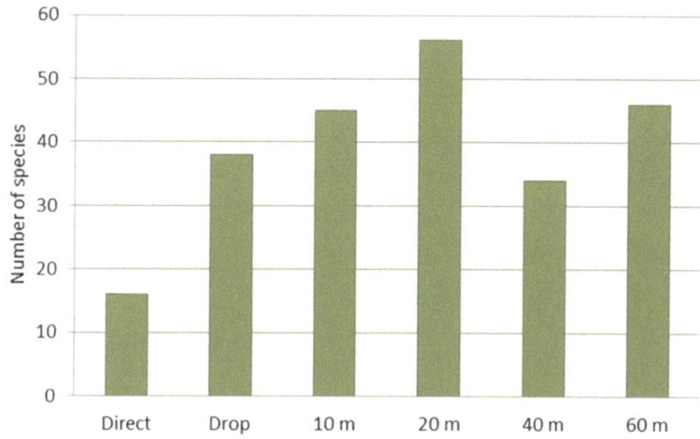

Figure 3.
Total number of species in the sampling points.

3.2 Physicochemical parameters

The values of the physicochemical parameters taken in the field and the nutrients of the samples analyzed in the laboratory are shown in **Table 1**. It is included the coefficient of variation, which showed that physicochemical parameters showed a low proportion of variation in most of them (2–18%), except in depth where it was obtained 37%. On the other hand, the nutrients also showed a low variation ranging from 14 to 20%.

The concentrations of NO_{2-} were found at an interval between 0.621 mg/L obtained at 20 meters and 0.751 mg/L in the sample taken directly under the dump; for NO_{3-} the concentrations interval was between 6.9 mg/L in the sampling point of 20 m, and 14 mg/L

	Turbidity (cm)	Depth (cm)	Temperature (°C)	Conductivity (μS/cm)	pH	TN (mg/L)	TP (mg/L)
Direct	ND	ND	21.5	711	8.3	14.6	7.6
Fall	70	70	21.7	772	8	11.5	9.4
10 m	70	150	21.6	772	7.8	11.1	7.3
20 m	85	95	21.8	775	8	10.3	6.4
40 m	72	92	22.1	784	7.9	10.4	6.9
60 m	50	60	22.5	787	7.7	11	5
Standard deviation	12.52	34.90	0.37	28.07	0.21	1.60	1.50
Arithmetic mean	69.4	93.4	21.87	766.83	7.95	11.48	7.1
Variation coefficient	0.18	0.37	0.02	0.04	0.03	0.14	0.20
VC expressed in percentage	18%	37%	2%	4%	3%	14%	20%

The nutriments are shown as total nitrogen (TN) and total phosphorus (TP). Variation coefficient (VC) is expressed as the ratio between the standard deviation and the arithmetic mean.

Table 1.
Values of environmental factors (physical and chemical) in the Cerro de la Estrella treatment plant dump and in the different sampling points (at different distances).

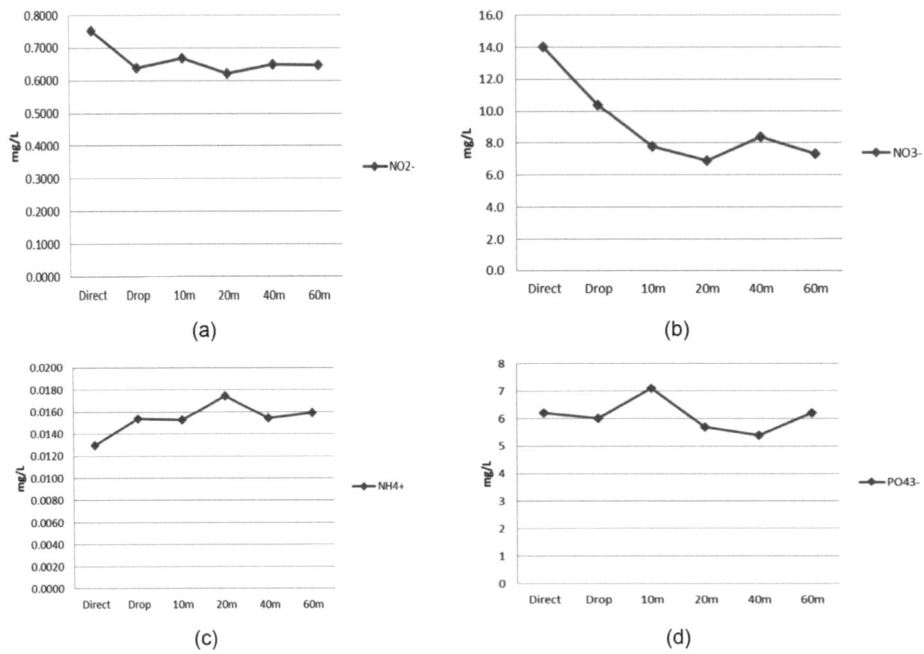

Figure 4.
(a-d). Nutrient concentrations (NO_2^-, NO_3^-, NH_4+ y PO_4^{3-}) at the different sampling points.

registered in the sample taken directly under the dump; moreover, the lowest concentration of NH_4+ was found in the sample taken directly under the dump with 0.0129 mg/L and the highest value was obtained in the sample of the spot 20 m with 0.0174 mg/L, observing that the water from the dump, was enriched by mixing with the water of the ecosystem, which is more rich in this compound in the further sampling points. The concentrations of ortho PO_4^{3-}were between 5.4 and 7.1 mg/L, having the lowest value in the sample from 40 m and the highest in the sample from 10 m (**Figure 4**).

Regarding the linear correlation analysis, it can be observed that NO_{2-}, NO_{3-} and NH_4+ have a higher correlation value compared to the parameters measured in field, this in Ref. to species richness, instead in terms of abundance it is observed that NO_{2-} and NH_4+ together with turbidity are those that have a higher correlation, nevertheless, the correlation was lower when comparing it to the one obtained with species richness (**Table 2**).

3.3 Isolation and growth of microalgae

As a result of the isolation methods (capillary pipetting and seeding in agar plates), it was possible to achieve the growth of three species of microalgae, two belonging to the Division Chlorophyta (*Chlamydomonas* sp. and *Chlorella* sp.) and one of the Division Bacillariophyta (*Nitzschia* cf. *amphibia*).

3.3.1 Growth curves

For the growth of isolated microalgae, the three strains were seeded in enriched liquid culture medium, based in the Bold Basal formula and cell counting were made for 10 days, having as a result the following:

Parameter	Species richness correlation	Abundance correlation
NO_{2-}	0.7584	0.6497
NO_{3-}	0.8485	0.5187
NH_4+	0.9059	0.7887
PO_4^{-3}	0.0009	0.0704
Temperature (°C)	0.1295	0.00001
Conductivity (μS/cm)	0.6184	0.3647
pH	0.4663	0.0848
Turbidity (cm)	0.6920	0.7997
Depth (cm)	0.4803	0.4484

Table 2.
Linear correlation analysis of the physicochemical parameters on the richness and abundance of the species.

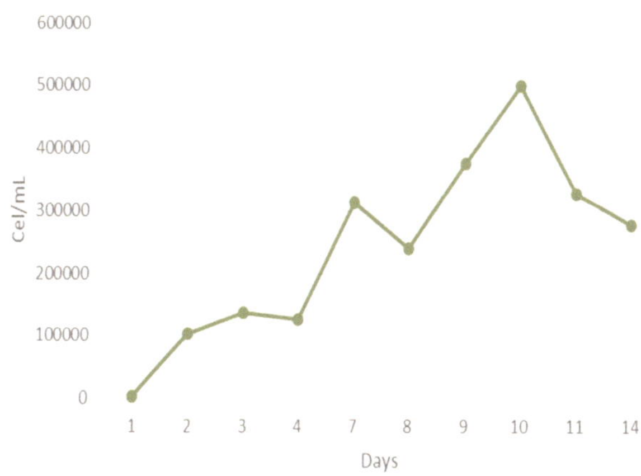

Figure 5.
Growth curve of Chlamydomonas *sp.*

Chlamydomonas sp.

In the following graph it is shown the growth behavior of microalga *Chlamydomonas* sp. in liquid culture medium presenting a maximum growth at ten days (**Figure 5**).

Chlorella sp.

Regarding *Chlorella* sp. it was observed that its growth was exponential even after 14 days (**Figure 6**).

Nitzschia cf. *amphibia.*

This microalga presented a heterogeneous growth in time, having its maximum growth point at thirteen days and then decrease (**Figure 7**).

3.4 Bioassays

3.4.1 Growth curves

As for the growth of microalga *Chlamydomonas* sp., it can be observed that there was an exponential increase of organisms during the first five days with a slight

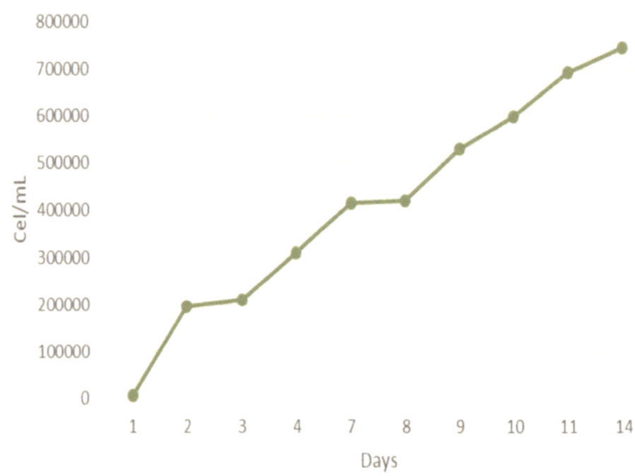

Figure 6.
Growth curve of chlorella *sp.*

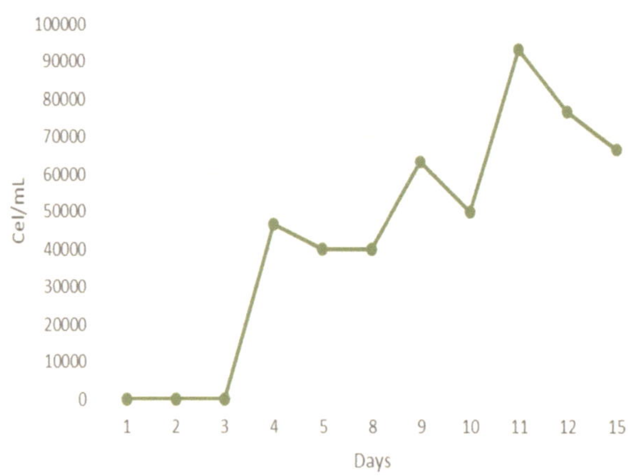

Figure 7.
Growth curve of Nitzschia *cf.* amphibia.

decrease at day six, recovering on the seventh day and from there presented marked fluctuations in the number of individuals until the end of the experiment. In the case of *Chlorella* sp. it was observed the same behavior as *Chlamydomonas* sp. because the maximum growth was reached at day five, presenting a decrease of organisms from day six until day ten, nevertheless, at day eleven there was an upturn of organisms maintaining it during three more days, to have a downbeat in the last two days.

On the other hand, the growth of *Nitzschia* cf. *amphibia* had an exponential growth in the first four days and a decease during the next two days, at seventh and eighth day it presented an upturn and then continue with ups and downs until the end of experiment (**Figure 8**).

3.4.2 Nutriments and heavy metals comparison between treatments

The concentrations of NO_2^- in water of bioassay were higher after incubation time, as opposed to expected, however this could be since not being an axenic culture

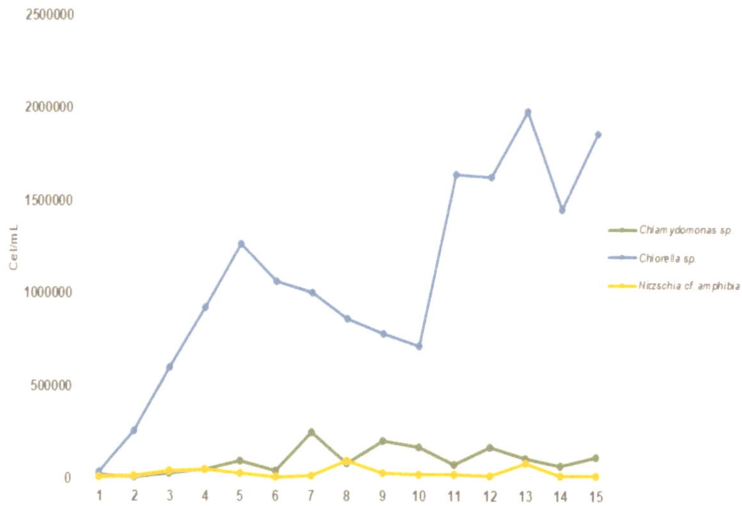

Figure 8.
Growth curves of the three microalgae in the bioassays.

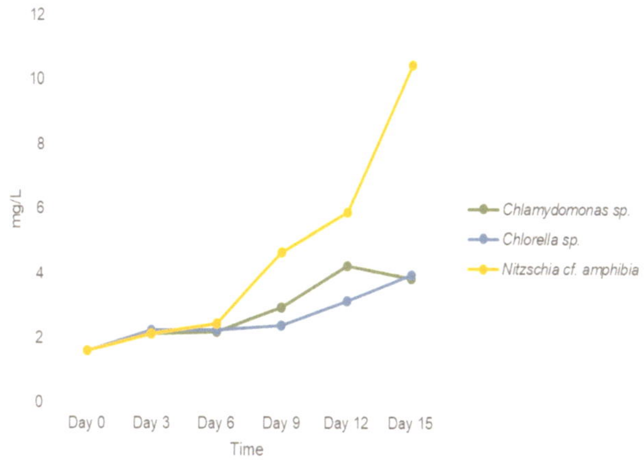

Figure 9.
Nitrites values of each treatment.

it could present nitrifying bacteria that can oxidate NH_4+, increasing the NO_2^- at the end of incubation (**Figure 9**).

In the case of NO_3-, it was observed that in treatments with *Chlamydomonas* sp. and *Chlorella* sp. there was a decrease in the quantities of this nutriment, contraire to the bioassays with *Nitzschia* cf. *amphibia* where its values increased after three days (**Figure 10**).

Moreover, in NH_4+ there was a decrease of this nutriment with the three microalgae from day three to the end of the experiment (**Figure 11**).

Regarding phosphate it did not show a continuous decrease in any of the treatments with the three microalgae, so there were ups and downs over the course of the days in all treatments, nevertheless, *Chlamydomonas* sp. and *Nitzschia* cf. *amphibia* presented the lowest quantities of this nutriment at day 12 (**Figure 12**).

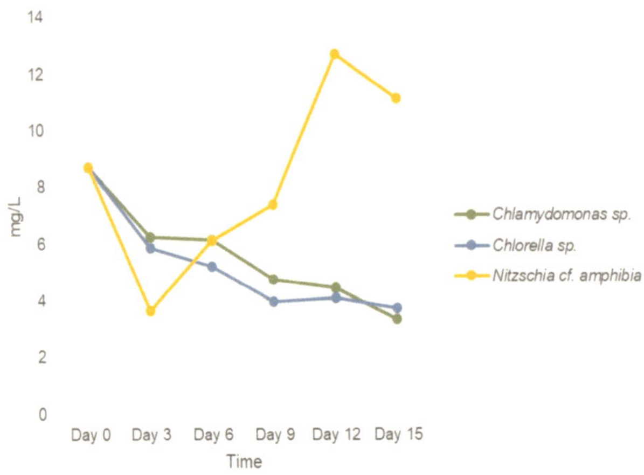

Figure 10.
Nitrates values of each treatment.

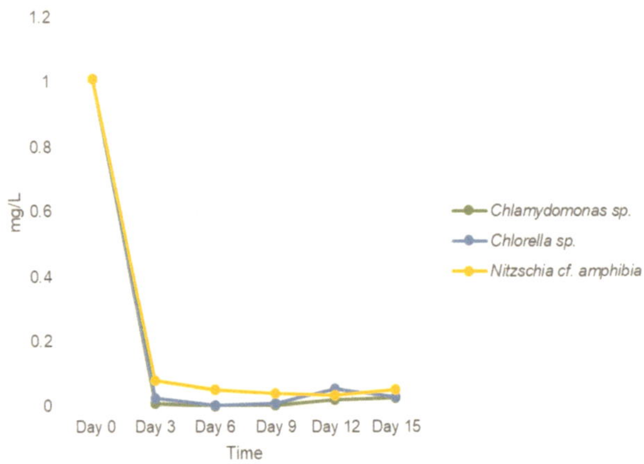

Figure 11.
Ammonium values of each treatment.

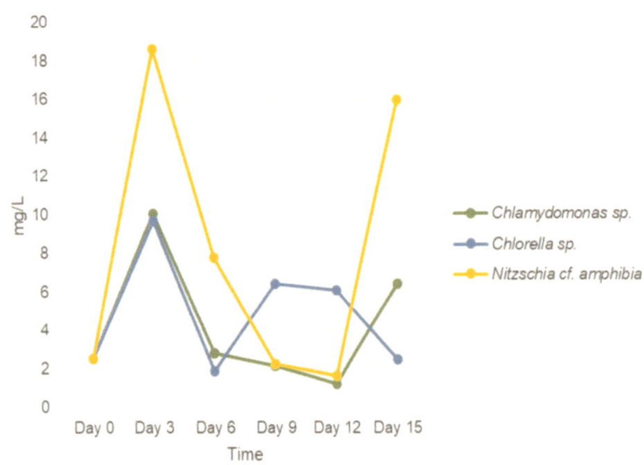

Figure 12.
Phosphates values of each treatment.

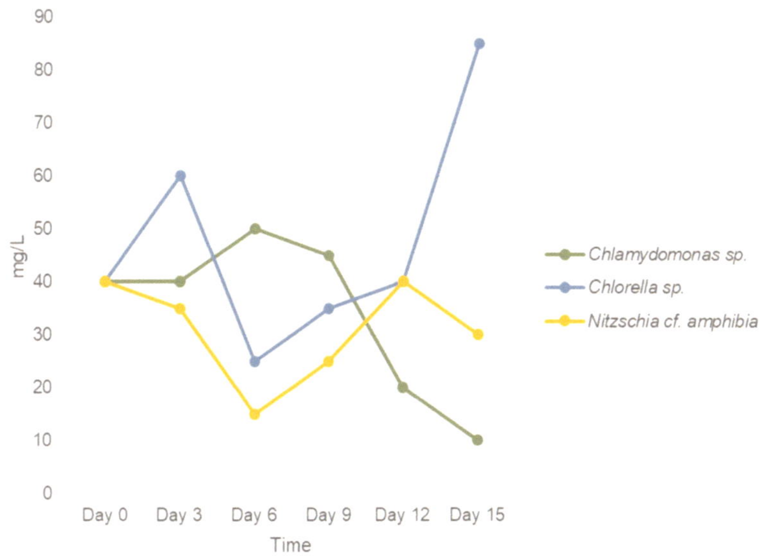

Figure 13.
Nickel values of each treatment.

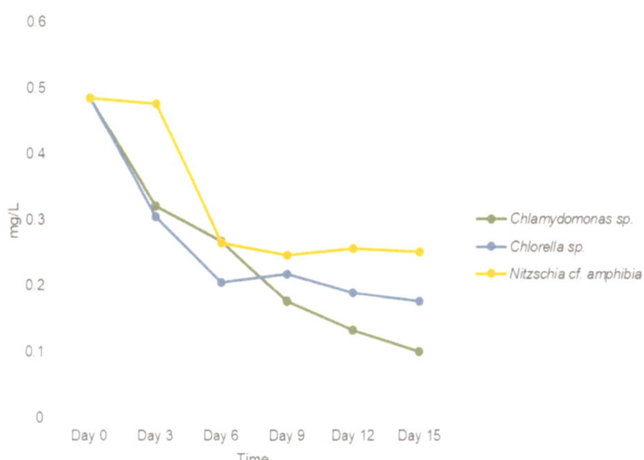

Figure 14.
Lead values of each treatment.

As regards heavy metals, in the case of copper a graph is not included because the values obtained were zero from the beginning of the experiment in the three treatments.

For nickel, it is noted that the quantities of this metal were not completely reduced, instead there were sharp fluctuations in values especially in *Chlorella* sp. and *Nitzschia* cf. *amphibia*, only in the bioassay with *Chlamydomonas* sp. it was observed a decrease of the metal from day nine (**Figure 13**).

In the case of lead treatments, it was observed that the three species of microalgae decreased the quantities of this metal, however, the bioassay with *Chlamydomonas* sp., was the one that obtained the lowest concentration of lead at the end of the 15 days of experimentation (**Figure 14**).

	Treatment	Day 0	Day 3	Day 6	Day 9	Day 12	Day 15
Nitrite	*Chlamydomonas* sp.	0	−34	−38	−84	−166	−141
	Chlorella sp.	0	−41	−41	−50	−97	−147
	Nitzschia cf. *amphibia*	0	−34	−53	−191	−269	−553
Nitrate	*Chlamydomonas* sp.	0	28	29	45	49	61
	Chlorella sp.	0	33	40	55	53	57
	Nitzschia cf. *amphibia*	0	58	30	15	−46	−28
Ammonium	*Chlamydomonas* sp.	0	99	100	99.5	98	98
	Chlorella sp.	0	98	100	99	95	97
	Nitzschia cf. *amphibia*	0	92	95	96	97	95
Phosphate	*Chlamydomonas* sp.	0	−302	−12	15	52	−157
	Chlorella sp.	0	−286	26	−157	−144	1
	Nitzschia cf. *amphibia*	0	−644	−212	12	36	−538
Nickel	*Chlamydomonas* sp.	0	0	−25	−13	50	75
	Chlorella sp.	0	−50	38	13	0	−113
	Nitzschia cf. *amphibia*	0	13	63	38	0	25
Lead	*Chlamydomonas* sp.	0	34	45	64	73	79
	Chlorella sp.	0	37	58	55	61	64
	Nitzschia cf. *amphibia*	0	2	45	49	47	48

Table 3.
Percentage of removal of each compound with each microalga.

Table 3 shows the removal percentage of each nutrient by the three microalgae in the different bioassays. Negative numbers indicate that there was no nutrient removal, but that its value increased.

4. Discussion

4.1 Microalgae

According to obtained results a total of 88 species of microalgae were determined, from which 55 have been previously reported by diverse authors, in different points in the Xochimilco channels and at different seasons. On the other hand, 24 species were new registers for the study zone, which indicates that nowadays there's no full knowledge of the species present in this place, this can be due to the fact that over time the conditions of the environment are changing, making changes in the microalgae community composition, as well as the introduction of new species, that come from treated water.

4.2 Physicochemical parameters

Regarding the parameters taken on field, it could be observed that most of them did not vary significantly, only depth had a variance higher than 30%, which decrease according to distance, one of the reasons can be the movement that the waterfall generates on the lake, because the greater the distance the movement of the water is less, which favors a higher deposit of sediments.

Nutriments values were compared with the NOM-001-ECOL-1996 [20], where the maximum permissible limits for basic pollutants in wastewater discharges are established. It was observed that the values of total nitrogen (TN) for use for agricultural irrigation do not exceed the maximum permitted limits (40–60 mg/L), because in the present investigation a maximum value of 14.6 mg/L was obtained. Nevertheless, this value is near the maximum limits permitted for urban public use which is of 15 mg/L.

Regarding to obtained total phosphorus values (TP), the values were in a range of 5 to 9.4 mg/L and they did not exceed the maximum permitted limits for use for agricultural irrigation which are of 20 mg/L daily average and 30 mg/L monthly. However, for urban public use, the obtained values are above the monthly average (5 mg/L) and very close to the permitted daily average which is of 10 mg/L [20].

As for nitrogen, it was observed a higher concentration in the form of Nitratos (NO_3^-) in the site of the dump (14 mg/L), observing a decrease as the sites were farther away from the waterfall, this can be due to the higher density of microalgae found in those sites, which could be using this nutrient, because NO_{3-} are one of the main forms of nitrogen that absorb microalgae [21].

The above is complemented with the linear correlation analysis because the higher correlation values were obtained in the NO_{3-} and NH_4+ on species richness, which points out that the presence of these nutrients is essential for the growth and formation of biomass, as microalgae absorb them directly [22].

4.3 Isolation and growth of microalgae

After work with isolation techniques, the species *Chlamydomonas* sp. and *Chlorella* sp., responded better to isolation in liquid medium, because both species are more of planktonic character, in addition to being species with wide ranges of tolerance regarding temperature and nutriments.

Nitzschia cf. *ampphibia* being a specie of benthic character, was isolated in solid medium with the Pasteur pipette spraying technique, so when growing it produced brown spots, which is a characteristic of the species of Bacillariophyta Division.

It should be mentioned that the species *Chlamydomonas* sp. and *Nitzschia* cf. *amphibia* did not were very abundant in in reviewing field samples, only *Chlorella* sp. was.

4.4 Bioassays

4.4.1 Growth curves

Regarding the growth of the three microalgae in the wastewater from the dump, it was observed that none of them presented a normal growth curve [23], in such a way that in the case of *Chlamydomonas* sp. and *Chlorella* sp. they had an exponential growth until the fifth day of experiment without having a stationary phase, but continued directly to death phase during the sixth day for *Chlamydomonas* sp. and tenth day for *Chlorella* sp., to then observe ups and downs in the number of cells until the end of the experiment. These results differ from the study conducted by [24] which worked with *Chlorella* sp. and obtained a more normal growth curve, and it was observed that the maximum growth value was obtained at 15 says, having its death phase between 17 and 18 days of their experimentation.

Nitzschia cf. *amphibia* also showed a different growth than conventional curves, because it had an exponential growth during the firsts days having its maximum growth on day four without having an stationary phase, but as in the other two species went directly to the death phase during the five and six days of the experiment, followed by various phases of exponential growth and death in few days, which could be explained due to the rapid growth that is given by this species, that according to Brennan and Owende [25], some species of microalgae can duplicate its biomass in less than 24 hours.

Additionally, it was observed that in some time the cells remained glued to the glass in the flask and despite the care a smaller number of cells was counted. When the flask was vigorously agitated it was not counted the same number of individuals as the days before.

4.4.2 Comparison between treatments

Regarding to comparison of the nutriment evaluation it was observed that *Chlamydomonas* sp. and *Chlorella* sp. use NO_3- and NH_4+ as source of nitrogen because they reduced the NO_{3-} in 61% for *Chlamydomonas* sp. and 57% for *Chlorella* sp. at the end of the experiment, and both species reduced up to 100% the NH_4+ at day six, in this regard. Oliveros and Wild [26] point out that *Chlorella* sp. was capable of removing up to 95% of NO_{3-} in wastewater and mentioned that this microalga is suitable for this type of treatment because it has a great ability to remove nutrients in wastewater. Meanwhile, Chacón et al. [27], mentioned in their study that the highest reductions by *Chlorella* sp. were of NH_4+, reducing it by 100% as in this study. However, their experimentation time was 27 days while in this investigation was 15 days.

In the case of *Nitzschia* cf. *amphibia*, it was observed that this specie use NH_4+ as source of nitrogen, because it reduced this nutriment in a 97% at 12 days of experimentation and according to Pérez [28], some microalgae prefer nitrogen in form of NH_4+, so when looking at the graphs of the nitrite and nitrate values an increase was noticed instead of decreasing.

On the other hand, in the case of phosphates, the method of using it was more dynamic, which was reflected in increases and decreases during the experiment by the three microalgae, so a removal of this nutriment could not be found. However, it was observed that *Nitzschia* cf. *amphibia* and *Chlamydomonas* sp., presented higher percentages of removal of this nutriment at 12 days of experimentation with 36% and 52% respectively. This is like the reported by Oliveros and Wild [26], which worked with *Chlorella* sp. and obtained a removal percentage of 20% at 8 and 12 hours of experimentation.

About heavy metals, as mentioned above the values of copper were zero, so they were discarded from the subsequent analyses. Meanwhile, the levels of nickel were reduced by *Nitzschia* cf. *amphibia,* after six days of experimentation by 63%. Nevertheless, after that time the values increased again, which might suggest that microalgae adsorb and retain it for some time, but when it dies, the metal is released again, and the value increase. Instead, *Chlamydomonas* sp. had a slight increase in the values on the first six days; however, from day nine it reduced the value of nickel by 75% at the end of experiment, being this specie that show the highest removal percentage of removal. On the other hand, *Chlorella* sp. at the third day of experimentation had an increase in nickel values, and at sixth day there was a slight decrease of 63%, nevertheless, after that it was observed that the values increased util the end of the experiment, contraire to the experiment made by Hammouda et al. [29] which reported a removal of 77.3% up to 81% of Ni by *Chlorella* sp.

The lead was the metal that presented highest removal by the three microalgae, with 49% by *Nitzschia* cf. *amphibia,* 64% with *Chlorella* sp. and 79% with *Chlamydomonas* sp., being this last one the one with the highest capacity of absorption of this metal. It is worth mentioning that the experiment lasted 15 days and it was not possible to decrease the totality of this metal by the microalgae. Perhaps, if the experimentation time was longer, the totality of this metal could have been removed, but would be necessary to remove the produced biomass, because the metal would remain in the microalgae and could be transferred to the next food chain levels, including reaching humans.

Of the three microalgae study, *Chlamydomonas* sp. appears to be the one that have better capacity of nutrients and heavy metals removal with 50%, except in NO_2-, so it is reiterated that apparently, they do not use this form of nitrogen for their growth. Secondly, there is *Chlorella* sp. which although with some nutrients such as phosphate had a low percentage of removal (26%), in other nutrients such as NH_4+ reached to remove from 95 to 100% throughout the experiment. This match with what was reported by Martínez et al. [30], which mention that some species of chlorophytes are capable of remove 98% of the phosphorus and up to 100% of nitrogen in wastewaters.

Meanwhile, *Nitzschia* cf. *amphibia* resulted being the microalgae with lowest efficiency regarding the quantity of removed compounds, because of not decreasing nitrite and nitrate values and had low removal of heavy metals, nevertheless, it can be considered as highly efficient in the removal of NH_4+ since it reached a 97%.

5. Conclusions

- The algal community was composed by a total of 88 species from the Divisions Bacillariophyta, Chlorophyta, Euglenophyta, Cyanoprocaryota and Pyrrophyta.
- 24 new records were found for the study area.
- Physicochemical parameters taken on field did not vary significantly over the sampling points.
- NO_3- and NH_4+ are the parameters that had highest relation with species richness.
- Valued nutriments in laboratory did not surpass the maximum permissible limits.
- The spraying and seeding techniques in agar were the most effective for the isolation of microalgae species.
- *Chlorella* sp. was the microalgae that best developed to the conditions of cultivation and the one that presented the greatest population growth.
- *Chlorella* sp. was the specie that had better growth in the treatments.
- *Chlamydomonas* sp. was the specie that presented highest percentages of nutrients and heavy metals removal.

- Regarding nutrients, the three species presented good efficiency in the removal of NH_4+.
- Regarding heavy metals, the three species were efficient in the removal of lead.
- It is recommended for future laboratory experiments, to work with axenic microalgal cultures, even though in natural conditions, microalgae coexist with bacteria.

Acknowledgements

We would like to express our gratitude to Universidad Autónoma Metropolitana and specially to Maestría en Ecología Aplicada for support to make the present research.

Author details

Saúl Almanza Encarnación[1*], María Guadalupe Figueroa Torres[2], María Jesús Ferrara Guerrero[2], Aída del Rosario Malpica Sánchez[2] and José Roberto Angeles Vázquez[2]

1 Universidad Autónoma Metropolitana-Unidad Xochimilco, Maestría en Ecología Aplicada, México

2 Universidad Autónoma Metropolitana-Unidad Xochimilco, Ciudad de México, México

*Address all correspondence to: saul20almanza@gmail.com

References

[1] Aranda M. Ficha Informativa de los Humedales de Ramsar (FIR). México: Secretaría de Medio Ambiente; 2004. p. 13

[2] Alatriste O. Xochimilco Aspectos histórico-culturales. México: CEPE-UNAM; 2010. p. 22

[3] Arcos R, González J, Ramos L, Rosas F. Evaluación del componente orgánico como un factor indicativo del estado trófico del lago de Xochimilco. México: Facultad de Estudios Superiores Zaragoza, UNAM; 2006. p. 10

[4] Figueroa M, Arana F, Almanza S, Ramos M, Ferrara M. Microflora de Xochimilco y San Gregorio Atlapulco. CienciaUAT. 2015;**9**(2):15-29

[5] Espinosa A, Mazari M. Pueblos indígenas de México y agua: Xochimilcas. Atlas de culturas del agua en américa latina y el caribe. México: Instituto de Ecología, UNAM; 2009. p. 31

[6] Figueroa M, Linares F, Ferrara M, Castro J. Manejo Integral de la Cuenca de Xochimilco y Sus Afluentes. Revista digital e-bios. México: Universidad Autónoma Metropolitana- Xochimilco; 2014. p. 124

[7] Bermeo L. Estudio del cosechado de cultivos de microalgas en agua residual mediante técnicas de centrifugado. España: Universidad Técnica Particular de Loja; 2011. p. 54

[8] Cañizares, R., Montes, M. & H. Perales. 2013. Las microalgas en la biorremediación acuática: Una alternativa biotecnológica. pp. 217-234. En Alarcón, A. y R. Ferrera. (eds). Biorremediación de Suelos y Aguas, Contaminadas con Compuestos Orgánicos e Inorgánicos. Editorial Trillas. México. 333 p.

[9] Codina M, García C, Barón J, da Silva S, Bosch J. Planta piloto de microalgas para mejoramiento del tratamiento efluentes urbanos en Catamarca, Argentina. Mendoza. Argentina: Facultad de Ingeniería, Universidad Nacional de Cuyo; 2012. p. 11

[10] González A. ¿Qué son las microalgas? Interés y uso. España: Fundación CAJAMAR–Grupo Cooperativo CAJAMAR; 2015. p. 11

[11] Jbari N. Utilización secuencial de microalgas en depuración y adsorción de Cr (VI). España: Universidad de Granada, Facultad de Ciencias, Departamento de Ingeniería Química; 2012. p. 413

[12] Carrión C, Ponce-de León C, Cram S, Sommer I, Hernández M, Vanegas C. Aprovechamiento potencial del lirio acuático (*Eichhornia crassipes*) en Xochimilco para fitorremediación de metales. Agrociencia. 2012;**46**(6): 609-620

[13] Abeja O. Biomarcadores de daño temprano en *Ankistrodesmus falcatus* y *Hyalella* azteca como indicadores de estrés en tres áreas (urbana, turística y agrícola), de la zona lacustre de Xochimilco. México: Instituto Politécnico Nacional, Escuela Nacional de Ciencias Biológicas; 2011. p. 110

[14] Arcos R, Díaz G, Domínguez A. Macrofitas acuáticas: ¿Contaminantes o soluciones de la contaminación por metales pesados? México: XXVIII Congreso Interamericano de Ingeniería Sanitaria y Ambiental Cancún; 2002. p. 8

[15] Gobierno del Distrito Federal (GDF). Gaceta Oficial del Distrito Federal (No. 1717 Tomo IV). México. 2013. p. 148.

Available from: https://data.consejeria.cdmx.gob.mx/portal_old/uploads/gacetas/5265e7b13b072.pdf

[16] Pedroza R. Productividad primaria y diversidad microbiana fotolitoautotrófica en los Canales de Xochimilco. México: Escuela Nacional de Ciencias Biológicas, IPN; 2007. p. 60

[17] Gobierno del Distrito Federal (GDF) Gaceta Oficial del Distrito Federal (No. 5). México. 2006. p. 164. Available from: https://paot.org.mx/centro/gaceta/2006/enero06/11enero06.pdf

[18] Schwöerbel J. Métodos de Hidrobiología. Madrid, España: H. Blume; 1975. p. 262

[19] American Public Health Association, American Water Works Association & Water Pollution Control Federation (APHA). Standard Methods for the Examination of Water and Wastewater. Washington, D. C. EUA: Decimoséptima edición. APHA; 1989. p. 1469

[20] Secretaria de Medio Ambiente y Recursos Naturales (SEMARNAT). Norma Oficial Mexicana NOM-001-Semarnat-1996, Que Establece Los Límites Máximos permisibles De Contaminantes En Las Descargas De Aguas Residuales En Aguas Y Bienes. México: Nacionales; 1997. p. 35

[21] Grobbelaar J. Algal nutrition, mineral nutrition. In: Richmond A, editor. Handbook of Microalgal Culture: Biotechnology and Applied Phycology. Hoboken, NJ: Blackwell Science; 2004. p. 576

[22] Tam, N, & Y. Wong. 1996. Effect of ammonia concentrations on growth of Chlorella vulgaris and nitrogen removal from media. Bioresource Technology 57: 45-50.

[23] Montes O. Efecto de la intensidad de la luz y de la tasa de inyección de aire en el crecimiento y la producción de la microalga *Nannochloropsis* sp., cultivada en un biorreactor plano. Tesis de posgrado en Ciencias en Acuicultura. Baja California. México: Centro de Investigación Científica y de Educación Superior de Ensenada; 2014. p. 70

[24] Vacca V, Angulo E, Puentes D, Torres J, Plaza M. Uso de la *Chlorella* sp. viva en suspensión en la decoloración del agua residual de una empresa textil. Prospect. 2017;**15**(1):93-99

[25] Brennan L, Owende P. Biofuels from microalgae-a review of technologies for production, processing, and extractions of biofuels and co-products. Renewable and Sustainable Energy Reviews. 2010; **14**:557-577

[26] Oliveros D, Wild J. Evaluación de la eficiencia de remoción de nutrientes presentes en aguas residuales municipales en un sistema de tratamiento terciario. Barranquilla, Colombia: Trabajo de grado en Ingeniero Ambiental; 2019. p. 112

[27] Chacón C, Andrade C, Cárdenas C, Araujo I, Morales E. Uso de Chlorella sp. y Scenedesmus sp. En la remoción de nitrógeno, fósforo y DQO de aguas residuales urbanas de Maracaibo, Venezuela. Boletín del centro de Investigaciones Biológicas. 2004;**38**(2): 94-108

[28] Pérez R. Eliminación de nutrientes para el tratamiento biológico de agua residual usando un sistema inmovilizado microalga-bacteria en crecimiento autotrófico, heterotrófico y mixotrófico. S. C. México: Tesis de maestría. Centro de Investigaciones Biológicas del Noroeste; 2009. p. 162

[29] Hammouda O, Gaber A, Abdel N. Microalgae and wastewater treatment. Ecotoxicology and Environmental Safety. 1995;**31**:205-210

[30] Martínez M, Sánchez S, Jiménez J, El Yousfi F, Muñoz L. Nitrogen and phosphorus removal from urban wastewater by the microalga Scenedesmus obliquus. Bioresource Technology. 2000;**73**:263-272

Chapter 10

A New Insight of Phycoremdiation Study: Using Filamentous Algae for the Treatment of Tertiary Municipal Wastewater

Yan Li, Ethan Wood, Gergely Kosa, Bushra Muzamil, Christian Vogelsang and Rune Holmstad

Abstract

This book chapter demonstrated that the filamentous algae could be used as a promising phycoremediation approach to purify municipal tertiary wastewater. Initial screening of 25 algae strains across multiple genera revealed that *Spirogyra* sp. and *Klebsormidium* sp. were suitable to treat the tertiary effluent from a modern wastewater treatment plant (WWTP), and their co-culture was validated in three consecutive outdoor pilot tests. In the first two pilot tests, the nutrient concentrations of phosphorous and ammonium were depleted close to zero within 24 hours, whereas the pH value increased from 7 to 9 in the wastewater. Therefore, CO_2 was added for pH control in the 3rd batch, but the nutrient removal efficacy indicated that fresh algae inoculum was critical to maintain treatment efficiency. The biomass accumulated notable amounts of Ca, Mg, K, Fe, Al, and heavy metals from the effluent, while the algae production increased by two to three times over 7 days with an average algae biomass productivity of 1.68 g m^2 d^{-1}. The derived biomass can be used for biogas production and biofertilizer applications based on the biochemical constituent. Given a great potential for further optimization and improvement, we provide a new insight to use phycoremediation approach to facilitate the green transition of wastewater treatment plants.

Keywords: filamentous algae, treated wastewater, phycoremediation, co-culture, *Klebsormidium*, *Spirogyra*

1. Introduction

Wastewater management has received increasing attention and interest in the context of circular economy. Different from the conventional perception, waste streams become emerging resource for valorisation instead of being treated as a problem [1]. This has gradually become a new consensus for the green transition of wastewater treatment plants (WWTPs), include the request to neutralize carbon emissions. Thereby, it is imperative to introduce new technologies and solutions to

IntechOpen

WWTPs to alleviate their environment footprints as part of the green transition. Algae represent a promising tool to recover and recycle the residual nutrients from wastewater into bioproducts, coupled with significant carbon abatement. However, due to several constraints of scalable microalgae cultivation, it is still debatable whether the algae-based treatment technique (or phycoremediation) is a viable approach to facilitate the circular economy development of WWTPs. Therefore, this book chapter will provide a new insight to evaluate the phycoremediation for wastewater treatment.

Based on a case study, this book chapter will elaborate the potentials of using filamentous algae to treat the municipal tertiary effluent from a modern Norwegian WWTP. Both laboratory research and pilot scale tests were employed for the demonstration, since transferring the results from lab-scale R&D to pilot scale is a critical process for phycoremediation studies. Different from the conventional monoculture approach, the filamentous co-culture was exploited on purpose in order to enhance the resilience and viability of proposed phycoremediation strategy. High Rate Algal Pond (raceway) was used for the pilot study as it is an efficient system for algae—wastewater treatment [2]. The depletion of nutrients was monitored to indicate the treatment efficiency, while the productivity of algae biomass was detected in the pilot tests. Finally, the biochemical constituent and elemental content of the produced algae biomass were characterized with an attempt to assess the potential valorisation. It is anticipated that this study will shed light on how to effectively deploy the phycoremediation technology to facilitate the green transition of WWTPs.

2. Emerging interest on filamentous algae

Research in the application of microalgae in wastewater treatment was initiated in the 1950s. It has been highlighted that phycoremediation can remove up to 99% of nitrogen (N) and phosphorous (P) and reduce these nutrients concentration below 1 mg/L [3]. Moreover, the produced algal biomass can be valorised from niche markets of special materials (e.g. biopolymers and coatings) to the large-scale uses of fertilizer [4]. Phycoremediation has been demonstrated to be technically possible at lab scale, pilot scale, and industrial scale, but not economically viable. Although the cost for microalgae cultivation can be subsidized to a large extent by recovering nutrients from wastewater and sequestrating CO_2 released by WWTPs [5], harvesting microalgae cells is actually a premier obstacle to impede the popularity and scalability of phycoremediation [1]. It is an energy-intensive and expensive process. For an operational facility, this process can occupy over 90% of capital expenses [6] and above 20–30% of the overall production cost [7]. Meanwhile, it cannot be ignored that most fast-growing microalgae are vulnerable to the fluctuation of biotic and abiotic conditions in wastewater [1]. This is another inevitable challenge for those delicate microalgae cultivations, especially for the commonly proposed monoculture [4]. To address these inherent problems on phycoremediation, research attention has gradually shifted to filamentous algae in recent years, as they possess some unique traits that singular microalgae does not have.

Species from the genera *Oedogonium, Cladophora, Spirogyra, Klebsormidium*, and *Stigeoclonium* have been demonstrated as good candidates for wastewater treatment applications [4]. These filamentous algae have several impressive advantages for wastewater treatment, such as robust ability to uptake nutrients from wastewater to achieve about 60% increase in dry biomass per day [8], simplicity of harvesting, stronger resistance to a variety of aquatic grazers and competing organisms, as well as better adaptation to dynamic conditions [9]. Even in the environment with varying N:P ratios, the filamentous algae still can remove 99% of N and P simultaneously [10]. Moreover, these filamentous algae can naturally grow and bloom in a broad spectrum of waste streams including ash dam water [11]. Although filamentous algae have been highly recommended for wastewater treatment [4], there is little information to investigate their potentials for the contemporary requirement of WWTPs. With the declined interest of algae bioenergy and increased knowledge on pragmatic algae potentials, there is a growing consensus that the appropriate phycoremediation should be implemented and aligned with realistic demand and specific wastewater conditions.

3. New demand for treating municipal tertiary wastewater

In the past decades, there have been numerous attempts to employ phycoremediation technologies for the treatment of primary or secondary effluents, while less attention has been given to tertiary treated effluents. Although the advanced physical, chemical, and/or biological techniques used in modern WWTPs can remove the most nutrients, the tertiary effluents still contain considerable loads of N and P contributing to eutrophication. With more concerns of environmental eutrophication and rapid loss of nonrenewable resource of P, how to effectively eliminate the residual nutrients in the final discharge becomes an emerging request for most WWTPs. In Norway, for example, it was reported that more than 900 tons of phosphorus and about 15,000 tons of nitrogen were discharged into the ecosystem via WWTPs per annum [12]. These nutrients are equivalent to approx. 5% and 15% of agriculture P and N fertilizer consumptions as reported at a national level in 2017 [13]. Instead of releasing them to the aquatic environment as pollutants, it is apparently beneficial to recover these nutrients from the released wastewater.

Compared to primary and secondary wastewaters, tertiary treated wastewater has relatively stable pH and less turbidity, and these are acceptable conditions for algae to grow. Furthermore, along with the biological oxidation and denitrification process, tertiary wastewater contains much less organic macronutrients and the major dissolved nutrients are inorganic forms and less bioavailable organic compounds. In fact, they are the preferable medium conditions for algae proliferation. With the potential benefits of biomass valorisation and carbon sequestration, it is conceived that the application of filamentous algae to purify tertiary wastewater could represent a new win-win strategy for WWTPs. However, there is a suspicion on the algae productivity when the filamentous algae are exposed to the treated municipal wastewater. After all, the low concentration of those nutrient residuals (normally in a level of mg L^{-1}) seems not optimal for the algae growth. In order to address this concern, the case study introduced in this book chapter will provide more details to demonstrate the possibility of employing filamentous phycoremediation for a Norwegian WWTP.

4. Experimental validation of filamentous algae co-culture

In the case study, a total of 25 freshwater algae strains from 11 genera (**Table 1**) were selected from the Norwegian Culture Collection of Algae (NORCCA, www.norcca.no) aiming to test the suitability of phytoremediation to local climate condition. According to the record from a local WWTP (VEAS: VeasSelvkost AS, Slemmestad, Norway), the temperature of municipal discharge primarily ranges from 10 to 15°C over the year. This range was thereby used a criterion in the case study for

No.	Phylum	Class	Species	Strain	Origin	Medium	Morphology
1	Chlorophyta	Chlorophyceae	Tetradesmusobliguus	NIVA-CHL6	Lake Årungen, Akershus, Norway, 1946	Z8	Single cell
2			Coelastrum sp.	NIVA-CHL86	Lake Malawi, Malawi, 1991	Z8	Single cell
3			Chlamydomonas reinhardtii	K-1016	Amherst, Massachusetts, USA, unknown	CW15	Single cell
4			C. reinhardtii	K-1017	Amherst, Massachusetts, USA, unknown	CW15	Single cell
5			Oedogonium vaucheri	K-0094	Store Magleby, Amager, Denmark, unknown	NF2	filamentous
6			Oedogonium cardiacum	K-1001	Dry Drayton, England, unknown	20% Z8 + vitamins + soil extract	filamentous
7			Oedogonium cardiacum	K-1002	Dry Drayton, England, unknown	20% Z8 + vitamins + soil extract	filamentous
8			Stigeoclonium sp.	K-0018	Avernakø, Denmark, unknown	NF2	filamentous
9			Stigeoclonium sp.	K-1030	unknown	20% Z8 + vitamins + soil extract	filamentous
10			Stigeoclonium sp.	K-1031	unknown	20% Z8 + vitamins + soil extract	filamentous
11			Stigeoclonium sp.	K-1032	unknown	20% Z8 + vitamins + soil extract	filamentous
12			Raphidocelissubcapitata	NIVA-CHL1	River Nitelva, Akershus, Norway, 1959	Z8	filamentous
13		Trebouxiophyceae	Chlorella vulgaris	K-1801	Revo (TN), garden soil, Italy, unknown	Z8	Single cell
14			Chlorella vulgaris	NIVA-CHL 108	Germany, unknown	Z8	Single cell
15			Chlorella sorokiniana	NIVA-CHL 176	Austin, Texas, USA, 1953	Z8	Single cell

DOI: http://dx.doi.org/10.5772/intechopen.104253

No.	Phylum	Class	Species	Strain	Origin	Medium	Morphology
16		Conjugatophyceae	Spirogyra singularis	K-1019	UtterslevMose, Denmark, unknown	20% Z8 + vitamins + soil extract	filamentous
17			Spirogyra sp.	K-1454	Samsø, Denmark, unknown	20% Z8 + vitamins + soil extract	filamentous
18			Spirogyra sp.	CHL-189	Pond, Kindrogan, Scotland, 2013	20% Z8 + vitamins + soil extract	filamentous
19		Klebsormidiophyceae	Klebsormidium sp.	K-0148	Little Island, Cork, Eire, unknown	NF2	filamentous
20			Klebsormidium sp.	NIVA-CHL 142	Dal, Akershun, Norway, 1993	Z8	filamentous
21			Klebsormidiumflaccidum	NIVA-CHL 80	Spruce nursery, 1990	Z8	filamentous
22	Cyanobacteria	Cyanophyceae	Arthrospira platensis	NIVA-VYA 428	Lake Lonar, Maharashtra, India, unknown	Z8	filamentous
23			Anabaena subcylindrica	NIVA-CYA323	Fuggdal, Rendalen, Hedmark, Norway, 1993	Z8	filamentous
24			Trichormus variabilis	NIVA-CYA 19	Lake Mendota, Madison, Wisconsin, USA, 1948	Z8	filamentous
25			Trichormus variabilis	NIVA-CYA 410	Mississippi, USA, 1964	Z8	filamentous

Table 1.
Algae selection for the experimental test. (unicellular species are included for a point of comparison).

the algae screening and tests. Based on the previous knowledge on local rivers ecosystem [e.g., 14, 15], inclusion of filamentous algae selection was to avoid the potential risk of introducing invasive species to the local environment. The VEAS tertiary wastewater was freshly sampled and filtered (0.2 μm) on the day for the laboratory test below. All the selection process was conducted following the NORCCA's standard protocols. There were two selection criteria, (1) being filamentous and (2) at least 100% of increase in chlorophyll fluorescence. The filamentous strains with the highest growth rates were selected for co-culture combination studies.

Among these candidates, only K-1454 (*Spirogyra* sp.) and NIVA-CHL142 (*Klebsormidium* sp.) passed the criteria, and their growth rates (109–137%) were comparable to the level of unicellular strains (129–195%) (**Figure 1**). Although *Trichormus* sp. was selected as the third candidate for the subsequent co-culture tests, the growth of monoculture was only detected on *Spirogyra sp*. and *Klebsormidium sp*. when these three strains were exposed directly to the VEAS effluent, rather than on *Trichormus sp*. (**Figure 2**). In the following combination model that simulating the co-culture conditions, the results showed that the co-culture of *Spirogyra sp*. and *Klebsormidium sp*. could grow in the effluent, based on the detection of fluorescence increment in the end. Actually, the different sized algae cells have different uptake and scaled uptake affinities for the nutrient utilization [16], so the algae mixture in different cell sizes will be better for the purpose of nutrient recovery from wastewater reclamation. As the cell of *Spirogyra sp*. was much bigger than that of *Klebsormidium sp*. (approx. 10 time of difference), their co-culture was supposed a matched combination. Moreover, mixed culture would have a better resilience to the variable conditions and complex microbial community in the wastewater. Therefore, the co-culture of *Klebsormidium sp*. and *Spirogyra sp*. was selected as a model filamentous combination to treat the tertiary VEAS effluent.

At NIVA's Solbergstrand Algae R&D Facility, these two filamentous algae of *Klebsormidium* sp. and *Spirogyra* sp. were scaled up to 100 liters separately. All the algae cells were harvested by filtration (35 mm plankton net) when the inoculum biomass reached to the level of 2 g L^{-1}, and then washed by clean water, and weighed (after 10 min of air-drying) for the pilot test. A few grams of biomass were sampled from each species and lyophilised for a benchmark study, and the rest was used for

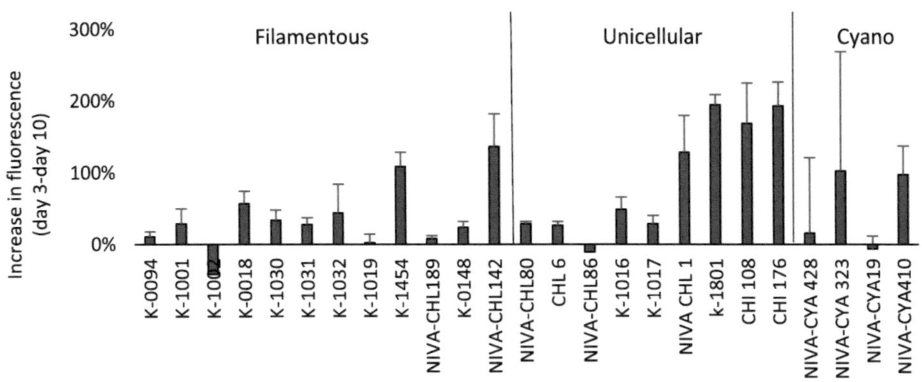

Figure 1.
The comparison of fluorescence increments among different selected algae in the screening test (mean ± SD). (note: The first 3 days were omitted to allow inoculated cells for acclimation, and thereby the results were obtained between days 3 and 10 during their exponential growth phase. The chlorophyll fluorescent was measured at a wavelength of 685 nm with excited emission at 450-550 nm.)

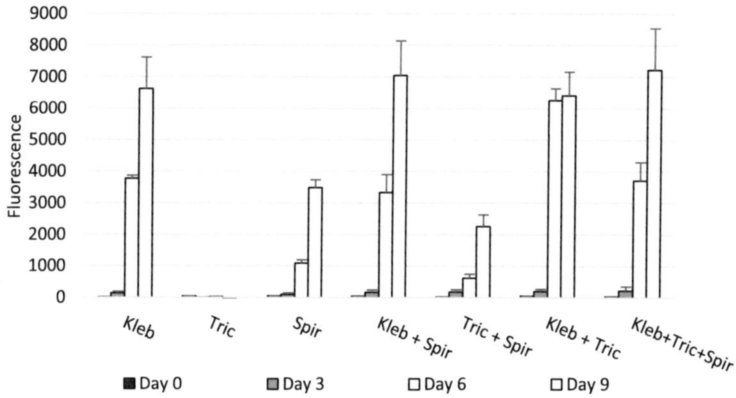

Figure 2.
Monoculture and co-culture flask experiments (mean ± SD; Kleb: *Klebsormidium sp.* Tric: *Trichormus sp.* Spir: *Spirogyra sp.)*

pilot tests. The pilot test was performed in three consecutive cycles with about 1500 L of municipal tertiary effluent (non-filtered, provided by VEAS) on each, which was conducted on 1:15 (v/v) with each indoor inoculum in a raceway. The 1st batch was inoculated with about 242 g wet *Klebsormidium* (equal to 22.3 g of dry weight, DW) and 547 g wet *Spirogyra* (about 39.9 g DW). With an attempt to continue algae cultivation, a similar amount (wet weight, WW) of co-culture was taken from the final produced biomass and subsequently used as the new inoculum for second batch, and so did on third cycle. The outdoor pilot test was carried out at 10–15°C (similar to the conditions at VEAS), with supplemental 24 hr. of LED light radiation (30–95 μE $m^{-2} s^{-1}$). The three batches were all monitored for 7 days, with 2 L of water sampling on each day. As the pH value in the raceway increased from 7 to 9 within 24 hr in the first 2 cycles (personal communication), the pH in the third batch was controlled at 7.5 with automatic CO_2 addition after day 2. In the end, the total co-culture biomass in the raceway was filtrated via a 80-mm plankton net, rinsed with tap water, and quantified for the yield measurement. The biomass was freeze-dried for various analytical analyses, including protein, starch, and lipids.

5. Nutrients depletion and biomass yield in the pilot test

This case study showed that the N and P residues (mg L^{-1} range) in the tertiary treated wastewater could be effectively removed by the inoculated co-culture (**Figure 3**). In the 1st and 2nd batches, most of NH_4^+ and P were depleted within 1 day. However, the depletion rate became slower in the 3rd batch. Moreover, the nutrient depletion followed a similar pattern. The depletion of NH_4^+ preceded the other inorganic N nutrients. Then, nitrate and nitrite started to deplete once NH_4^+ was close to 0. This is consistent with the previous reports that algae have a preference on inorganic N: NH_4^+ over other sources of NO_3^- [17, 18]. Regardless of the difference in 3rd batch, it seems that the total N could be reduced by 1.5 mg L^{-1} in 3 days. As this rate was achieved by the experimental amount of inoculated co-culture, perhaps the nutrient depletion could be accelerated with more co-culture inoculum.

Albeit of a similar amount of algae inoculum used in each batch (**Table 2**), their production was different. As there was a low level of nutrients (especially P) in the

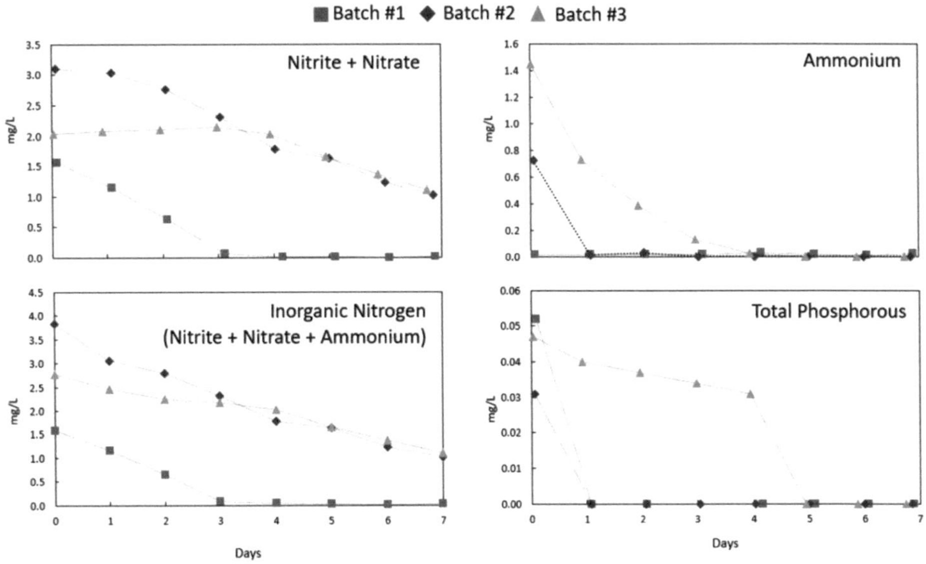

Figure 3.
Macronutrients during WWTP effluent treatment with algal co-culture.

Batches	Batch #1	Batch #2	Batch #3
Initial biomass inoculum (g L^{-1}, DW)	0.04	0.04	0.05
Harvested biomass on day 7 (g L^{-1}, DW)	0.13	0.10	0.10
Algae productivity over 7 days (DW g m^2 d^{-1})	2.27	1.51	1.26
Biomass yield vs. total N consumed in 7 days (DW g g^{-1})	55.59	19.93	19.13
Biomass yield vs. total P consumed in 7 days (DW g g^{-1})	1668	1801	641
Initial N:P ratio in the wastewater	30.38	123.29	58.94

Table 2.
The summary of algae production and nutrient consumed in the pilot tests.

treated wastewater, the exposed biomass was likely to confront prolonged P starvation with each subsequent batch. This is contradictive to findings that initial P-starvation can be implemented on microalgae to maximize the P uptake in wastewater [19]. This could happen in the 2nd pilot test but did not occurr in the 3rd batch. Although it was attempted to recycle the produced algal biomass for continuous effluent purification, the results support that the efficiency turned unsustainable after 14 days (duration of first two batches) even with CO_2 supplement. It is thereby deduced that the fresh algae inoculum is imperative to the rapid nutrient recovery in municipal tertiary wastewater.

Overall, the biomass of co-culture increased by 330, 250, and 200%, respectively, in three consecutive pilot tests (**Table 2**). The obtained algae productivity was consistent with the reported range of 0.8– 50.0 g m^2 d^{-1} on filamentous algae [4]. The results showed that the algae grew slower after day 3 in the first two pilot tests, when the N and P became deficient afterward (**Figure 3**). It is indicated that the real productivity of co-culture would be above the average level over 7 days. The effective

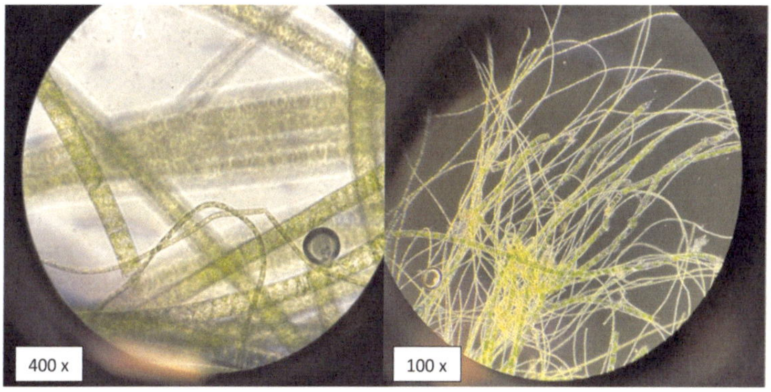

Figure 4.
*Microscopy picture of co-culture: A taken from the 1st batch pilot test on day 6. B taken from the 3rd batch pilot test on day 7. (*Spirogyra *sp.—Wider, spiral chlorophyll;* Klebsormidium *sp.—Thinner filaments; and air bubble—Empty round cycle).*

retention time for the wastewater reclamation could be less than 3 days as well. Obviously, the biomass yield can be elevated purposely by using an optimal amount of algae inoculum. Given the more biomass yield (vs both N and P consumed) in the first two batches (especially in the 1st cycle), this also underpins the above suggestion on fresh inoculum preparation for tertiary wastewater reclamation. Therefore, how to optimize the algae inoculum for the pilot treatment will be vital for further (semi-) continuous treatment process.

During the pilot test, both algae species grew well in the municipal treated wastewater. There was no sign of growth inhibition (dead cells or faded color) identified in the microscopy examinations (**Figure 4**). Interestingly, these two filamentous species clumped together. The co-culture formed numerous small algae colonies (about 1 cm) in the raceway, making biomass harvested much easier and quicker via a simple filtration on a 35 mm plankton net. Apparently, this can benefit to the practicability and scalability of selected co-culture cultivation for wastewater treatment. Only a few ciliates *Vorticella* were visualized to attach to the filamentous algae colonies in the 3rd batch, and they were possibly the "carryover" with the consecutive cultivation. However, they were not a predator to the co-culture as filamentous algae were too big to be the prey.

In this case study, big variations were obtained for algae growth measurements between sampling days (data not shown) by filtering 2 L of water samples. So, the biomass quantification was only based on the total algae production at the end of each pilot batch. The cell density of *Klebsormidium sp*. appeared gradually increase in the colonies along with the tests according to the microscope observation. In order to identify the change of combination ratio in the co-culture, the sensitive gene sequencing techniques (e.g. QPCR) or laser scanning confocal microscope (LSCM) with appropriate probes could be considered for future study.

6. Other discoveries in the phycoremediation pilot tests

Three consecutive pilot batches showed that the pH value increased from 7 to 9 within a day and stabilized at 9 if CO_2 was not supplied. At this point, it needs to clarify that the treated wastewater used in this study was collected after the denitrification treatment, and the water thereby could be CO_2 saturated. Apparently, the 3rd batch needed CO2 addition to maintain pH level. Albeit its algae growth was not as good as previous two batches (with the reasons mentioned above), it is believed that the fresh filamentous co-culture can rapidly deplete CO_2 in the treated wastewater. With this regard, a large amount of CO_2 supplement will be needed for the proposed filamentous co-culture in the process of wastewater purification. As a return, it is foreseen that this can significantly reduce the carbon footprint of WWTPs.

A total of 14 elements were measured in this case study, but only the mercury (Hg) was undetectable (**Table 3**). As a background study, these elements were also analysed on the indoor monocultures (inoculum), to exclude the influence of indoor cultivation medium. The differences between indoor inoculums of *Klebsormidium sp.* and *Spirogyra sp.* showed that they had variable update affinities to these elements. In contrast, the detected discrepancies from the outdoor co-cultures indicated the effectiveness of accumulation/absorption of these chemicals and heavy metals from the wastewater treatment. As discussed above, it was conceived that the biomass produced in pilot test could have less P. Interestingly, the indoor and outdoor biomass contained similar amount of Mg and K, while Ca content was more in the outdoor biomass. As showed in the results, these mineral chemicals in the municipal treated wastewater can be assimilated effectively by the tested co-culture to constitute the produced algae biomass at a level of g kg^{-1}.

It is worth noticing that there was a substantial accumulation of AI and Fe residues in the outdoor samples (g kg^{-1}). Although these two metal ions also existed in the algal cultivation medium, indoor algae samples contained much less than that of outdoor samples. One side, it is verified that there were still certain amounts of AI^+ and Fe^+ ions in the treated wastewater. It is believed that they were derived from the WWTP's chemical treatment process, as cationic coagulants/flocculants (e.g. ferric chloride,

Parameter	Unit	Spirogyra#	Klebsormidium	Batch #1	Batch #2	Batch #3
K	g kg^{-1}	6.10	11.2 ± 0.1	7.00 ± 0.09	6.50 ± 0.04	6.10 ± 0.05
Mg	g kg^{-1}	1.45	2.35 ± 0.02	2.82 ± 0.04	3.21 ± 0.02	1.70 ± 0.01
Ca	g kg^{-1}	9.14	2.74 ± 0.07	84 ± 1	77.3 ± 0.4	41.3 ± 0.6
Tot-P	g kg^{-1}	3.24	8.31 ± 0.05	2.31 ± 0.03	1.46 ± 0.00	1.79 ± 0.01
Fe	g kg^{-1}	0.13	0.34 ± 1.69	0.97 ± 0.04	1.29 ± 0.02	2.65 ± 0.03
Al	g kg^{-1}	<0.07	<0.07	0.82 ± 0.00	1.34 ± 0.01	2.08 ± 0.03
Cr	mg kg^{-1}	<2	<2	6.4 ± 0.2	5.14 ± 0.05	7.79 ± 0.02
Cu	mg kg^{-1}	<17	23.1 ± 1.69	19.2 ± 0.3	18.02 ± 0.09	45 ± 1
Ni	mg kg^{-1}	<2	<2	5.1 ± 0.3	3.5 ± 0.1	3.1 ± 0.1
Mn	mg kg^{-1}	187	34.0 ± 0.18	176 ± 4	334 ± 3	357 ± 5
Pb	mg kg^{-1}	<2	<2	2.6 ± 0.3	2.50 ± 0.06	3.6 ± 0.1
Zn	mg kg^{-1}	102	56.9 ± 0.06	153 ± 1	121.8 ± 0.4	194 ± 2
Cd*	mg kg^{-1}	<0.06	<0.06	0.07	0.07	0.08
Hg	mg kg^{-1}	<0.04	<0.04	<0.04	<0.04	<0.04

#*Spirogyra is the mean value of 2 replicates because of an accident with one of the replicates.*
**Cd is the result of only one replicate.*

Table 3.
Chemical elements analysis in the experimental algae biomass (mean ± SE).

aluminum chloride, and polymers) are usually used for P precipitation at the secondary treatment process of municipal wastewater. Although there is little information on their scavenge in water, using the filamentous algae can act a new approach to reduce these cationic residues from the municipal tertiary wastewater. Apparently, this can further diminish the environmental burden of the chemical treatment for the WWTP.

Heavy metals are hazardous substances and persistent pollutants in municipal wastewater [4]. Although they are close to undetectable levels at ug L^{-1} or ng L^{-1} [20], their appearance is a recalcitrant problem. As indicated in this case study, these trace elements were encapsulated in the algae biomass at a level of mg kg^{-1} from tons of municipal wastewater. This process is normally accomplished through a robust combination of non-active biosorption and active metabolism-dependent mechanisms [21, 22] because of algae's high binding affinity, abundance of binding sites, and large surface area [23]. Numerous studies in recent years have approved the existence of heavy metals and emerging contaminants in aquatic system and pointed out wastewater discharge as one of the main pollution sources [24, 25]. Therefore, using filamentous algae co-culture can be an effective and pragmatic approach to purify the tertiary wastewater in an environmental-friendly manner.

7. Potential valorisations of produced biomass

Since Fourier-transform infrared spectroscopy (FTIR) represents a rapid, simple, and reproducible method to identify the different compositions in the different biomass [26], it was employed to firstly examine the quality of algae samples in this study. FTIR spectroscopy revealed different proximate biochemical composition (lipids, carbohydrates, phosphates, and proteins) of indoor and outdoor cultivated algae (**Figure 5A**). In comparison, *Klebsormidium* biomass had more lipid, protein, and phospholipid, but *Spirogyra* contained more carbohydrates (**Figure 5B**). For the co-culture, the biochemical profile was a bit consistent among three batches, and the proportion of those biochemicals seemed to between the levels of inoculums. In the principal component analysis (PCA) analysis (**Figure 5C**), it seemed that the *Klebsormidium* could take over *Spirogyra* during consecutive pilot tests. This result was coincident with the microscopy observation. This is probably because that *Spirogyra* prefers growing in warm temperature [28].

FTIR results were further validated by the following analytical analysis. About 50% of *Klebsormidium* biomass was protein, but *Spirogyra* biomass was only 21% (**Figure 6**). However, *Spirogyra* had more starch (7% of DW) than *Klebsormidium* (3.3%). In the outdoor pilot test, the protein content of co-culture was increased gradually from 1st to 3rd batch with 20–30% of DW. However, the starch content became less from 5–4%. Like the PCA analysis, this result also suggested that the proportion of *Klebsormidium* in the co-culture increased. The lipid content was detected below 8% of DW with a small variation across different samples. Despite microalgae could increase lipid content in a condition of nutrient starvation [6], this is not applicable to the biological response of experimental filamentous algae in the outdoor pilot tests. Maybe it is why that filamentous algae are not compelling to the research attention as did on most of microalgae for typical algae economy values (e.g. biofuel and omega-3 oil). Moreover, with the notable protein content, the potential impact on anaerobic digestion (AD) process shall be investigated if the biomass is used for AD biogas production.

Another consideration is the ash content in the produced biomass. It was high in the first two WWT batches (26.6% –23.3%), lower in 3rd batch (13.8%). However,

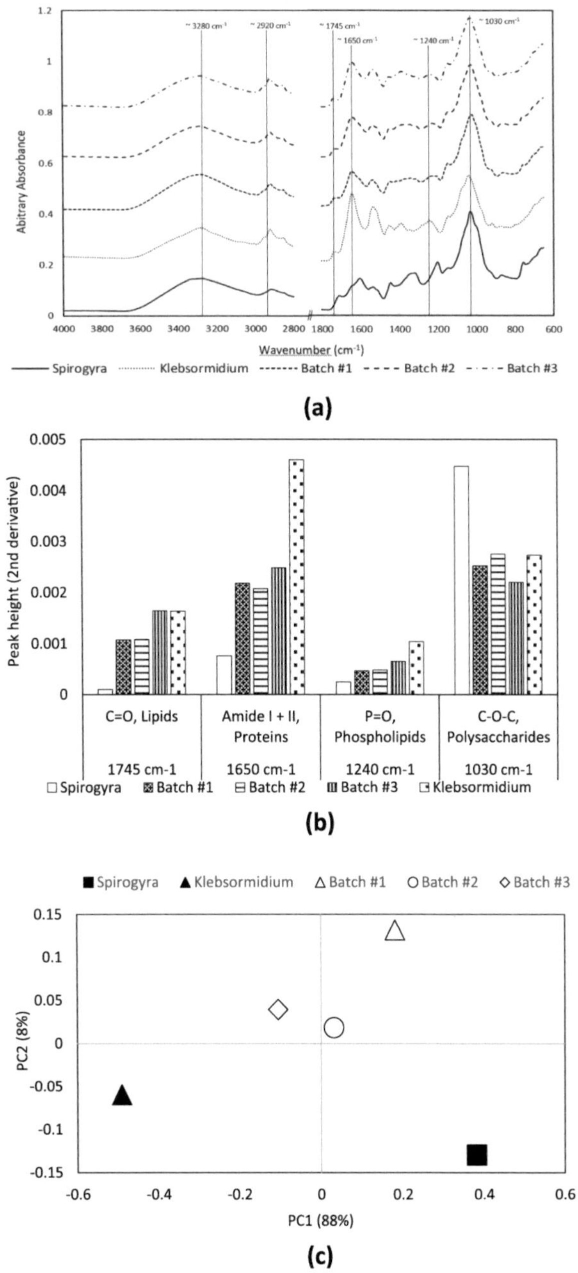

Figure 5.
FTIR analysis of microalgae biomass. (A) FTIR spectra with characteristic bands noted (note: The moisture condition was similar between different algae samples (3400–3200 cm^{-1}). As the variation between 3000 and 2500 cm^{-1} was not correlated to the changes in biochemical composition [27], the major differences between 1800 and 800 cm^{-1} were used for assessment.), (B) peak height of characteristic bands, and (C) scores plot of principal component analysis (PCA).

they were all more than the content in the indoor inoculums of *Klebsormidium* (6.4%) and *Spirogyra* (9.4%). Since the ash content is almost associated with the minerals content in biomass [29], those higher values of ash content in the pilot test also can

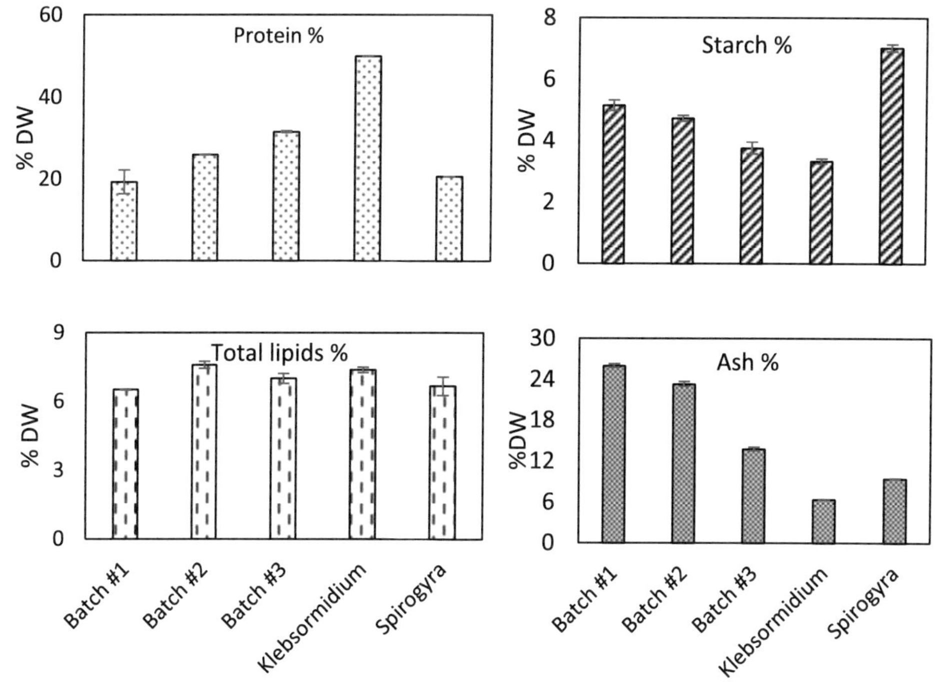

Figure 6.
Proximate biochemical analyses of microalgae biomass from indoor monocultures and outdoor co-cultures on wastewater.

evidence that filamentous algae removed certain amounts of mineral chemicals and heavy metals along with the treatment of tertiary wastewater in this case study. Regardless of heavy metals and/or other hazardous substances accumulated in the biomass, high ash content will also affect the algae inclusion level for food and feed utilization [30] and increase problems in combustion for energy conversion [31].

Given these concerns, there is still a great potential to utilize the produced biomass for biofertilizer, soil ameliorator, or new material development. For example, the heavy metals content in the produced biomass was below the maximum limit for permissible content in the organic fertilizers, according to the Norwegian regulations on organic fertilizers (FOR-2003-2007-04-951). There is no doubt that a comprehensive evaluation will be needed prior to this viable application, such as to match the restrictions of hygiene conditions, pesticides, and requirements for soil mixtures, as well as public perception. Apparently, it is going to be an inclusive question to evaluate the potential usage of produced biomass from this case study. However, it is undeniable that this represents a new value creation, as an authentic solution to facilitate the green transition of WWTP. Therefore, this case study provides a new paradigm for WWTPs to integrate the management of tertiary wastewater with emerging circular economy requirement.

8. Conclusions

Based on a realistic case study, this book chapter reveals that the mixed culture of filamentous algae *Klebsormidium* and *Spirogyra* can act as an effective tool to treat the municipal tertiary wastewater, with notable algae productivity. The co-culture can effectively recover macronutrients, mineral elements, and heavy metals from the wastewater, and their cultivation potentially can consume a considerable amount of CO2 for biomass production. Thereby, this phycoremediation process could significantly reduce or eliminate the environmental footprint of municipal tertiary wastewater in the ecosystem. Future research will focus on the remaining questions derived from the case study, which require proper optimization in areas of algae inoculum preparations, the nutrient depletion vs. carbon supplementation, hydraulic retention time reduction, and new cultivation strategy (e.g. two-stage) for continuous process. In order to improve the viability of proposed concept, the associated techno-economic analysis and environmental impact assessment will need to be deployed prior to the full-scale implementation.

Although the produced filamentous algae biomass will not be suitable for some typical algae economy purpose, their nutrient profiles and the easy scalable production can bring new hopes to other different value-added applications. As highlighted in the case study, the produced bulk biomass can become an optimal feedstock for new green fertilizer production. Although the case study was performed in pilot test, the outcomes endorse the feasibility of extrapolation to a full-scale wastewater purification or deployment. Thus, the proposed wastewater algae will represent a win-win strategy for WWTP and agriculture enterprise, as a typical model of circular economy. Apart from the contribution to the green transition of WWTP, this approach also can alleviate the pressure of soil deterioration and environment pollution due to the vast usage of chemical fertilizers. With a better understanding on the filamentous algae for municipal tertiary wastewater treatment, it is anticipated that this new phycoremediation approach can shape future investment plans of WWTPs or other new business consideration. Overall, this book chapter sheds lights on a new approach for the green transition of wastewater management and provides a new insight on the potentials of phycoremediation technology for WWTPs' sustainable development.

Acknowledgements

We would like to thank the industry partner VEAS for supporting the tertiary wastewater transportation, water analyses, and biomass heavy metals/phosphorous analyses, as well as the CO2 supplementation used in the 3rd batch of pilot test. Also we thank the Labtek for the analytical services at Norwegian University of Life Sciences.

Competing interests

The authors declare that they have no competing interests.

Funding

The authors thank the Regional Research Fund scheme (316690—RFFVIKEN), the NordForskNCoE programme "NordAqua" (no. 82845) and the Norwegian Research Council (320079—ALEGCO) for funding the study.

Author details

Yan Li[1*], Ethan Wood[1], Gergely Kosa[1], Bushra Muzamil[1], Christian Vogelsang[1] and Rune Holmstad[2]

1 Norwegian Institute for Water Research (NIVA), Oslo, Norway

2 VeasSelvkost AS, Slemmestad, Norway

*Address all correspondence to: michael.li@niva.no

References

[1] Liu R et al. Capabilities and mechanisms of microalgae on removing micropollutants from wastewater: A review. Journal of Environmental Management. 2021;**285**:112149

[2] Park JBK, Craggs RJ, Shilton AN. Wastewater treatment high rate algal ponds for biofuel production. Bioresource Technology. 2011;**102**(1): 35-42

[3] Silkina A et al. Large-scale waste bio-remediation using microalgae cultivation as a platform. Energies. 2019;**12**(14):2772

[4] Liu JJ et al. Wastewater treatment using filamentous algae—A review. Bioresource Technology. 2020;**298**: 122556

[5] Whitton R et al. Microalgae for municipal wastewater nutrient remediation: Mechanisms, reactors and outlook for tertiary treatment. Environmental Technology Reviews. 2015;**4**(1):133-148

[6] Amer L, Adhikari B, Pellegrino J. Technoeconomic analysis of five microalgae-to-biofuels processes of varying complexity. Bioresource Technology. 2011;**102**(20):9350-9359

[7] Rawat I et al. Dual role of microalgae: Phycoremediation of domestic wastewater and biomass production for sustainable biofuels production. Applied Energy. 2011;**88**(10):3411-3424

[8] Auer MT, Canale RP. Ecological studies and mathematical modeling of Cladophora in Lake Huron: 2. Phosphorus uptake kinetics. Journal of Great Lakes Research. 1982;**8**(1):84-92

[9] Grayburn WS et al. Harvesting, oil extraction, and conversion of local filamentous algae growing in wastewater into biodiesel. International Journal of Energy and Environmental Engineering. 2013;**4**:185-190

[10] Liu JZ, Vyverman W. Differences in nutrient uptake capacity of the benthic filamentous algae Cladophora sp., Klebsormidium sp. and Pseudanabaena sp. under varying N/P conditions. Bioresource Technology. 2015;**179**: 234-242

[11] Sternberg SPK, Dorn RW. Cadmium removal using Cladophora in batch, semi-batch and flow reactors. Bioresource Technology. 2002;**81**(3): 249-255

[12] Norskeutslipp. Wastewater Treatment Plant. 2022. Available from: https://www.norskeutslipp.no/no/Avlopsannlegg/?SectorID=100

[13] FAOSTAT. 2017. Available from: http://www.fao.org/faostat/en/#data/RFN

[14] Schneider SC, Lindstrøm E-A. The periphyton index of trophic status PIT: A new eutrophication metric based on non-diatomaceous benthic algae in Nordic rivers. Hydrobiologia. 2011; **665**(1):143-155

[15] Schneider S, Lindstrøm EA. Bioindication in Norwegian rivers using non-diatomaceous benthic algae: The acidification index periphyton (AIP). Ecological Indicators. 2009;**9**(6):1206-1211

[16] Edwards KF et al. Allometric scaling and taxonomic variation in nutrient utilization traits and maximum growth rate of phytoplankton. Limnology and Oceanography. 2012;**57**(2):554-566

[17] Van Den Hende S. Microalgal bacterial flocs for wastewater treatment :

From concept to pilot scale. In: Faculty of Bioscience Engineering. Ghent, Belgium: Ghent University; 2014. p. 324

[18] Lopez-Serna R et al. Removal of contaminants of emerging concern from urban wastewater in novel algal-bacterial photobioreactors. Science of the Total Environment. 2019;**662**: 32-40

[19] Solovchenko AE et al. Luxury phosphorus uptake in microalgae. Journal of Applied Phycology. 2019; **31**(5):2755-2770

[20] Kõrgmaa V et al. Removal of hazardous substances in municipal wastewater treatment plants. Water Science and Technology. 2020;**81**(9): 2011-2022

[21] Chekroun KB, Baghour M. The role of algae in phytoremediation of heavy metals: A review. Journal of Materials and Environmental Science. 2013;**4**(6): 873-880

[22] Matagi SV, Swai D, Mugabe R. A review of heavy metal removal mechanisms in wetlands. African Journal of Tropical Hydrobiology and Fisheries. 1998;**8**:23-25

[23] Cameron H, Mata MT, Riquelme C. The effect of heavy metals on the viability of Tetraselmis marina AC16-MESO and an evaluation of the potential use of this microalga in bioremediation. PeerJ. 2018;**6**:e5295

[24] Barra CA, Topp E, Grenni P. Pharmaceuticals in the environment: Biodegradation and effects on natural microbial communities. A review. Journal of Pharmaceutical and Biomedical Analysis. 2015;**106**:25-36

[25] Villar-Navarro E et al. Removal of pharmaceuticals in urban wastewater: High rate algae pond (HRAP) based technologies as an alternative to activated sludge based processes. Water Research. 2018;**139**:19-29

[26] Stehfest K, Toepel J, Wilhelm C. The application of micro-FTIR spectroscopy to analyze nutrient stress-related changes in biomass composition of phytoplankton algae. Plant Physiology and Biochemistry. 2005;**43**(7):717-726

[27] Mayers JJ, Flynn KJ, Shields RJ. Rapid determination of bulk microalgal biochemical composition by Fourier-transform infrared spectroscopy. Bioresource Technology. 2013;**148**: 215-220

[28] Graham JM et al. Physiological responses to temperature and irradiance in Spirogyra (Zygnematales, Charophyceae). Journal of Phycology. 1995;**31**(4):531-540

[29] Liu KS. Effects of sample size, dry ashing temperature and duration on determination of ash content in algae and other biomass. Algal Research. 2019; **40**:101486

[30] Austic RE et al. Potential and limitation of a new defatted diatom microalgal biomass in replacing soybean meal and corn in diets for broiler chickens. Journal of Agricultural and Food Chemistry. 2013;**61**(30):7341-7348

[31] Hupa M. Ash-related issues in fluidized-bed combustion of biomasses: Recent research highlights. Energy and Fuels. 2012;**26**(1):4-14

Chapter 11

Anaerobic Co-Digestion of Microalgae and Industrial Wastes: A Critical and Bibliometric Review

David de la Lama-Calvente, Juan Cubero, María José Fernández-Rodríguez, Antonia Jiménez-Rodríguez and Rafael Borja

Abstract

Microalgae are photosynthetic organisms able to grow faster than land plants and produce biomass with relatively high energy potential. Accumulated high-value compounds like lipids, minerals, or proteins have focused the attention of scientists due to the potential production of biofuels and other value-added products. However, several drawbacks regarding both the biochemical structure of these organisms and technological difficulties have prevented the industry for implementing a comprehensive low-cost process regarding energy and environmental contamination. Among these technologies, anaerobic digestion (AD) has greatly increased research attention because of its simplicity and the ability to produce easily recycle by-products. Moreover, anaerobic co-digestion (AcoD) has shown promising results as a method to bypass the AD problems of microalgae as a sole substrate. This review is focused on the recent trends and comparison of the AcoD process to maximize energy recovery from microalgae biomass and agro-industrial wastes. The yield of methane gas among the studied bibliography is compared and a critical review of published data and methods used is included.

Keywords: anaerobic co-digestion, microalgae, methane production, review methodology, agro-industrial wastes

1. Introduction

Microalgae are a wide family of photosynthetic organisms able to increase their biomass by using CO_2 and sunlight as energy sources by a rate 100 times faster than plants [1]. Moreover, the water and nutrients consumption is lesser than the needed for the same amount of biomass of terrestrial crops and it does not compete with other biomass from land areas [1, 2].

For the above-mentioned reasons, microalgae have been studied for decades for their potential conversion to energy. However, it was not until recently that microalgae have been considered feasible biomass to be used as a feedstock to produce biofuels (e.g. biohydrogen, biodiesel, biomethane, bio-oil, bioethanol, etc.) [1–4]. This is mainly due to several drawbacks (e.g. the need of solvents to produce biodiesel contributing to greenhouse gas emission, the use of expensive enzymes for bioethanol production, etc.) that have been overpassed thanks to technology and research efforts over the years focusing on the understanding and optimizing those factors that affect the different systems along with a better understanding of the biomass itself (e.g. the effect of the cell wall, the algae growth requirements, etc.) [1–5].

Among these technologies, anaerobic digestion (AD) has shown promising results. AD is a biological process where the organic compounds from certain biomass are degraded in the absence of oxygen (O_2) by a microbial consortium. The main effluents of this process are biogas (i.e. a gas composed primarily of methane and CO_2) and a nutrient-rich digestate [3]. While the produced biogas is considered a renewable energy source to produce electricity and heat in cogeneration plants, the nutrient-rich digestate could be used as a fertilizer [3, 4].

Microalgae are not only a viable AD feedstock, but can also serve as a means of biogas upgrading and its cultivation in the digestate can reduce the excess of nutrients and mitigate its potential toxicity [5]. However, the mono-digestion of microalgae has shown some concerns regarding its viability at industrial scales. Briefly, these concerns are related to the presence of long-chain organic compounds, mainly in the cell wall, the low carbon to nitrogen (C/N) ratio, and the high retention time needed in the reactors, which led to low methane yields, undigested organic matter in the digestate and more importantly to the inhibition of the AD process [2, 3].

In order to overpass these problems, anaerobic co-digestion (AcoD) of microalgae along with a wide range of co-substrates has been the focus of several research groups recently [1–7]. AcoD has several benefits due to its capacity to enhance the C/N ratio, the buffer capacity, the nutrient balance, and to dilute inhibitory compounds [4]. These improvements produce higher methane yields which in most cases are higher than the theoretical values obtained from the sole digestion of each co-substrate showing a synergetic effect [1–5].

Nevertheless, AcoD presents some drawbacks such as a higher organic load in the digestate, the usual need of pretreatments, or the difficulties to maintain a stable feedstock along the seasons [3]. This chapter aims to summarize the state of the art of microalgae used as co-substrate in AcoD processes and to highlight knowledge gaps and potential future developments. Moreover, a comprehensive analysis of the parameters affecting AcoD of microalgae in order to enhance methane yield is included in detail. Lastly, the energetic viability of several scenarios is discussed and future trends proposed.

2. Review methodology

For this bibliographic review, 92 articles have been selected from the 137 articles were carried out on the Scopus with the keywords "microalgae" and "anaerobic co-digestion" during the periods of 2011–2021 [8–100]. The bibliometric analysis was realized using the VOSviewer software using these 92-original articles.

Thanks to this software, it has been verified that there are few groups dedicated to the study of AcoD with microalgae. These investigations focus mainly on countries such as Spain (30 articles), the United States (17 articles), China (8 articles), Mexico (8 articles), and Brazil (5 articles). In addition, only eight authors have five or more articles to 321 authors (**Figure 1a**). In **Figure 1b**, as can see the connection between

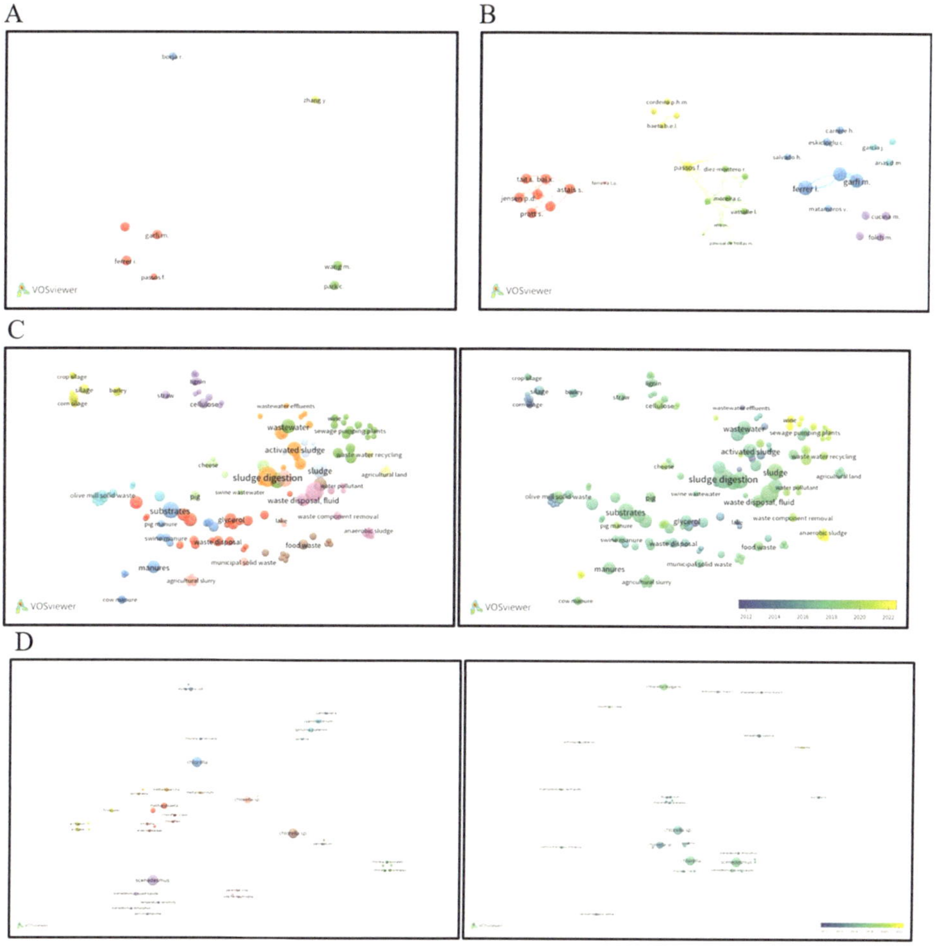

Figure 1.
Bibliometric analysis with VOSviewer: (a) authors with five or more articles, (b) connexion between authors, (c) substrates and (d) microalgae.

different authors and groups. The color and size indicate the group and the citations of these eight authors.

3. Microalgae and growth mediums

Since the first study on biogas production with microalgae by Golueke [101], a wide variety of microalgae genera have been studied. Due to the composition of the microalgae, it has been seen that each strain has a very specific biogas production, and has very diverse productions. One of the factors that most affect the AD of microalgae is the structure of its cell wall, which is why the selection of the type of microalgae is important. One of the main characteristics for microalgae to have a good methane production potential is to have a thin or null cell wall, large cytoplasmic components, a high growth rate, and a high tolerance to stress [102]. Other important aspects are whether it has a low content of hollocellulose in the cell walls, metabolic, and growth

Nutrient	BG11	BBM	Conway	Jaworki's	Zarrouk
Nitrate	$NaNO_3$	$NaNO_3$	KNO_3	$Ca(NO_3)_2 \cdot 4H_2O$ $NaNO_3$	$NaNO_3$
Potassium	K_2HPO_4	K_2HPO_4 KH_2PO_4	Na_3PO_4	KH_2PO_4 $Na_2HPO_4 \cdot 12H_2O$	K_2HPO_4 K_2SO_4
Sodium	Na_2CO_2			$NaHCO_3$	$NaHCO_3$ and Na_2CO_3
Magnesium	$MgSO_4 \cdot 7H_2O$	$MgSO_4 \cdot 7H_2O$		$MgSO_4 \cdot 7H_2O$	$MgSO_4 \cdot 7H_2O$
Chloride	$CaCl_2 \cdot 2H_2O$	NaCl $CaCl_2 \cdot 2H_2O$			NaCl $CaCl_2$
Boron	H_3BO_3	H_3BO_3	H_3BO_3	H_3BO_3	
Zinc	$ZnSO_4 \cdot 7H_2O$	$ZnSO_4 \cdot 7H_2O$	$ZnCl_2$		
Manganese	$MnCl_2 \cdot 4H_2O$	$MnCl_2 \cdot 4H_2O$	$MnCl_2 \cdot 4H_2O$	$MnCl_2 \cdot 4H_2O$	
Molybdenum	$Na_2MoO_4 \cdot 2H_2O$	MoO_3	$(NH_4)_6Mo_7O_{24} \cdot 4H_2O$	$(NH_4)_6Mo_7O_{24} \cdot 4H_2O$	
Copper	$CuSO_4 \cdot 6H_2O$	$CuSO_4 \cdot 6H_2O$	$CuSO_4 \cdot 6H_2O$		
Cobalt	$Co(NO_3)_2 \cdot 6H_2O$	$Co(NO_3)_2 \cdot 6H_2O$	$CoCl_2 \cdot 6H_2O$		
Iron	Ferric ammonium citrate	$FeSO_4 \cdot 7H_2O$	$FeCl_3 \cdot 6H_2O$	NaFeEDTA	$FeSO_4 \cdot 7H_2O$
EDTA	Na_2EDTA	Na_2EDTA (KOH)	$Na_2H_2EDTA \cdot 2H_2O$	Na_2EDTA	EDTA
Vitamin	Citric acid		Thiamin HCl		
Cyanocobalamin	Thiamin HCl				
Cyanocobalamin Biotin					

Table 1.
Synthetic culture medium for the growth of microalgae.

conditions are favorable, the morphological traits of the microalgae strains [27]. In addition, the selected strains offer a feasible genetic manipulation to control metabolic activities and improve tolerance to nutrient and ecological stress [102].

Nineteen genera of microorganisms have been investigated in the 92 articles found on the AcoD of microalgae. Of these 92 articles, 51 of these works have been studied with the genus *Chlorella* and 24 of these works with the genus *Scenedesmus*. The rest of the work carried out was with Chlorophyta as *Nannochloropsis* (5 studies), *Micractinium* (3 studies), *Dunaliella* (2 studies), *Dictyosphaerium*, *Closteriopsis*, *Desmodesmus*, *Chlamydomonas*, *Stigeoclonium*, *Botryococcus* and *Tetraselmis* genus. Other genus as *Tribonema* (Ochrophyta), *Phaeodactylum* (Bacillariophyta) and *Tisochrysis* (Haptophyta) have been also studied as co-substrate. In addition, the following cyanobacteria have been studied: *Arthrospira* (4 studies), *Spirulina* (2 studies), *Merismopedia*, *Oscillatoria*.

Another factor that should be considered is the culture medium for microalgae growth. Depending on the medium, it could favor the production of biogas in a later step due to the nutrient requirement of the microorganisms in the AD process [96]. According to the reviewed bibliography, the microalgae used for AcoD are obtained through other research groups or are cultivated using three different types of medium for growing it.

1. Synthetic medium: BG11, Bold's basal (BBM), Conway enriched medium, Jaworki's medium, modified Zarrouk medium, BlueBIOTech Ltf. (**Table 1**).

2. Digestate or effluent anaerobic: Anaerobic sludge, continuous stirred tank reactor (CSTR) digestate, up-flow anaerobic sludge blanket (UASB) digestate, pri-

Algae	$N\text{-}NH_4$		$P\text{-}PO_4$		Ref.
	[Conc] mg/L	%Remove	[Conc] mg/L	%Remove	
Scenedesmus sp. + *Chlorella* sp.	7.9	97	2.5	93	[10]
Mix culture	35.5	61	nd*	nd*	[80]
Arthrospira platensis	375	88.4	158	97.01	[42]
Chlorella vulgaris	285	99.6	117	91.2	[13]
Oscillatoria tenius	10.2	96.1	0.8	82.9	[18]
Chlorella 1067	202.68	94.33	7.18	29.67	[38]
Chlorella sp.	1144	58	188	98	[84]
Micractinium nov	2.2	94	4	95	[82]
Chlorella sp.	1.5	96	3.5	95	
Micractinium nov	27	92	0.7	51	[88]
Chlorella sp.					
Spirulina platensis	120	68	55	30	[92]
Chlorella sp.	225	85	120	20	
Cholerella sp.	45	88	9	58	[56]

nd: not determined.

Table 2.
Removal of $N\text{-}NH_4$ and $P\text{-}PO_4$ in microalgae culture.

mary effluent and sludge, chicken manure digestate, swine digestate, Anaerobic membrane bioreactor (AnMBR) digestate.

3. Wastewater: Tannery effluents, piggery WW, industry, domestic, municipal, fresh waters, lake waters, natural seawater enriched, soft drink. WW, Winery WW.

As shown in **Table 1**, microalgae are microorganisms that need certain nutrients to perform vital functions. Therefore, synthetic culture media contain macronutrients such as calcium, sodium, potassium, magnesium, and chloride. In addition to adding micronutrients such as iron, cobalt, molybdenum, manganese, copper, zinc, and vitamins. Finally, ethylenediaminetetraacetic acid (EDTA) is also added to form a complex ring (a chelate) with the trace elements, which, when used in low concentrations, stimulates the growth of microalgae by making this element available in low quantities. These nutrients are found in nature and are bioavailable in the other two natural culture media such as anaerobic digestate and wastewater, but may be found in lower concentrations than necessary [10].

In addition, using wastewater for the growth of microalgae could prevent eutrophication of the water due to the consumption of excess nutrient of these type of waters. In this sense, microalgae have been succesfully used for removing nitrogen and phosphorous from various wastes as shown in **Table 2**. Significant amounts of removal of nutrients and biomass production achieved in these studies demonstrate the feasibility of coupled wastewater treatment and microalgae cultivation processes.

4. Co-substrates

AcoD is the use of a mixture of biomasses to obtain a relatively higher methane yield [10]. This modified technique is considered economically more viable and easier to control mixed biomass compared to traditional mono-digestion systems. Depending on the species, microalgae contain significantly high or low amounts of protein, carbohydrates, and lipids. To balance the nutritional requirements of microorganisms in anaerobic reactors [13]. In addition, it would be possible to improve the stabilization of the process with well-balanced mixtures. With this, it could be possible to increase the organic load capacity, reduce the concentration of possible inhibitors, and increase the buffering capacity of the digestates. Apart from the nutrient balances and net synergistic effects that would occur when co-digesting microalgae with an efficient substrate. In the revised bibliography, synthetic co-substrates, agri-food residues and slurry, and liquid residues can be found.

1. Synthetic substrate: Cellulose and glycerol, and synthetic food wastes.

2. Solid waste: Cattle manure (chicken, cow, pig), cheese whey powder, chicken litter, green willow, chromium tanned leather shavings (SHA), coffee wastes, corn silage, fat, oil and grease (FOG), food waste, millet grass (*Pennisetum glaucum*), model kitchen waste, olive mill solid waste (OMSW), oil palm empty fruit bunches, *Opuntia maxima*, papaya waste, potato wastes, rice straw, *Sida hermaphrodita* (L.), silage and seaweed, sugarcane leaves, teak leaves (*Tectona grandis*), used cooking oil.

3. Wastewater: Bacteria biomass from anaerobic sludge, biosolids of water resource recovery facilities, catering waste leachate and raw sludge, cow rumen

liquid, deproteinated cheese whey, dewatered ww, food waste leachate, mill residue, municipal, paper sludge, palm oil mill effluent, septic tank sludge (STS), sewage sludge, swine wastewaters, waste activated sludge (WAS), and thickened WAS.

The production and processing of food and feed results in the generation of a large amount of waste. AD stands out as a suitable technology to reduce the environmental impact of agro-industrial waste and increase the energy self-sufficiency of these industries [29]. However, agro-waste is characterized by the lack of nutrients in its composition necessary for AD. In addition, having a high C/N ratio can affect the performance of AD. Another waste used as a co-substrate is animal manure (i.e. pig, cattle, and poultry). Contrary to agri-food residues, it has a relatively low C/N ratio, which increases the risk of ammonia inhibition [102]. Finally, one of the most studied co-substrates is sewage sludge, which was studied for the first time in 1983 [103], when it was co-digested with *Spirulina maxima*. Sludge of various varieties can be used as co-substrate, although like the rest, the interactions between microalgae-substrate must be studied to obtain a good balance between the different parameters that can affect AD [12].

5. Factors influencing anaerobic co-digestion

AD process is affected by several factors which led to a higher or lower biogas yield; these factors can be split into two main sources: operational conditions and substrates composition. Although both sources are widely related, they can be studied and controlled separately. The main operational conditions are temperature, pH, configuration of bioreactors, acclimation of inoculum, hydraulic retention time (HRT), organic load rate (OLR), and inoculum to substrate ratio (ISR) [1, 104]. The main factors related to substrates composition are the C/N ratio and macro and micronutrients [1, 5, 6, 104].

In this sense, AcoD is proposed as a feasible alternative in order to balance these factors and allow a better performance and a higher biogas production.

5.1 Effect of initial conditions

Most studies agree that the best operational temperatures for AD are under mesophilic (20–45°C) or thermophilic (>45°C) conditions [104]. Temperature affects, either directly or indirectly, the solubility of substrate compounds and the specific growth rate of the microorganisms involved, provoking a change in the HRT, the pH, and the methane yield [104]. Among the literature, the most common range is the mesophilic conditions due to its lower energy cost requirements and the similar methane yield when compared with higher temperature conditions [105]. However, temperature variations during AD performance have shown significant reductions in methane yield and the kinetics of the process [104, 106].

The effect of pH is mainly related to the optimum pH of the microorganism performance during AD. Based on that, a pH between 4.6 and 6.0 favored the hydrolysis, acidogenesis, and acetogenesis stages, while a pH between 6.0 and 8.0 favored the methanogenesis phase [1]. Literature shows that the most suitable pH range is between 6.5 and 7.5, where the methane production is most benefited [1, 104]. Furthermore, it has been reported that the initial pH of the substrate had a significant impact on methane yield, being the optimum value in the range of 7.0–7.5 for the co-digestion of swine manure and maize stalk [107]. However, pH is highly affected during the AD process, lowering its value if the buffer system is not strong enough due to the accumulation of volatile fatty acids (VFA) (e.g. acetic, propionic, butyric, and valeric acids) [104] or increasing it if ammonium nitrogen is accumulated (around 5.0 g NH_3-N L^{-1}) [33].

Temperature and pH have been linked to free ammonium nitrogen and ammonium ions equilibrium, showing that when one parameter is fixed there is a linear increase in methane yield when the other two independent variables increased up to a certain limit, after which the methane production decrease [104, 108]. A recent study showed optimal conditions for the mono-digestion of chicken manure of 34.0°C, 5.0 g NH_3-N L^{-1}, and pH 7.5 [108]. Moreover, in an earlier study, an increase of pH from 7 to 8 and above, enhanced the biogas production with similar methane proportion when temperature conditions increased from 37 to 55°C during the AD of buffalo manure [109].

As shown above, temperature and pH have a significant impact on AD performance and biogas production. Thus, co-digestion is presented as a suitable technique able to enhance the buffer system, the substrate pH, and free ammonium nitrogen values, allowing a higher methane yield and a more stable process [67, 77, 78, 99, 100]. Meneses-Reyes et al. showed that increasing the C/N ratio of the substrate by reducing the microalgae ratio provoked an increase in pH, however, the pH of digestate is similar among the different co-digestion ratios studied (7.33–7.51) [41]. AcoD of *Chlorella* sp. and food waste in batch mode (35°C) showed that methane yield was related to the initial pH of the substrates (7.3–8.7), being the optimum value 8.0 and reporting that the methane yield decreased almost linearly when pH differs from the optimum value, although these results were not conclusive as other variables were also different (e.g. $VS_{feeding}$ from 8.0 to 9.2 g) [35]. However, a mixture of microalgae biomass with thermally treated wheat straw presented a pH of 12 with no significant effect on AcoD performance when compared with the AcoD of microalgae biomass and untreated wheat straw (pH: 6.82) [71]. Another study assessed the effect of temperature within the mesophilic range (25°C, 30°C, 35°C, 40°C) in biogas production, reporting a significant increase of biogas production (45%) when the temperature was increased up to 35°C, but a reduction of production or no significant improvement (depending on the C/N ratio of substrates) when the temperature was set up at 40°C [24]. This result is in accordance with other studies reporting a decrease of methane yield for the AcoD of microalgae with undigested sewage sludge in batch mode when increasing the temperature from mesophilic (37°C) to thermophilic (55°C) conditions, even though the pH was not affected (6.91–7.03) and the processes were stable [45]. AcoD of corn silage and *Nannochloropsis salina* in a semi-continuous mode (38°C; C/N: 21.2) do not present any stability deviation by the increasing OLR (2–4.7 g VS L^{-1} d^{-1}), while the AcoD of two microalgae (*Scenedesmus* sp. and *Opuntia maxima*) in semi-continuous mode (37°C; C/N: 15.6) showed that pH was affected by OLR (2–6.67 g VS L^{-1} d^{-1}) and ISR (6:8% VS basis), is the most stable conditions OLR: 2 g VS L^{-1} d^{-1} and ISR: 8% VS basis [60, 110]. Furthermore, a study assessing the effect of alkali, acid, and thermal pretreatment of *Oscillatoria tenuis*, before the AcoD with pig manure, reported that pH control affected the biogas production rather than physical or chemical pretreatments [18]. Similar results were reported during the AcoD of alkali pretreated microalgae consortium with swine wastewater, where the negative effect of ammonia inhibition at high pH (11) was stronger than the positive effect of the destruction of microalgae's cell walls [47].

pH is also being widely used as a control parameter able to indicate the stability of the reactors since is strongly linked to VFA accumulation [40, 43]. AcoD of *Arthrospira platensis* with several carbon-rich co-substrates proved to be stable at low OLR (1 g VS L^{-1} d^{-1}) in a semi-continuous mode but unstable at higher OLR (2–5 g VS L^{-1} d^{-1}) due to VFA accumulation and pH dropped, this study is in accordance with the AcoD of *Chlorella* sp. and glycerol that presented a stable pH range (6.6–7.32) when the HRT was above 5 days, at which point volatile fatty acids (VFA) accumulation inhibit the biogas production [24, 33]. However, AcoD of naturally grown microalgae consortium with WAS in a semi-continuous mode (37°C) showed that when the HRT was increased from 1 to 3 to 4–6, pH was reduced (from 7.51 to 7.04), although VFA accumulation was not observed and the system remained stable with no difference in methane production except at HRT of 6 were slightly dropped [46].

Another important factor is alkalinity, which helps to prevent large changes in pH due to the accumulation of volatile fatty acids, or the generation of ammonia due to protein hydrolysis. Alkalinity provides the necessary buffering capacity to counteract possible changes in pH, produced by the balance between carbonate and bicarbonate. The ideal alkalinity values for AD would be between 2000 and 4000 mg $CaCO_3$ L^{-1} [99, 100]. A study assessed the relation between pH and alkalinity, where the initial pH was fixed at 7.0 while the initial alkalinity changed (70–3200 mg $CaCO_3$ L^{-1}), however, pH remained around neutral values (6.9–7.2) during the AcoD process suggesting that initial alkalinity has no impact avoiding an ammonium concentration from nitrogen-rich substrates as microalgae [30].

The C/N ratio is another factor that influences the AD process [29]. A good substrate C/N ratio can range between 20 and 30, with an optimal value of 25 [102]. With a C/N ratio below 20, there is an imbalance between C and N in the reactor, which ultimately releases a large amount of NH_3, which usually happens with the degradation of microalgae [102]. The high concentration of NH_3 in the digester affects the growth and metabolism of microorganisms and produces an accumulation of volatile fatty acids, which results in a decrease in biogas yield. This factor can also be supplied by choosing a good co-substrate and optimizing this C/N ratio [94].

The ISR is another key parameter that influences AD and methane production [13]. To find the maximum methane potential, a proper balance between the substrate and the microorganisms is necessary so that limitations and inhibitions do not occur due to the loading of the substrate. For biochemical methane potential (BMP) assays, a ISR $\geq$ 2 is suggested as the default value [111].

5.2 Effect of pretreatments

As a common way to improve the methane yield of AcoD of microalgae, several studies have reported the effect of pretreatments on microalgae biomass. **Table 3** shows some of these pretreatments and the effect on AcoD calculated as the increase of biomethane production in percentage over the non-pretreated AcoD.

It has been reported that although some pretreatments can improve the methane yield of microalgae as a sole substrate, these have a negative effect during the AcoD, mainly due to the high organic matter consumption or the inefficacy of pretreatments breaking down the cell wall [47, 86, 97].

From **Table 3** it can be seen that low thermal pretreatments (60–55°C; 1–2d) have none or very little effect on AcoD [19, 86], however, when the temperature increased to 120°C, the biomethane production improve greatly (up to 43%) [18, 49]. Other successfully tested pretreatments are ultrasonication, hot water, and a

Pretreatment	Conditions	Microalgae	Co-substrate	Improvement	Reference
Enzymatic	0.1% v/v; 0.5 h	*Scenedesmus* sp. + *Chlorella* sp.	WAS	-2%	[97]
Ultrasonication	n/d	*Nannochloropsis oculata*	Cow manure	16.7%	[57]
Hot water	121°C; 15 psi	*Nannochloropsis oculata*	Cow manure	36.7%	[57]
Ultrasonication:hot water	121°C; 15 psi	*Nannochloropsis oculata*	Cow manure	16.7%	[57]
Thermal	60°C, 24 h	*Scenedesmus* sp. + *Chlorella* sp.	WAS	−33.8%	[86]
Thermal	55°C, 2d	*Chlorella vulgaris*	WW	8.3%	[19]
Thermal	120°C, 20 min	*Oscillatoria tenius*	Pig manure	43%	[18]
Alkaline	NaOH 3% w/v	*Scenedesmus* sp. + *Chlorella* sp.	Swine WW	−10.4%	[47]
Thermal	120°C, 1 h	*Chlorella* sp.	Coffee wastes	13.9%	[49]
Thermo-alkaline	10%CaO, 75°C, 24 h	*Chlorella* sp.	Wheat straw	9.0%	[71]
Thermo-alkaline	10%CaO, 75°C, 24 h	*Chlorella* sp.	Wheat Straw	15%	[71].

Table 3.
Effect of pretreatment on AcoD.

combination of thermal and alkaline pretreatments [57, 71]. Nevertheless, based on the higher improvement on biomethane production, thermal pretreatment at 120°C is the most effective process, where time and pressure would be the variables to analyze [49, 57].

6. Biomethane potential (BMP) performance

Table 4 summarizes all the BMP assay results up to date complying with the following prerequisites:

- The co-substrates ratio is well described as VS or C/N.
- ISR has been considered.
- Methane yield have been reported under standard conditions.

The above prerequisites have been selected as they are crucial for comparison purposes. The authors of this review acknowledge the lack of homogenization regarding BMP performance, which difficult the task to evaluate scientific data and provide reliable conclusions. The authors of this review also acknowledge the lack of a widely approved and used standard methodology for BMP tests, although, some are published and proved to be reliable enough [111, 112]. The authors of this

Co-substrates	Co-substrate ratio (VS)	ISR (VS)	Temperature (°C)	Methane yield ($NL_{CH_4}\ kg_{VS}^{-1}$)	Synergetic effect (%)	Ref.
Dictyosphaerium sp.:synthetic food wastes	1:3	2	35	514	27.6	[29]
Tribonema sp.:pig manure	2:8	2	35	580.4	16.9	[34]
Chlorella sp.:sludge:FOG	4:4:2	2	35	334	3.7	[72]
Scenedesmus quadricauda:OMSW	1:3, C/N: 25.3	2	35	461	36.4	[26]
Chlamydomonas reinhardtii 6145:OMSW	1:1, C/N: 18.3	2	35	542	37.2	[27]
Chlamydomonas reinhardtii cw15:OMSW	1:1, C/N: 18.3	2	35	451	16.5	[27]
Nannochloropsis limnetica:piggery slurry	4:6	5	53	355	12	[77]
Microalgae consortium:WW	1:3	2	35	339	5.9	[74]
Scenedesmus sp.:deproteinated cheese whey	17:83	3.3	35	302	−7.6	[16]
Scenedesmus sp.:cellulose	16:84	3.3	35	272	−2.0	[16]
Microalgae consortium:WW	37:63	2	35	237.1	2.1	[46]
Chlorella vulgaris:cattle manure	4:1	4	55	431	8–15	[40]
Chlorella sp.:swine manure	6:94, C/N: 33.9	2	35	348	11.2	[84]
Scenedesmus sp. + *Chlorella* sp.:food waste	1:4, C/N: 26.4	2	Mesophilic	639.8	32.8	[99]
Nannochloropsis gaditana:cellulose	1:3, C/N: 20.3	2	37	268	1.5	[11]
Scenedesmus sp and *C. vulgaris*:sewage sludge	12:88, C/N: 9.4	2	37	388	0	[45]
Dunaliella salina:Olive mill solid waste	1:3, C/N: 26.7	2	37	330	28.7	[28]
Nannochloropsis salina:corn silage	1:6, C/N: 21.2	2.7	40	660	15	[60]

Table 4.
Methane yield of co-digested microalgae with different substrates at different ratios in BMP assays.

review would also like to highlight that only 20 out of 120 reports complies with the above prerequisites, and if other crucial factors were included within the prerequisites (e.g. use of positive control or details on inoculum acclimation) no reports could have been included in this review. Additionally, when more than one co-substrate ratio was measured in the same study, only the one producing the highest methane yield was included in **Table 4**.

As can be observed in **Table 4**, methane yield ranged from 237.1 to 639.8 $NL_{CH_4}\ kg_{VS}^{-1}$. AcoD of microalgae with industrial wastes showed positive synergetic effects in most cases, being the improvement against the theoretical values up to 37%. However, some studies had reported negative effects. This could be due to the limits of batch methodologies and could lead to higher methane productions during continuous tests as pointed out by several authors [16, 33] or could be related to the low C/N ratio [45].

The optimum C/N ratio produced higher synergetic effects than those studies where the C/N ratio was above 30 or below 20. The most common and successful temperature is within the mesophilic range with an ISR of 2–3, although, some studies have reported improvements when applying thermophilic temperature at a higher ISR (4–5) [40, 77].

Regarding the co-substrates ratio, when microalgae were used as a nitrogen source in order to balance the C/N ratio, it was added commonly as a fourth part of the whole influent, or even less. When microalgae were co-digested with other low C/N substrates, it was added at higher concentrations. Nevertheless, the optimum ratio between microalgae and other co-substrates is unique and needs to be assessed through experimentation.

7. Scaling up the AcoD process

Table 5 summarizes all the results obtained by semi-continuous or continuous assays at lab or pilot scale results up to date complying with the following prerequisites:

- The co-substrates ratio is well described as VS or C/N.
- Temperature, OLR, and HRT/ISR has been considered and reported.
- Methane yield have been reported under standard conditions (when values were reported under normal conditions a factor of 0.8871 was applied to obtain the methane yield under standard conditions).

Semi-continuous or continuous processes could overpass some drawbacks from batch assays as reported by several authors [71, 77, 84]. This could be due to the acclimation of the inoculum during the experiments. To this sense, these processes had shown successful results in the co-digestion of substrates with low C/N ratios, low pH, and higher ammonium content, allowing a higher concentration of microalgae in the influent. However, some studies had shown negative results when compared with the AD of sole substrates [32, 46, 77]. This low methane production had been related to the difficulty of microalgae cell-wall digestion, accumulation over time of ammonium, or high OLR [32, 46, 77].

Regarding operational parameters, the most common used temperature was within the mesophilic range due to its low energy cost and good performance. OLR ranged from 0.5 to 4 g COD $L^{-1}\ d^{-1}$, although, most studies showed a higher methane

Co-substrate	Co-substrate ratio (VS)	System	Acclimated inoculum	Conditions	Methane yield	Improvement over control	Ref.
Microalgae biomass:WW	26.4:1 (VSS)	UASB (pilot scale)	Yes	25 C OLR: 1.0 g COD L^{-1} d^{-1} HRT: 8.1 h	235 NL kg_{VS}^{-1}	−20% (wastewater)	[32]
Chlorella sp.:WW	38:62, C/N: 7.08	AnMBR (lab scale)	Yes	25 C OLR: 0.5 g COD L^{-1} d^{-1} HRT: 30 days HRT: 70 days	391 NL kg_{VS}^{-1}	nd*	[62]
Scenedesmus sp. *Chlorella* sp.:WW	38:62, C/N: 7.08	AnMBR (pilot scale)	Yes	35°C OLR: 0.5 g COD L^{-1} d^{-1} HRT: 30 days HRT: 70 days	370 NL kg_{VS}^{-1}	nd*	[62]
Chlorella sp.:WW	38:62, C/N: 7.08	CSTR (lab scale)	Yes	55°C OLR: 0.5 g COD L^{-1} d^{-1} HRT: 30 days	242 NL kg $_{VS}^{-1}$	nd*	[61]
Microalgae biomass:sewage	(Flow rate) algae: 0.5 L h^{-1}, sewage: 49 L h^{-1}	UASB (pilot scale)	No	23°C OLR: 0.7 g VS L^{-1} d^{-1} HRT: 7 h	211 NL kg_{VS}^{-1}	34.8% (Sewage only)	[79]
Chlorella sp.:WW	38:62, C/N: 7.08	Two stage: AnMBR, AnR (lab scale)	No	AnMBR 35°C OLR: 0.52 g COD L^{-1} d^{-1} HRT: 30 d, AnR 35°C OLR: 0.15 g COD L^{-1} d^{-1} HRT: 100 days	291 NL kg_{COD}^{-1}	nd*	[65]

Co-substrate	Co-substrate ratio (VS)	System	Acclimated inoculum	Conditions	Methane yield	Improvement over control	Ref.
Scenedesmus sp.:primary sludge	38:62	AnMBR (lab scale)	Yes	35 C OLR: 0.5 g COD L^{-1} d^{-1} HRT: 30 d	241 NL kg_{COD}^{-1}	40.1% (algae only)	[93]
Chlorella sp.:primary sludge	38:62	AnMBR (lab scale)	Yes	35 C OLR: 0.5 g COD L^{-1} d^{-1} HRT: 30 d	228 NL kg_{COD}^{-1}	6.5% (algae only)	[93]
Chlorella vulgaris:chicken litter:glycerol	30:67:3, C/N: 8.15	Semi-continuous (lab scale)	Yes	37 C OLR: 0.716 g VS L^{-1} d^{-1} HRT: 30 d	240 NL kg_{VS}^{-1}	39% (chicken litter only)	[43]
Nannochloropsis limnetica:piggery slurry	2:3	CSTR (lab scale)	nd*	53 C OLR: 1.4 g VS L^{-1} d^{-1} HRT: 15 d	216 NL kg_{VS}^{-1}	22.7% (pig slurry only); −31% (algae only)	[77]
Microalgae biomass:WW	1:3	Semi-continuous (lab scale)	No	37 C OLR: 1.17 g VS L^{-1} d^{-1} HRT: 30 days	460 NL kg_{VS}^{-1}	187.5% (pretreated algae only)	[74]
Stigeoclonium Scenedesmus:papaya waste	1:1 (w/w)	Semi-continuous (lab scale)	Yes	35 C OLR: 1.1 g VS L^{-1} d^{-1} HRT: 31 days	230 NL kg_{COD}^{-1}	59.7% (algae only); 12.2% (theoretical)	[17]
Microalgae biomass:primary sludge:WAS	2:3, C/N: 8.49	Semi-continuous (lab scale)	nd*	37 C OLR: 2.4 g COD L^{-1} d^{-1} HRT: 15 days	168 NL kg_{VS}^{-1}	−15.8% (WWTP only)	[46]
Chlorella sp.:wheat straw	1:1, C/N: 13.1	Semi-continuous (lab scale)	nd*	37 C OLR: 1 g VS L^{-1} d^{-1} HRT: 20 days	240 NL kg_{VS}^{-1}	15% (no pretreated); 75% (algae only)	[71]

Co-substrate	Co-substrate ratio (VS)	System	Acclimated inoculum	Conditions	Methane yield	Improvement over control	Ref.
Chlorella sp.:swine manure	6:94	Semi-continuous (lab scale)	nd*	35 C OLR: 1.16–1.68 g VS L^{-1} d^{-1} HRT: 21 days	348 NL kg_{VS}^{-1}	9.8% (swine only)	[84]
Chlorella vulgaris (MACC-755):used cooking oil	1:1, C/N: 477	Semi-continuous (lab scale)	No	38 C OLR: 4.01 g VS L^{-1} d^{-1} HRT: 90 days	880 NL kg_{VS}^{-1} 2.86 L L^{-1} d^{-1}	6.0% (algae only)	[54]
Chlorella vulgaris (MACC-755):maize silage	1:1, C/N: 16	Semi-continuous (lab scale)	No	38 C OLR: 4.01 g VS L^{-1} d^{-1} HRT: 90 days	1.99 L L^{-1} d^{-1}	nd*	[54]
Chlorella vulgaris (MACC-755):mill residue	1:1, C/N: 12	Semi-continuous (lab scale)	No	38 C OLR: 4.01 g VS L^{-1} d^{-1} HRT: 90 days	1.96 L L^{-1} d^{-1}	nd*	[54]

**nd: not determined.*

Table 5.
Methane yield of co-digested microalgae with different substrates at different ratios in semi-continuous or continuous assays.

production when the OLR is below 1.5 g COD L^{-1} d^{-1}. HRT was around 30 days, with some studies reporting 90 days when the OLR is around 4 g COD L^{-1} d^{-1}. OLR and HRT had been discussed as key factors for a complete energy-effective harnessing of microalgae AcoD, to this sense, high OLR and low HRT would be the most effective way for achieving this. However, pH decreases due to VFA accumulation, and ammonium release are the main factors affecting the OLR.

8. Conclusions

The status and current trends of AcoD of microalgae and industrial wastes were reviewed in this chapter. AcoD performance improvements still need further research on varied co-substrates and optimal mix ratios. Operational parameters and their control are key to achieving optimal biogas. Pretreatments of microalgae biomass are a promising way to enhance biogas production. The majority of research investigations are done by a few research groups and centered on biomethane potential tests lacking a common methodology, thus, further research and the application of common criteria need to be implemented. Moreover, pilot-scale assays have shown promising results, however, very few research groups have the ability to implement these studies.

Acknowledgements

The authors wish to express their gratitude to the regional government of Andalucía, Junta de Andalucía, Consejería de Transformación Económica, Industria, Conocimiento y Universidades (Project FEDER UPO-1380782) and to the Spanish national R&D plan (PID2020-114975RB-100/AEI/10.13039/5011000/11033) for providing financial support.

Conflict of interest

The authors declare no conflict of interest.

Author details

David de la Lama-Calvente[1*], Juan Cubero[1,2], María José Fernández-Rodríguez[3], Antonia Jiménez-Rodríguez[3] and Rafael Borja[1]

1 Spanish National Research Council (CSIC), Seville, Spain

2 University of Huelva, Huelva, Spain

3 University of Pablo de Olavide, Seville, Spain

*Address all correspondence to: dlama@ig.csic.es

References

[1] Vargas-Estrada L, Longoria A, Arenas E, Moreira J, Okoye PU, Bustos-Terrones Y, et al. A review on current trends in biogas production from miroalgae biomass and microalgae waste by anaerobic digestion and co-digestion. Bioenergy Research. 2021;**15**:77-92

[2] Thorin E, Olsson J, Schwede S, Nehrenheim E. Biogas from co-digestion of sewage sludge and microalgae. Energy Procedia. 2017;**105**:1037-1042

[3] Ganesh Saratale R, Kumar G, Banu R, Xia A, Periyasamy S, Dattatraya SG. A critical review on anaerobic digestion of microalgae and macroalgae and co-digestion of biomass for enhanced methane generation. Bioresource Technology. 2018;**262**:319-332

[4] Choudhary P, Assemany PP, Naaz F, Bhattacharya A, Castro J, Couto E, et al. A review of biochemical and thermochemical energy conversion routes of wastewater grown algal biomass. Science of the Total Environment. 2020;**726**:137961

[5] Wall DM, McDonagh S, Murphy JD. Cascading biomethane energy systems for sustainable green gas production in a circular economy. Bioresource Technology. 2017;**243**:1207-1215

[6] Thorin E, Olsson J, Schwede S, Nehrenheim E. Co-digestion of sewage sludge and microalgae—Biogas production investigations. Applied Energy. 2018;**227**:64-72

[7] Calicioglu O, Demirer GN, Ganesh Saratale R, Kumar G, Banu R, Xia A, et al. Recent achievements in the production of biogas from microalgae. Waste and Biomass Valorization. 2017;**8**:129-139

[8] Agustini CB, da Fontoura JT, Mella B, Gutterres M. Evaluating co-substrates to supplement biogas production from tannery solid waste treatment—Cattle hair, microalgae biomass, and silicone. Biofuels, Bioproducts and Biorefining. 2018;**12**:1095-1102

[9] Ahmad A, Shah SMU, Buang A, Othman MF, Abdullah MA. Evaluation of aerobic and anaerobic co-digestion of *Tetraselmis suecica* and oil palm empty fruit bunches by response surface methodology. Advanced Materials Research. 2014;**925**:243-247

[10] Avila R, Justo Á, Carrero E, Crivillés E, Vicent T, Blánquez P. Water resource recovery coupling microalgae wastewater treatment and sludge co-digestion for bio-wastes valorisation at industrial pilot-scale. Bioresource Technology. 2022;**343**:126080

[11] Barontini F, Biagini E, Dragoni F, Corneli E, Ragaglini G, Bonari E, et al. Anaerobic digestion and co-digestion of oleaginous microalgae residues for biogas production. Chemical Engineering Transactions. 2016;**50**:91-96

[12] Beltrán C, Jeison D, Fermoso FG, Borja R. Batch anaerobic co-digestion of waste activated sludge and microalgae (*Chlorella sorokiniana*) at mesophilic temperature. Journal of Environmental Science and Health—Part A Toxic/Hazardous Substances and Environmental Engineering. 2016;**51**:847-850

[13] Calicioglu O, Demirer GN. Carbon-to-nitrogen and substrate-to-inoculum ratio adjustments can improve co-digestion performance of microalgal biomass obtained from domestic wastewater treatment. Environmental Technology (United Kingdom). 2019;**40**:614-624

[14] Caporgno MP, Clavero E, Torras C, Salvadó J, Lepine O, Pruvost J, et al. Energy and nutrients recovery from lipid-extracted nannochloropsis via anaerobic digestion and hydrothermal liquefaction. ACS Sustainable Chemistry and Engineering. 2016;**4**:3133-3139

[15] Caporgno MP, Olkiewicz M, Torras C, Salvadó J, Clavero E, Bengoa C. Effect of pre-treatments on the production of biofuels from *Phaeodactylum tricornutum*. Journal of Environmental Management. 2016;**177**:240-246

[16] Carminati P, Gusmini D, Pizzera A, Catenacci A, Parati K, Ficara E. Biogas from mono- and co-digestion of microalgal biomass grown on piggery wastewater. Water Science and Technology. 2018;**78**:103-113

[17] Cea-Barcia G, Pérez J, Buitrón G. Co-digestion of microalga-bacteria biomass with papaya waste for methane production. Water Science and Technology. 2018;**78**:125-131

[18] Cheng Q, Deng F, Li H, Qin Z, Wang M, Li J. Nutrients removal from the secondary effluents of municipal domestic wastewater by *Oscillatoria tenuis* and subsequent co-digestion with pig manure. Environmental Technology (United Kingdom). 2018;**39**:3127-3134

[19] Damtie MM, Shin J, Jang HM, Kim YM. Synergistic co-digestion of microalgae and primary sludge to enhance methane yield from temperature-phased anaerobic digestion. Energies. 2020;**13**:4547

[20] Ahmad A, Shah SMU, Othman MF, Abdullah MA. Enhanced palm oil mill effluent treatment and biomethane production by co-digestion of oil palm empty fruit bunches with *Chlorella* sp. Canadian Journal of Chemical Engineering. 2014;**92**:1636-1642

[21] Dębowski M, Zieliński M, Kisielewska M, Krzemieniewski M. Anaerobic co-digestion of the energy crop *Sida hermaphrodita* and microalgae biomass for enhanced biogas production. International Journal of Environmental Research. 2017;**11**:243-250

[22] Du X, Tao Y, Li H, Liu Y, Feng K. Synergistic methane production from the anaerobic co-digestion of *Spirulina platensis* with food waste and sewage sludge at high solid concentrations. Renewable Energy. 2019;**142**:55-61

[23] Du X, Tao Y, Liu Y, Li H. Stimulating methane production from microalgae by alkaline pretreatment and co-digestion with sludge. Environmental Technology (United Kingdom). 2020;**41**:1546-1553

[24] Ehimen EA, Sun ZF, Carrington CG, Birch EJ, Eaton-Rye JJ. Anaerobic digestion of microalgae residues resulting from the biodiesel production process. Applied Energy. 2011;**88**:3454-3463

[25] Fernández-Rodríguez MJ, de la Lama-Calvente D, Jiménez-Rodríguez A, Borja R, Rincón B. Evolution of control parameters in biochemical methane potential tests of olive mill solid waste (OMSW), thermal pre-treated OMSW, and its co-digestion with *Dunaliella salina*. Journal of Applied Phycology. 2021;**33**:419-429

[26] Fernández-Rodríguez MJ, de la Lama-Calvente D, Jiménez-Rodríguez A, Borja R, Rincón-Llorente B. Anaerobic co-digestion of olive mill solid waste and microalga *Scenedesmus quadricauda*: Effect of different carbon to nitrogen ratios on process performance and kinetics. Journal of Applied Phycology. 2019;**31**:3583-3591

[27] Fernández-Rodríguez MJ, de la Lama-Calvente D, Jiménez-Rodríguez A, Borja R, Rincón-Llorente B. Influence

of the cell wall of *Chlamydomonas reinhardtii* on anaerobic digestion yield and on its anaerobic co-digestion with a carbon-rich substrate. Process Safety and Environmental Protection. 2019;**128**:167-175

[28] Fernández-Rodríguez MJ, Rincón B, Fermoso FG, Jiménez AM, Borja R. Assessment of two-phase olive mill solid waste and microalgae co-digestion to improve methane production and process kinetics. Bioresource Technology. 2014;**157**:263-269

[29] Ferreira LO, Astals S, Passos F. Anaerobic co-digestion of food waste and microalgae in an integrated treatment plant. Journal of Chemical Technology and Biotechnology. 2021

[30] Garoma T, Nguyen D. Anaerobic co-digestion of microalgae *Scenedesmus* sp. and TWAS for biomethane production. Water Environment Research. 2016;**88**:13-20

[31] Ahmed D, Wagdy R, Said N. Evaluation of biogas production from anaerobic co-digestion of sewage sludge with microalgae and agriculture wastes. BioResources. 2019;**14**:8405-8412

[32] Gonçalves RF, Assis TI, Maciel GB, Borges RM, Cassini STA. Co-digestion of municipal wastewater and microalgae biomass in an upflow anaerobic sludge blanket reactor. Algal Research. 2020;**52**:102117

[33] Herrmann C, Kalita N, Wall D, Xia A, Murphy JD, Rétfalvi T, et al. Optimised biogas production from microalgae through co-digestion with carbon-rich co-substrates. Bioresource Technology. 2016;**28**:2741-2752

[34] Hu Y, Kumar M, Wang Z, Zhan X, Stengel DB. Filamentous microalgae as an advantageous co-substrate for enhanced methane production and digestate dewaterability in anaerobic co-digestion of pig manure. Waste Management. 2021;**119**:399-407

[35] Kim J, Kang C-M. Increased anaerobic production of methane by co-digestion of sludge with microalgal biomass and food waste leachate. Bioresource Technology. 2015;**189**:409-412

[36] Kumar P, Bhattacharya A, Prajapati SK, Malik A, Vijay VK. Anaerobic co-digestion of waste microalgal biomass with cattle dung in a pilot-scale reactor: Effect of seasonal variations and long-term stability assessment. Biomass Conversion and Biorefinery. 2020;**12**:1203-1215

[37] Li R, Duan N, Zhang Y, Liu Z, Li B, Zhang D, et al. Anaerobic co-digestion of chicken manure and microalgae *Chlorella* sp.: Methane potential, microbial diversity and synergistic impact evaluation. Waste Management. 2017;**68**:120-127

[38] Li R, Duan N, Zhang Y, Liu Z, Li B, Zhang D, et al. Co-digestion of chicken manure and microalgae *Chlorella* 1067 grown in the recycled digestate: Nutrients reuse and biogas enhancement. Waste Management. 2017;**70**:247-254

[39] Lu D, Liu X, Apul OG, Zhang L, Ryan DK, Zhang X. Optimization of biomethane production from anaerobic co-digestion of microalgae and septic tank sludge. Biomass and Bioenergy. 2019;**127**:105266

[40] Mahdy A, Fotidis IA, Mancini E, Ballesteros M, González-Fernández C, Angelidaki I. Ammonia tolerant inocula provide a good base for anaerobic digestion of microalgae in third generation biogas process. Bioresource Technology. 2017;**225**:272-278

[41] Meneses-Reyes JC, Hernández-Eugenio G, Huber DH, Balagurusamy N, Espinosa-Solares T. Biochemical methane potential of oil-extracted microalgae

and glycerol in co-digestion with chicken litter. Bioresource Technology. 2017;**224**:373-379

[42] Álvarez X, Arévalo O, Salvador M, Mercado I, Velázquez-Martí B. Cyanobacterial biomass produced in the wastewater of the dairy industry and its evaluation in anaerobic co-digestion with cattle manure for enhanced methane production. PRO. 2020;**8**:1-16

[43] Meneses-Reyes JC, Hernández-Eugenio G, Huber DH, Balagurusamy N, Espinosa-Solares T. Oil-extracted *Chlorella vulgaris* biomass and glycerol bioconversion to methane via continuous anaerobic co-digestion with chicken litter. Renewable Energy. 2018;**128**:223-229

[44] Neumann P, Torres A, Fermoso FG, Borja R, Jeison D. Anaerobic co-digestion of lipid-spent microalgae with waste activated sludge and glycerol in batch mode. International Biodeterioration and Biodegradation. 2015;**100**:85-88

[45] Olsson J, Feng XM, Ascue J, Gentili FG, Shabiimam MA, Nehrenheim E, et al. Co-digestion of cultivated microalgae and sewage sludge from municipal waste water treatment. Bioresource Technology. 2014;**171**:203-210. DOI: 10.1016/j.biortech.2014.08.069

[46] Olsson J, Forkman T, Gentili FG, Zambrano J, Schwede S, Thorin E, et al. Anaerobic co-digestion of sludge and microalgae grown in municipal wastewater—A feasibility study. Water Science and Technology. 2018;**77**:682-694

[47] Panyaping K, Khiewwijit R, Wongpankamol P. Enhanced biogas production potential of microalgae and swine wastewater using co-digestion and alkaline pretreatment. Water Science and Technology. 2018;**78**:92-102

[48] Park S, Li Y. Evaluation of methane production and macronutrient degradation in the anaerobic co-digestion of algae biomass residue and lipid waste. Bioresource Technology. 2012;**111**:42-48

[49] Passos F, Cordeiro PHM, Baeta BEL, de Aquino SF, Perez-Elvira SI. Anaerobic co-digestion of coffee husks and microalgal biomass after thermal hydrolysis. Bioresource Technology. 2018;**253**:49-54

[50] Peng S, Colosi LM. Anaerobic digestion of algae biomass to produce energy during wastewater treatment. Water Environment Research. 2016;**88**:29-39

[51] Quintana-Najera J, Blacker AJ, Fletcher LA, Ross AB. Influence of augmentation of biochar during anaerobic co-digestion of *Chlorella vulgaris* and cellulose. Bioresource Technology. 2022;**343**:126086

[52] Ramos-Suárez JL, Carreras N. Use of microalgae residues for biogas production. Chemical Engineering Journal. 2014;**242**:86-95

[53] Ambarsari H, Adrian R, Manurung BS. Anaerobic biogas production using microalgae *Chlorella* sp. as biomass co-digested by cow manure and cow rumen fluid as inoculum. IOP Conference Series: Earth and Environmental Science. 2018

[54] Rétfalvi T, Szabó P, Hájos AT, Albert L, Kovács A, Milics G, et al. Effect of co-substrate feeding on methane yield of anaerobic digestion of *Chlorella vulgaris*. Journal of Applied Phycology. 2016;**28**:2741-2752

[55] Rincón-Pérez J, Celis LB, Morales M, Alatriste-Mondragón F, Tapia-Rodríguez A, Razo-Flores E. Improvement of methane production at alkaline and neutral pH from anaerobic

co-digestion of microalgal biomass and cheese whey. Biochemical Engineering Journal. 2021;**169**:107972

[56] Rusten B, Sahu AK. Microalgae growth for nutrient recovery from sludge liquor and production of renewable bioenergy. Water Science and Technology. 2011;**64**:1195-1201

[57] Saleem M, Hanif MU, Bahada A, Iqbal H, Capareda SC, Waqas A. The effects of hotwater and ultrasonication pretreatment of microalgae (*Nannochloropsis oculata*) on biogas production in anaerobic co-digestion with cow manure. PRO. 2020;**8**:1-10

[58] Santos-Ballardo DU, Font-Segura X, Ferrer AS, Barrena R, Rossi S, Valdez-Ortiz A. Valorisation of biodiesel production wastes: Anaerobic digestion of residual *Tetraselmis suecica* biomass and co-digestion with glycerol. Waste Management and Research. 2015;**33**:250-257

[59] Scarcelli PG, Serejo ML, Paulo PL, Boncz MÁ. Evaluation of biomethanization during co-digestion of thermally pretreated microalgae and waste activated sludge, and estimation of its kinetic parameters. Science of the Total Environment. 2020;**706**:135745

[60] Schwede S, Kowalczyk A, Gerber M, Span R. Anaerobic co-digestion of the marine microalga *Nannochloropsis salina* with energy crops. Bioresource Technology. 2013;**148**:428-435. DOI: 10.1016/j.biortech.2013.08.157

[61] Serna-García R, Borrás L, Bouzas A, Seco A. Insights into the biological process performance and microbial diversity during thermophilic microalgae co-digestion in an anaerobic membrane bioreactor (AnMBR). Algal Research. 2020;**50**:101981

[62] Serna-García R, Mora-Sánchez JF, Sanchis-Perucho P, Bouzas A, Seco A. Anaerobic membrane bioreactor (AnMBR) scale-up from laboratory to pilot-scale for microalgae and primary sludge co-digestion: Biological and filtration assessment. Bioresource Technology. 2020;**316**:123930

[63] Serna-García R, Ruiz-Barriga P, Noriega-Hevia G, Serralta J, Pachés M, Bouzas A. Maximising resource recovery from wastewater grown microalgae and primary sludge in an anaerobic membrane co-digestion pilot plant coupled to a composting process. Journal of Environmental Management. 2021;**281**:111890

[64] Arias DM, Solé-Bundó M, Garfí M, Ferrer I, García J, Uggetti E. Integrating microalgae tertiary treatment into activated sludge systems for energy and nutrients recovery from wastewater. Bioresource Technology. 2018;**247**:513-519

[65] Serna-García R, Zamorano-López N, Seco A, Bouzas A. Co-digestion of harvested microalgae and primary sludge in a mesophilic anaerobic membrane bioreactor (AnMBR): Methane potential and microbial diversity. Bioresource Technology. 2020;**298**

[66] Sialve B, Bernet N, Bernard O. Anaerobic digestion of microalgae as a necessary step to make microalgal biodiesel sustainable. Biotechnology Advances. 2009;**27**:409-416

[67] Siddique MNI, Wahid ZBA. Enhanced methane yield by codigestion of sewage sludge with microalgae and catering waste leachate. Water Environment Research. 2018;**90**:835-839

[68] Sitthikitpanya N, Sittijunda S, Khamtib S, Reungsang A. Co-generation of biohydrogen and biochemicals from co-digestion of *Chlorella* sp. biomass

hydrolysate with sugarcane leaf hydrolysate in an integrated circular biorefinery concept. Biotechnology for Biofuels. 2021;**14**:197

[69] Sittijunda S, Reungsang A. Methane production from the co-digestion of algal biomass with crude glycerol by anaerobic mixed cultures. Waste and Biomass Valorization. 2020;**11**:1873-1881

[70] Solé-Bundó M, Cucina M, Folch M, Tàpias J, Gigliotti G, Garfí M, et al. Assessing the agricultural reuse of the digestate from microalgae anaerobic digestion and co-digestion with sewage sludge. Science of the Total Environment. 2017;**586**:1-9

[71] Solé-Bundó M, Eskicioglu C, Garfí M, Carrère H, Ferrer I. Anaerobic co-digestion of microalgal biomass and wheat straw with and without thermo-alkaline pretreatment. Bioresource Technology. 2017;**237**:89-98

[72] Solé-Bundó M, Garfí M, Ferrer I. Pretreatment and co-digestion of microalgae, sludge and fat oil and grease (FOG) from microalgae-based wastewater treatment plants. Bioresource Technology. 2020;**298**:122563

[73] Solé-Bundó M, Garfí M, Matamoros V, Ferrer I. Co-digestion of microalgae and primary sludge: Effect on biogas production and microcontaminants removal. Science of the Total Environment. 2019;**660**:974-981

[74] Solé-Bundó M, Salvadó H, Passos F, Garfí M, Ferrer I. Strategies to optimize microalgae conversion to biogas: Co-digestion, pretreatment and hydraulic retention time. Molecules. 2018;**23**:2096

[75] Astals S, Musenze RS, Bai X, Tannock S, Tait S, Pratt S, et al. Anaerobic co-digestion of pig manure and algae: Impact of intracellular algal products recovery on co-digestion performance. Bioresource Technology. 2015;**181**:97-104

[76] Srivastava G, Kumar V, Tiwari R, Patil R, Kalamdhad A, Goud V. Anaerobic co-digestion of defatted microalgae residue and rice straw as an emerging trend for waste utilization and sustainable biorefinery development. Biomass Conversion and Biorefinery. 2020;**12**:1193-1202

[77] Tsapekos P, Kougias PG, Alvarado-Morales M, Kovalovszki A, Corbière M, Angelidaki I. Energy recovery from wastewater microalgae through anaerobic digestion process: Methane potential, continuous reactor operation and modelling aspects. Biochemical Engineering Journal. 2018;**139**:1-7

[78] Varol A, Ugurlu A. Biogas production from microalgae (*Spirulina platensis*) in a two stage anaerobic system. Waste and Biomass Valorization. 2016;**7**:193-200

[79] Vassalle L, Díez-Montero R, Machado ATR, Moreira C, Ferrer I, Mota CR, et al. Upflow anaerobic sludge blanket in microalgae-based sewage treatment: Co-digestion for improving biogas production. Bioresource Technology. 2020;**300**:122677

[80] Vassalle L, Passos F, Rosa-Machado AT, Moreira C, Reis M, Pascoal de Freitas M, et al. The use of solar pre-treatment as a strategy to improve the anaerobic biodegradability of microalgal biomass in co-digestion with sewage. Chemosphere. 2022;**286**:131929

[81] Wágner DS, Radovici M, Smets BF, Angelidaki I, Valverde-Pérez B, Plósz BG. Harvesting microalgae using activated sludge can decrease polymer dosing and enhance methane production via co-digestion in a bacterial-microalgal process. Algal Research. 2016;**20**:197-204

[82] Wang M, Kuo-Dahab WC, Park C. Investigation of characteristics of microalgae grown in different wastewater and their enhancing anaerobic digestibility of waste activated sludge. 86th Annual Water Environment Federation Technical Exhibition and Conference, WEFTEC 2013. 2013. p. 2071-2080

[83] Wang M, Lee E, Zhang Q, Ergas S. Energy production from anaerobic co-digestion of swine manure and microalgae *Chlorella* sp. 87th Annual Water Environment Federation Technical Exhibition and Conference, WEFTEC 2014. 2014. p. 3806-3816

[84] Wang M, Lee E, Zhang Q, Ergas SJ. Anaerobic co-digestion of swine manure and microalgae *Chlorella* sp.: Experimental studies and energy analysis. Bioenergy Research. 2016;**9**:1204-1215

[85] Wang M, Park C. Investigation of anaerobic digestion of *Chlorella* sp. and *Micractinium* sp. grown in high-nitrogen wastewater and their co-digestion with waste activated sludge. Biomass and Bioenergy. 2015;**80**:30-37

[86] Avila R, Carrero E, Crivillés E, Mercader M, Vicent T, Blánquez P. Effects of low temperature thermal pretreatments in solubility and co-digestion of waste activated sludge and microalgae mixtures. Algal Research. 2020;**50**:101965

[87] Wang M, Sahu AK, Rusten B, Park C. Anaerobic co-digestion of microalgae *Chlorella* sp. and waste activated sludge. Bioresource Technology. 2013;**142**:585-590

[88] Wang M, Zhu Z, Dolan S, Park C. Investigation of algal cultivation and anaerobic co-digestion of sewage sludge and algae at wastewater treatment plant (WWTP). WEFTEC 2012—85th Annual Technical Exhibition and Conference. 2012. p. 3586-3599

[89] Wirth R, Lakatos G, Böjti T, Maróti G, Bagi Z, Kis M, et al. Metagenome changes in the mesophilic biogas-producing community during fermentation of the green alga *Scenedesmus obliquus*. Journal of Biotechnology. 2015;**215**:52-61

[90] Wirth R, Pap B, Dudits D, Kakuk B, Bagi Z, Shetty P, et al. Genome-centric investigation of anaerobic digestion using sustainable second and third generation substrates. Journal of Biotechnology. 2021;**339**:53-64

[91] Woinaroschy A, Matei CB, Gălan A-M, Stepan E. Anaerobic co-digestion of delipidized microalgae biomass and food residues. UPB Scientific Bulletin, Series B: Chemistry and Materials Science. 2018;**80**:41-52

[92] Yuan X, Wang M, Park C, Sahu AK, Ergas SJ. Microalgae growth using high-strength wastewater followed by anaerobic co-digestion. Water Environment Research. 2012;**84**:396-404

[93] Zamorano-López N, Borrás L, Seco A, Aguado D. Unveiling microbial structures during raw microalgae digestion and co-digestion with primary sludge to produce biogas using semi-continuous AnMBR systems. Science of the Total Environment. 2020;**699**:134365

[94] Zhang Y, Caldwell GS, Blythe PT, Zealand AM, Li S, Edwards S, et al. Co-digestion of microalgae with potato processing waste and glycerol: Effect of glycerol addition on methane production and the microbial community. RSC Advances. 2020;**10**:37391-37408

[95] Zhang Y, Caldwell GS, Sallis PJ. Semi-continuous anaerobic co-digestion of marine microalgae with potato processing waste for methane production. Journal of Environmental Chemical Engineering. 2019;**7**:102917

[96] Zhang Y, Caldwell GS, Zealand AM, Sallis PJ. Anaerobic co-digestion of microalgae *Chlorella vulgaris* and potato processing waste: Effect of mixing ratio, waste type and substrate to inoculum ratio. Biochemical Engineering Journal. 2019;**143**:91-100

[97] Avila R, Carrero E, Vicent T, Blánquez P. Integration of enzymatic pretreatment and sludge co-digestion in biogas production from microalgae. Waste Management. 2021;**124**:254-263

[98] Zhang Y, Kang X, Wang Z, Kong X, Li L, Sun Y, et al. Enhancement of the energy yield from microalgae via enzymatic pretreatment and anaerobic co-digestion. Energy. 2018;**164**:400-407

[99] Zhen G, Lu X, Kobayashi T, Kumar G, Xu K. Anaerobic co-digestion on improving methane production from mixed microalgae (*Scenedesmus* sp., *Chlorella* sp.) and food waste: Kinetic modeling and synergistic impact evaluation. Chemical Engineering Journal. 2016;**299**:332-341

[100] Wannapokin A, Ramaraj R, Whangchai K, Unpaprom Y. Potential improvement of biogas production from fallen teak leaves with co-digestion of microalgae. 3 Biotech. 2018;**8**:123

[101] Golueke CG, Oswald WJ, Gotaas HB. Anaerobic Digestion of Algae. Applied Microbiology. 1957;**5**(1):47-55

[102] Zabed HM, Akter S, Yun J, Zhang G, Zhang Y, Qi X. Biogas from microalgae: Technologies, challenges and opportunities. Renewable and Sustainable Energy Reviews. 2020;**117**:109503

[103] Samson R, Leduy A. Improved performance of anaerobic digestion of *Spirulina maxima* algal biomass by addition of carbon-rich subtrates. Biotechnology Letters. 1983;**5**(10): 677-682

[104] Selormey GK, Barnes B, Kemausuor F, Darkwah L. A review of anaerobic digestion of slaughterhouse waste: effect of selected operational and environmental parameters on anaerobic biodegradability. Reviews in Environmental Science and Biotechnology. 2021;**20**:1073-1086

[105] Kabouris JC, Tezel U, Pavlostathis SG, Engelmann M, Dulaney JA, Todd AC, et al. Mesophilic and thermophilic anaerobic digestion of municipal sludge and fat, oil, and grease. Water Environment Research. 2009;**81**:476-485

[106] Younus Bhuiyan Sabbir ASM, Saha CK, Nandi R, Zaman MFU, Alam MM, Sarker S. Effects of seasonal temperature variation on slurry temperature and biogas composition of a commercial fixed-dome anaerobic digester used in Bangladesh. Sustainability (Switzerland). 2021;**13**:11096

[107] Zhang T, Mao C, Zhai N, Wang X, Yang G. Influence of initial pH on thermophilic anaerobic co-digestion of swine manure and maize stalk. Waste Management. 2015;**35**:119-126

[108] Cai Y, Gallegos D, Zheng Z, Stinner W, Wang X, Pröter J, et al. Exploring the combined effect of total ammonia nitrogen, pH and temperature on anaerobic digestion of chicken manure using response surface methodology and two kinetic models. Bioresource Technology. 2021;**337**:125328

[109] Carotenuto C, Guarino G, Morrone B, Minale M. Temperature and pH effect on methane production from buffalo manure anaerobic digestion. International Journal of Heat and Technology. 2016;**34**:S425-S429

[110] Ramos-Suárez JL, Martínez A, Carreras N. Optimization of the digestion process of *Scenedesmus* sp. and *Opuntia maxima* for biogas production. Energy Conversion and Management. 2014;**88**:1263-1270

[111] Raposo F, Borja R, Ibelli-Bianco C. Predictive regression models for biochemical methane potential tests of biomass samples: Pitfalls and challenges of laboratory measurements. Renewable and Sustainable Energy Reviews. 2020;**127**:109890

[112] Holliger C, Alves M, Andrade D, Angelidaki I, Astals S, Baier U, et al. Towards a standardization of biomethane potential tests. Water Science and Technology. 2016;**74**:2515-2522

Chapter 12

Removal of Divalent Nickel from Aqueous Solution Using Blue Green Marine Algae: Adsorption Modelling and Applicability of Various Isotherm Models

Ramsenthil Ramadoss, Durai Gunasekaran and Dhanasekaran Subramanian

Abstract

The adsorption of Ni(II) onto blue green marine algae (BGMA) in batch conditions is being investigated. The highest adsorption capacity of BGMA was found to be 42.056 mg/g under ideal testing conditions, where the initial Ni(II) metal ion concentration was adjusted from 25 ppm to 250 ppm. The optimal pH, biomass loading, and agitation rate for maximum Cu(II) ion removal have been determined to be 6, 2 g and 120 rpm, respectively. For the equilibrium condition, 24 hours of contact time is allowed. At room temperature, all of the experiments are conducted. The isotherm has a L shape, based on the equilibrium experimental data. It indicates that there is no considerable competition for active sites between the solvent and Ni(II). There is no strong competition between the solvent and Ni(II) for the active sites of BGMA, indicating that there is no strong competition between the two. It also suggests that the BGMA's Ni sorption ability is restricted (II). The experimental data is validated using multiple isotherm models, and the mechanism of adsorption is then discovered, as well as the process design parameters. The Fritz-Schlunder-V isotherm model is particularly relevant in defining the mechanism of Ni(II) adsorption under the conditions used in this study, according to modelling studies. This model's q_{max} of 41.89 mg/g shows that it matches experimental data more closely.

Keywords: divalent nickel, adsorption, isotherm models, blue green marine algae, modelling

1. Introduction

Metal-processing industries would undoubtedly have a challenge in disposing of metal-bearing effluents. Most industries dishonestly discharge their effluents into surrounding drains and water streams, either untreated or partially treated [1] due to increase in overall industrial cost. Industries such as alloys, pigments, electroplating, mining, metallurgical activities, nuclear power plant operations, aerospace industries, electrical contacts, printing, and the manufacture of paper, rubber, plastics, and batteries play a major role in water pollution by releasing heavy metal ions in their effluents [2].

Nickel, a non-biodegradable hazardous metal ion, is the heavy metal ion studied in this work [3]. Ni(II) ions in drinking water are allowed to be at a concentration of 0.02 mg/L. Anaemia, diarrhoea, encephalopathy, hepatitis, lung and kidney damage, gastrointestinal distress, pulmonary fibrosis, renal edoema, skin dermatitis, and central nervous system dysfunction are just some of the negative health impacts of exceeding the permissible limit [4]. As a result, before being discharged into the environment, industrial effluents containing Ni(II) must be treated [5].

Traditional methods for removing Ni(II) metal ions include coagulation, electro dialysis, flotation, ion exchange, precipitation, reverse osmosis, and others [6]. Low competency performances, especially when using these methods on very small concentrations of metal ions [7], are some of the limitations of these traditional approaches.

Adsorption with a low-cost sorbent is a frequently used approach in the treatment of industrial effluents [8]. However, it is still necessary to develop a low-cost, readily available, high-adsorption-capacity waste water treatment material that can address the aforementioned environmental concerns [9]. Because it is efficient, avoids secondary wastes, and utilises low-cost resources, biosorption onto live or non-living biomass, such as fungi, bacteria, yeast, moss, aquatic plants, and algae, can be a viable approach for removing heavy metals from their source [10]. Marine algae in coastal locations play a significant role in world ecology, are exceedingly efficient, and are taxonomically varied [11]. Many sections of the world gather or produce marine macroalgae, making them easily accessible in huge amount in the manufacture of very efficient bio sorbent resources [12]. The goal of this study is to utilise Blue Green Marine Algae (BGMA) as an adsorbent to adsorb the metal ions Ni(II) present in an artificial aqueous solution [13].

There is a scarcity of relevant literature on Ni(II) adsorption with three, four, and five parameter models [14, 15]. The constraints of the simple, one- and two-parameter models would be overcome by the high-parameter models. The use of large parameter models to describe the adsorption process under equilibrium conditions can provide highly clear and accurate information. The purpose of this work is to determine the biosorption capacity of BGMA for the removal of Ni(II) metal ions from a synthetically generated stock solution under optimal experimental circumstances of pH 6, 2 g biomass loading, and 120 rpm agitation speed. In addition, the experimental data is examined using one, two, three, four, and five parameter isotherm models. For the purpose of modelling, the experimental data is examined using one, two, three, four, and five parameter isotherm models.

2. Materials and methods

The BGMA was collected along the coast of Tamil Nadu, India, near Chidambaram. It is cleaned and dried at room temperature with distilled water. It is then pulverised to 150–200 microns in size. For IR spectrum investigations of dried biomass and Ni(II)-sorbed biomass in the range of 4000–400 cm^{-1}, a Fourier transform infrared (FT-IR) spectrometer (BRUKER FT-IR, ALPHA-T, GERMANY) was employed.

The synthetic Ni(II) solution is made with an analytical grade salt, Nickel(II) sulfate heptahydrate ($NiSO_4{\cdot}7H_2O$). To make 1 L of solution for the stock purpose of

1000 ppm, exactly 4.7852 g of $NiSO_4{\cdot}7H_2O$ is weighed and utilised. For stock solution preparation, double distilled water is employed. This stock solution is diluted to a concentration of 25–250 parts per million.

Each solution's pH is changed from 2 to 7. pH is adjusted with 0.1 N nitric acid (HNO_3) and 0.1 N sodium hydroxide (NaOH) solutions. The amount of biomass put to the conical flask varies from 0.5 to 2.5 g (0.5, 1.0, 1.5, 2.0 and 2.5 g).

A 500 mL conical flask is used for the batch adsorption experiment. The starting concentrations of metal ions are 25, 50, 75, 100, 125, 150, 175, 200, 225, and 250 parts per million. Each 500 mL Erlenmeyer flask contains 400 mL of 25 ppm metal ion solution. Each Erlenmeyer flask receives 0.5, 1.0, 1.5, 2.0, and 2.5 g of BGMA, respectively. In all five flasks, the pH of the solution is kept at 2. The flasks are agitated at 120 rpm in a rotary shaker. There is a total of 24 hours of contact (shaking) time given. This is more than enough to attain the desired equilibrium (maximum adsorption). The starting and ultimate concentrations of the solution are determined using a double-beam Atomic Adsorption Spectrophotometer (AAS SL176-Elico Limited India). The % removal of metal ions is computed using Eq. (1) from the starting (C_{in}) and equilibrium final concentration (C_{eq}).

$$\%\text{Removal} = \frac{C_{in} - C_{eq}}{C_{in}} \times 100 \qquad (1)$$

Eq. (2) is used to compute the equilibrium metal uptake, q_{eq}, using the starting (C_i) and equilibrium (C_{eq}) concentrations of the metal ion solution.

$$q_{eq} = \frac{V}{M}\left(C_{in} - C_{eq}\right) \qquad (2)$$

where V is the litre volume of the liquid sample and M is the gram weight of the adsorbent. In order to improve the pH value and biomass loading, the same operation is repeated with adjusting the solution pH as 3, 4, 5, 6 and 7. Similarly, 50 ppm, 75 ppm, 100 ppm, 125 ppm, 150 ppm, 175 ppm, 200 ppm, 225 ppm, and 250 ppm metal ion concentrations are used in the studies. For concordant results, experiments are repeated (**Table 1**).

S. no.	C_{in} (mg/L)	C_{eq} (mg/L)	q_{eq} (mg/g)	Removal (%)
1	25	1.52	4.696	93.92
2	50	3.25	9.350	93.50
3	75	5.23	13.954	93.03
4	100	6.54	18.692	93.46
5	125	15.39	21.922	87.69
6	150	21.58	25.684	85.61
7	175	27.64	29.472	84.21
8	200	29.39	34.122	85.31
9	225	35.79	37.842	84.09
10	250	39.72	42.056	84.11

Table 1.
Experimental values of adsorption of Ni(II) onto BGMA.

One parameter model (Henry's law), two parameter models (Henry's law with intercept, Langmuir, Freundlich, Dubinin-Radushkevich, Temkin, Hill-de Boer, Fowler-Guggenheim, Flory-Huggins, Halsey, Harkin-Jura, Jovanovic, Elovich and Kiselev), three parameter models (Hill, Redlich-Peterson, Sips, Langmuir-Freundlich, Fritz-Schlunder-III, Radke-Prausnits-I, Radke-Prausnits-II, Radke-Prausnits-III, Toth, Khan, Koble-Corrigan, Jossens, Jovanovic-Freundlich, Brouers-Sotolongo, Vieth-Sladek, Unilan, Holl-Krich and Langmuir-Jovanovic), four parameter models (Fritz-Schlunder-IV, Baudu Weber-van Vliet and Marczewski-Jaroniec) and five parameter model (Fritz-Schlunder-5) are used to examine the experimental facts and to discover its applicability for modelling purpose. The parameter values are predicted using the cftool kit available in MATLAB R2010a software. This toolkit aids in the estimation of model parameters, including the non-linear regression coefficient (R^2), Sum of Squares due to Error (SSE), and Root Mean Squared Error (RMSE).

3. Isotherm models: theoretical knowledge

Isotherm models are used to determine an adsorbent's maximal sorption capacity and is represented in terms of the amount of metal absorbed per unit mass of adsorbent.

3.1 One parameter model

The simple adsorption isotherm model has only one parameter to explain the adsorption mechanism. The most basic adsorption isotherm is one in which the amount of solute adsorbed is proportional to the equilibrium effluent concentration [16]. Eq. (3) is the model.

$$q_{eq} = KC_{eq} \tag{3}$$

3.2 Two parameter models

This section focuses on gaining theoretical insight into models with two parameters that explain the adsorption mechanism. Henry's law with constant, Langmuir, Freundlich, Dubinin-Radushkevich, Temkin, Hill-de Boer, Fowler-Guggenheim, Flory-Huggins, Halsey, Harkin-Jura, Jovanovic, Elovich, and Kiselev are among the 13 models covered.

3.2.1 Henry's law with intercept isotherm model

This model was created to address the contradiction highlighted by the one parameter model and to be applicable over a wide range of metal ion concentrations [16]. Eq. (4) describes the model.

$$q_{eq} = KC_{eq} + C \tag{4}$$

3.2.2 Langmuir isotherm model

The Langmuir model assumes a homogenous surface and explains how the adsorbate forms monolayer coverage on the adsorbent's outer surface [17, 18]. The rate of adsorption is proportional to the percentage of open adsorbent surface, and the rate of desorption is related to the fraction of covered adsorbent surface. The model is described in Eq. (5).

$$q_{eq} = \frac{q_{max} b_L C_{eq}}{1 + b_L C_{eq}} \tag{5}$$

b_L is the Langmuir constant, which links the fluctuation of the appropriate area and porosity of the adsorbent with its adsorption capacity (mg/g). A dimensionless constant called the Langmuir separation factor R_L, which is computed as Eq. (6), helps explain the key properties of the Langmuir isotherm.

$$R_L = \frac{1}{1 + b_L q_{max}} \tag{6}$$

When $R_L > 1$, adsorption is unfavourable, when $R_L = 1$, linear when $R_L = 1$, favourable when $R_L = 0$, and irreversible when $R_L = 0$.

3.2.3 Freundlich isotherm model

The Freundlich adsorption isotherm model depicts the adsorbent surface heterogeneity. The adsorptive sites are made up of tiny homogenous heterogeneous adsorption sites [19]. Eq. (7) is the model.

$$q_{eq} = a_F C_{eq}^{\frac{1}{n_F}} \tag{7}$$

Freundlich adsorption capacity (L/mg) is denoted by a_F, while adsorption intensity is denoted by n_F. The higher the adsorption capacity, the larger the a_F value. The magnitude of $1/n_F$, which varies from 0 to 1, indicates favourable adsorption and becomes more heterogeneous as it approaches zero [18–21].

3.2.4 Dubinin-Radushkevich isotherm model

This empirical model implies that physical adsorption processes are multilayered and involve Van Der Waal's forces [22], it is frequently used to estimate the characteristic porosity. This model [23] gives insight into gas and vapour adsorption on micro porous sorbents.

The Dubinin-Radushkevich isotherm's temperature dependence is another distinguishing trait [24, 25]. Eq. (8) represents the Dubinin-Radushkevich isotherm.

$$q_{eq} = K_{DR} \exp\left[-B_{DR}\left(RT\ln\left(1 + \frac{1}{C_{eq}}\right)\right)^2\right] \tag{8}$$

$$\varepsilon = RT\ln\left[\frac{1}{C_{eq}}\right] \tag{9}$$

ε is known as the Polanyi potential, as seen in Eq. (9). The activation energy or mean free energy E (kJ/mol) of adsorption per molecule of adsorbate when it is transported from infinity in the solution to the surface of the solid may be computed using Eq. (10).

$$E = \frac{1}{\sqrt{2K_{DR}}} \tag{10}$$

The value of E is used to forecast whether an adsorption is physical or chemical. In nature, physisorption occurs when E = 8 KJ/mol, whereas chemisorption occurs when E = 8–16 KJ/mol [23].

3.2.5 Temkin isotherm model

For forecasting the gas phase adsorption equilibrium, the Temkin adsorption isotherm model is quite useful [25, 26]. Eq. (11) illustrates the Temkin adsorption isotherm model.

$$q_{eq} = \frac{RT}{b_T}\left(\ln A_T C_{eq}\right) \tag{11}$$

The heat of adsorption is represented by the equation RT/b_T, and the equilibrium binding constant (L/mg) corresponding to the maximal binding energy is represented by A_T.

3.2.6 Hill-de Boer isotherm model

This Hill-Deboer isotherm model accurately describes mobile adsorptions as well as lateral interactions between adsorbed molecules [27, 28]. Eq. (12) depicts the Hill-Deboer isotherm model.

$$K_1 C_{eq} = \frac{\theta}{1-\theta} \exp\left(\frac{\theta}{1-\theta} - \frac{K_2\theta}{RT}\right) \tag{12}$$

A positive value of K_2 implies attraction between adsorbed species, whereas a negative value of K_2 suggests repulsion. If K_2 is equal to zero, there is no interaction between adsorbed molecules, and the Volmer equation [29] is used

3.2.7 Fowler-Guggenheim isotherm model

The Fowler-Guggenheim adsorption isotherm model describes how adsorbed molecules interact laterally. Eq. (13) represents the Fowler-Guggenheim adsorption isotherm.

$$K_{FG} C_{eq} = \frac{\theta}{1-\theta} \exp\left[\frac{2\theta W}{RT}\right] \tag{13}$$

The presence of a positive W indicates that the contact between the adsorbed molecules is attractive. In contrast, if W = 0, the contact between adsorbed molecules is repulsive, and the heat of adsorption decreases with loading, the Fowler-Guggenheim equation reduces to the Langmuir model. W = 0 when there is no contact between adsorbed molecules. Furthermore, this model is only viable when the surface coverage is less than 0.6 [30, 31].

3.2.8 Flory-Huggins isotherm model

The degree of surface coverage of the adsorbate onto the adsorbent is discussed by the Flory-Huggins isotherm [32–35]. Eq. (14) shows the Flory-Huggins adsorption isotherm.

$$\frac{\theta}{C_{in}} = K_{FH}\,[1-\theta]^{n_{FH}} \tag{14}$$

K_{FH} is used to calculate the spontaneity Gibbs free energy.

3.2.9 Halsey isotherm model

In multi layer adsorption, the hetero porous nature of the adsorbent is demonstrated by the fitting of experimental data to this equation [36]. Eq. (15) denotes the Halsey adsorption isotherm model.

$$q_{eq} = \exp\left[\frac{\ln K_{Ha} - \ln C_{eq}}{n_{Ha}}\right] \tag{15}$$

3.2.10 Harkin-Jura isotherm model

The Hurkin-Jura adsorption isotherm is used to explain multilayer adsorption on the surface of absorbents with heterogeneous pore distribution [37]. Eq. (16) describes the Hurkin-Jura adsorption isotherm model.

$$q_{eq} = \sqrt{\frac{A_{HJ}}{B_{HJ} - \log C_{eq}}} \tag{16}$$

3.2.11 Jovanovic isotherm model

With the approximation of monolayer localised adsorption without lateral contacts, the Jovanovic adsorption isotherm model is analogous to the Langmuir model [38]. Eq. (17) shows the Jovanovic adsorption isotherm model.

$$q_{eq} = q_{max}\left[1 - e^{-(K_J C_{eq})}\right] \tag{17}$$

It forms the Langmuir isotherm at large adsorbate concentrations but does not obey Henry's rule.

3.2.12 Elovich isotherm model

Adsorption sites expand exponentially with adsorption, demonstrating multilayer adsorption, according to the Elovich isotherm model [39]. The Elovich adsorption isotherm model is depicted in Eq. (18).

$$\frac{q_{eq}}{q_{max}} = K_E C_{eq} \exp^{\frac{q_{eq}}{q_{max}}} \tag{18}$$

3.2.13 Kiselev isotherm model

The Kiselev adsorption isotherm model is also known as localised monomolecular layer model [40]. This model is valid only for surface coverage $\theta > 0.68$. The Kiselev adsorption isotherm model is given in Eq. (19).

$$K_{eqK}C_{eq} = \frac{\theta}{(1-\theta)\,(1+K_{nK}\theta)} \tag{19}$$

3.3 Three parameter models

Models containing three parameters to explain the mechanism of adsorption are discussed using 16 models viz., Hill, Redlich-Peterson, Sips, Langmuir-Freundlich, Fritz-Schlunder-III, Radke-Prausnits, Toth, Khan, Koble-Corrigan, Jossens, Jovanovic-Freundlich, Brouers-Sotolongo, Vieth-Sladek, Unilan, Holl-Krich and Langmuir-Jovanovic.

3.3.1 Hill isotherm model

To characterise the adhesion of diverse species to homogeneous substrates, the Hill isotherm model is developed [41]. Eq. (20) depicts the Hill adsorption isotherm model.

$$q_{eq} = \frac{q_{max}C_{eq}^{n_H}}{K_H + C_{eq}^{n_H}} \tag{20}$$

If n_H is greater than 1, this isotherm indicates positive co-operativity in binding, n_H is equal to 1, it indicates non-cooperative or hyperbolic binding and n_H is less than 1, indicating negative co-operativity in binding.

3.3.2 Redlich-Peterson isotherm model

The Redlich-Peterson isotherm model is created by combining elements of the Langmuir and Freundlich isotherms [42]. Eq. (21) depicts the Redlich-Peterson isotherm model.

$$q_{eq} = \frac{A_{RP}C_{eq}}{1 + B_{RP}C_{eq}^{\beta}} \tag{21}$$

When the liquid phase concentration is low, this model approaches Henrys Law. β_{RP}, the exponent, is usually between 0 and 1. When $\beta_{RP} = 1$, this model is similar to the Langmuir model, and when $\beta_{RP} = 0$, this isotherm is similar to the Freundlich model.

3.3.3 Sips isotherm model

The Sips adsorption isotherm model [43] was designed to represent localised adsorption without adsorbate-adsorbate interactions [44] at high adsorbate concentrations. Eq. (22) is the Sips adsorption isotherm model.

$$q_{eq} = \frac{q_m K_s C_{eq}^{\beta_S}}{1 + K_s C_{eq}^{\beta_S}} \tag{22}$$

When β_S equal to 1 this isotherm approaches Langmuir isotherm and β_S equal to 0, this isotherm approaches Freundlich isotherm.

3.3.4 Langmuir-Freundlich isotherm model

Adsorption in heterogeneous surfaces is described by the Langmuir-Freundlich isotherm model [17, 18]. Eq. (23) depicts the Langmuir-Freundlich isotherm model.

$$q_{eq} = \frac{q_{max}\left(K_{LF} C_{eq}\right)^{m_{LF}}}{1 + \left(K_{LF} C_{eq}\right)^{m_{LF}}} \tag{23}$$

m_{LF} is a heterogeneous parameter with a value between 0 and 1. This value rises when the degree of surface heterogeneity decreases. For m_{LF} is equal to 1, this model covert to Langmuir model.

3.3.5 Fritz-Schlunder-III isotherm model

Because of the large number of coefficients in their isotherm, the Fritz-Schunder three parameter isotherm model was constructed to suit a wide variety of experimental findings [45]. Eq. (24) has this expression.

$$q_{eq} = \frac{q_{max} K_{FS3} C_{eq}}{1 + q_{max} C_{eq}^{m_{FS3}}} \tag{24}$$

If m_{FS3} is equal to 1, the Fritz-Schlunder-III model becomes the Langmuir model but for high concentrations of adsorbate, the Fritz-Schlunder-III reduces to the Freundlich model.

3.3.6 Radke-Prausnits isotherm model

At low adsorbate concentrations, the Radke-Prausnits isotherm model has numerous essential qualities that make it preferable in most adsorption systems [46]. When the Radke-Prausnits model exponent mRaP3 is equal to zero. Eqs. (25)–(27) show Radke-Prausnits isotherm models.

$$\text{Model 1}: q_{eq} = \frac{q_{max} K_{RaP1} C_{eq}}{\left[1 + K_{RP1} C_{eq}\right]^{m_{RaP1}}} \tag{25}$$

$$\text{Model 2}: q_{eq} = \frac{q_{max} K_{RaP2} C_{eq}}{1 + K_{RP2} C_{eq}{}^{m_{RaP2}}} \tag{26}$$

$$\text{Model 3}: q_{eq} = \frac{q_{max} K_{RaP3} C_{eq}^{m_{RaP3}}}{1 + K_{RP3} C_{eq}{}^{m_{RaP3-1}}} \tag{27}$$

When both m_{RaP1} and m_{RaP2} are equal to 1, the Radke-Prausnitz 1, 2 models decrease to the Langmuir model; nevertheless, at low concentrations, the models become Henry's law; but, at high adsorbate concentrations, the Radke-Prausnitz 1 and

2 models become the Freundlich model. When the exponent m_{RaP3} is equal to 1, the Radke-Prausnitz-3 equation simplifies to Henry's law, and when the exponent mRaP3 is equal to 0, it becomes the Langmuir isotherm.

3.3.7 Toth isotherm model

The Toth adsorption isotherm model is used to explain heterogeneous adsorption systems that fulfil both the low and high end boundaries of adsorbate concentration [47]. Eq. (28) represents the Toth isotherm model.

$$q_{eq} = \frac{q_{max} C_{eq}}{\left(\frac{1}{K_T} + C_{eq}^{n_T}\right)^{\frac{1}{n_T}}} \tag{28}$$

When n = 1, this equation simplifies to the Langmuir isotherm equation, indicating that the process is approaching the homogenous surface. As a result, the value n describes the adsorption system's heterogeneity. The system is considered to be heterogeneous if it deviates farther from unity.

3.3.8 Khan isotherm model

For adsorption of bi-adsorbate from pure dilute equations solutions, the Kahn isotherm model is devised [48]. Eq. (29) presents the Kahn isotherm model.

$$q_{eq} = \frac{q_m b_K C_{eq}}{\left(1 + b_K C_{eq}\right)^{a_K}} \tag{29}$$

When a_K is equal to one, the Toth model approaches the Langmuir isotherm model, and when a_K is more than one, the Toth model simplifies to the Freundlich isotherm model.

3.3.9 Koble-Corrigan isotherm model

The Sips isotherm model is similar to the Koble-Carrigan isotherm model. This model includes both the Langmuir and the Freundlich isotherms [49]. Eq. (30) depicts the Koble-Carrigan isotherm model.

$$q_{eq} = \frac{A_{KC} C_{eq}^{n_{KC}}}{1 + B_{KC} {C_{eq}}^{n_{KC}}} \tag{30}$$

This model reduces to the Freundlich isotherm at large adsorbate concentrations. It is only acceptable when n is higher than or equal to 1. When n is lesser than one, it indicates that the model, despite a high concentration coefficient or a low error value, is incapable of describing the experimental data.

3.3.10 Jossens isotherm model

The Jossens isotherm model is based on the energy distribution of adsorbate-adsorbent interactions at adsorption sites [50]. At low concentrations, this model is reduced to Henry's law. Eq. (29) depicts the Jossens isotherm model.

$$q_{eq} = \frac{K_J C_{eq}}{1 + J\, C_{eq}^{b_J}} \tag{31}$$

At low capacity, J equates to Henry's constant. b_J is the Jossens isotherm constant, which is constant regardless of temperature or the composition of the adsorbent.

3.3.11 Jovanovic-Freundlich isotherm model

To depict single-component adsorption equilibrium on heterogeneous surfaces, the Jovanovic-Freundlich isotherm model is developed [51]. Eq. (32) represents the Jovanovic-Freundlich isotherm model.

$$q_{eq} = q_{max}\left[1 - e^{-\left(K_{JF}\, C_{eq}{}^{n_{JF}}\right)}\right] \tag{32}$$

3.3.12 Brouers-Sotolongo isotherm model

This isotherm is built in the form of a deformed exponential function for adsorption onto a heterogeneous surface [52]. Eq. (33) depicts the Brouers-Sotolongo model.

$$q_{eq} = q_{max}\left[1 - e^{\left(-K_{BS} C_{eq}^{\alpha_{BS}}\right)}\right] \tag{33}$$

The parameter α_{BS} is related with distribution of adsorption energy and the energy of heterogeneity of the adsorbent surfaces at the given temperature [53].

3.3.13 Vieth-Sladek isotherm model

This model includes two independent parts for calculating transient adsorption diffusion rates in solid adsorbents [54]. Eq. (34) represents the Vieth-Sladek isotherm model.

$$q_{eq} = K_{VS} C_{eq} + \frac{q_{max} \beta_{VS} C_{eq}}{1 + \beta_{VS} C_{eq}} \tag{34}$$

3.3.14 Unilan isotherm model

The application of the local Langmuir isotherm and uniform energy distribution is assumed for the Unilan isotherm model [44]. Eq. (35) presents the Unilan isotherm model.

$$q_{eq} = \frac{q_{max}}{2\beta_U} \ln\left[\frac{1 + K_U C_{eq} e^{\beta_U}}{1 + K_U C_{eq} e^{-\beta_U}}\right] \tag{35}$$

The higher the model exponent β_U, the system is more heterogeneous. If β_U is equal to 0, the Unilan isotherm model becomes the classical Langmuir model as the range of energy distribution is zero in this limit [50, 55, 56].

3.3.15 Holl-Krich isotherm model

The Langmuir Isotherm [57] is a version of the Holl-Krich Isotherm Model. The Freundlich isotherm is formed when the concentration of the solvent is low [22]. The Holl-Krich Isotherm Model may be seen in Eq. (36).

$$q_{eq} = \frac{q_{max} K_{HK} C_{eq}^{n_{HK}}}{1 + K_{HK} C_{eq}^{n_{HK}}} \tag{36}$$

3.3.16 Langmuir-Jovanovic isotherm model

This empirical model is the combined form of both Langmuir and Jovanovic isotherm [58]. The Langmuir-Jovanovic model is given in Eq. (37).

$$q_{eq} = \frac{q_{max} C_{eq} \left[1 - e^{\left(K_{LJ} C_{eq}^{n_{LJ}}\right)}\right]}{1 + C_{eq}} \tag{37}$$

3.4 Four parameter models

The four parameter models discussed in this study are Fritz-Schlunder-IV, Baudu, Weber-van Vliet and Marczewski-Jaroniec models.

3.4.1 Fritz-Schlunder-IV isotherm model

Fritz-Schlunder IV model is another model comprised of four-parameter with combine features of Langmuir-Freundlich isotherm [45]. The model is given in Eq. (38).

$$q_{eq} = \frac{A_{FS5}\ C_{eq}^{\alpha_{FS5}}}{1 + B_{FS5} C_{eq}^{\beta_{FS5}}} \tag{38}$$

When the values of α_{FS5} and β_{FS5} are less than or equal to one, this isotherm is true. The Fritz-Schlunder-IV isotherm transforms into the Freundlich equation at high adsorbate concentrations. If both α_{FS5} and β_{FS5} are equal to one, the isotherm is reduced to the Langmuir isotherm. This isotherm model becomes the Freundlich at large concentrations of adsorbate in the liquid-phase.

3.4.2 Baudu isotherm model

The Baudu isotherm model was created in response to a disagreement in computing the Langmuir constant and coefficient from slope and tangent across a wide range of concentrations [59]. The Langmuir isotherm model has been modified into the Baudu isotherm model. Eq. (39) explains it.

$$q_{eq} = \frac{q_{max}\ b_o\ C_{eq}^{(1+x+y)}}{1 + b_o C_{eq}^{(1+x)}} \tag{39}$$

This model is only applicable in the range of (1 + x + y) < 1 and (1 + x) < 1. For lower surface coverage, Baudu model reduces to the Freundlich equation [58], i.e.:

$$q_{eq} = \frac{q_{m0}\ b_o C_{eq}^{(1+x+y)}}{1 + b_0} \quad (40)$$

3.4.3 Weber-van Vliet isotherm model

With four parameters, the Weber and van Vliet isotherm model is used to represent equilibrium adsorption data [60–62]. Eq. (41) depicts the model.

$$C_{eq} = P_1 q_{eq}^{\left(P_2 q_{eq}^{P_3} + P_4\right)} \quad (41)$$

Multiple nonlinear curve fitting approaches based on the reduction of the sum of squares of residuals can be used to define the isotherm parameters P_1, P_2, P_3, and P_4.

3.4.4 Marczewski-Jaroniec isotherm model

The Marczewski-Jaroniec isotherm model is analogous to the Langmuir isotherm model [62, 63]. Eq. (42) represents the Marczewski-Jaroniec isotherm model.

$$q_{eq} = q_{max}\left[\frac{\left(K_{MJ} C_{eq}\right)^{n_{MJ}}}{1 + \left(K_{MJ} C_{eq}\right)^{n_{MJ}}}\right] \quad (42)$$

The spreading of distribution along the route of increasing adsorption energy is described by K_{MJ}.

3.5 Five parameter model

Accounting for the high parameter models offers unmistakable information on the process of adsorption under equilibrium conditions. Only one five-parameter model, the Fritz-Schlunder-V isotherm model, is used in this section.

3.5.1 Fritz-Schlunder-V isotherm model

The Fritz-Schlunder adsorption isotherm model was created with the goal of more precisely reproducing model modifications for applicability over a wide range of equilibrium data [45]. Eq. (43) represents the Fritz-Schlunder adsorption isotherm model.

$$q_{eq} = \frac{q_{max} K_{1FS5} C_{eq}^{\alpha_{FS5}}}{1 + K_{2FS5} C_{eq}^{\beta_{FS5}}} \quad (43)$$

4. Results and discussion

The findings of the experiments suggest that the highest Ni(II) adsorption by BGMA may be attained at a pH of 6 and a biomass loading of 2 g of BGMA. BGMA has an adsorption capability of 42.056 mg/g. During the continuous 24 hours of contact time, the agitation rate of 120 rpm is maintained.

Figure 1 depicts the visual effect of Ni(II) starting metal ion concentration on equilibrium metal absorption and % clearance. The largest percent elimination of Ni (II) metal ions is observed at low starting metal ion concentrations. The diminishing trend in Ni(II) metal ion removal is observed with an increase in initial metal ion concentration due to an increase in the ratio of the initial number of metal ions to the fixed number of active sites. Furthermore, for a certain number of active sites, the amount of substrate metal ions accommodated in the interlayer gap rises, resulting in decreased metal ion removal. An increase in the initial metal ion concentration causes a decrease in the ionic strength of the solution, which helps to improve metal absorption. As a result of the lowering ionic strength, a rise in initial metal ion concentration raises the equilibrium metal uptake (q_{eq}).

4.1 FTIR-characterisation of BGMA biomass

Figure 2 depicts an FTIR spectroscopic investigation of BGMA prior to Ni(II) adsorption. The —NH stretching is shown by the wide adsorption bands at 3696.36 cm^{-1}, 3620.77 cm^{-1}, and 3408.94 cm^{-1}. The —CH_2 stretching is measured at 2928.18 cm^{-1}. The wide adsorption band at 1643.81 cm^{-1} might be attributed to the carboxylic C=O group, whereas the carboxylate group is represented by the adsorption band at 1427.97 cm^{-1}. Furthermore, the band at 1039.87 cm^{-1} shows C—N amide stretching. **Figure 3** depicts an FTIR spectroscopic investigation of BGMA following Ni(II) adsorption. The shifts of peaks at 3695.90–34696.36 cm^{-1}, 1643.81–1644.54 cm^{-1}, and 1041.07–1039.87 cm^{-1} after Ni(II) adsorption indicate that the amide —NH bonding, CH stretching, carboxylic acid, and hydroxyl groups are the main functional groups involved in the adsorption of Ni(II) metal ions.

4.2 Adsorption isotherms

Figure 4 depicts the experimental adsorption behaviour of Ni(II) from its synthetic aqueous solutions onto BGMA, which is particularly important in distinguishing

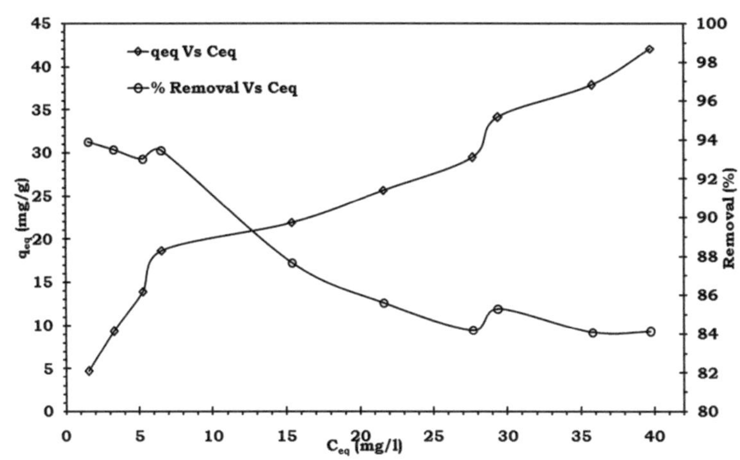

Figure 1.
Effect of initial metal ion concentration of equilibrium metal uptake and % removal for adsorption of Ni(II) onto BGMA.

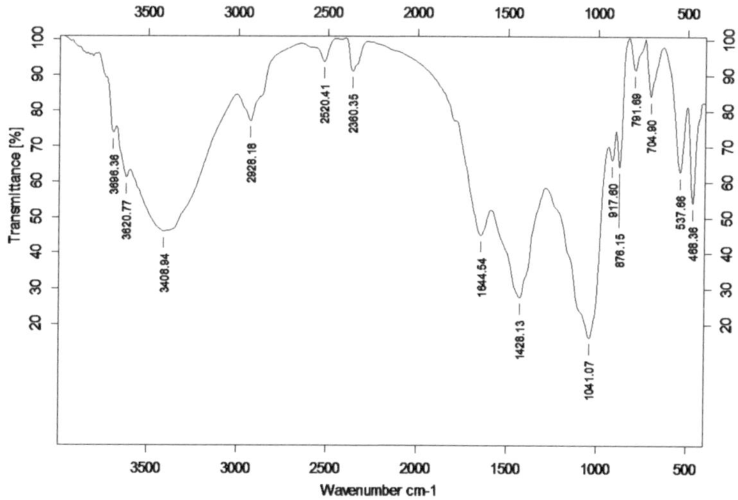

Figure 2.
FTIR spectra of BGMA biomass after adsorption of Ni(II).

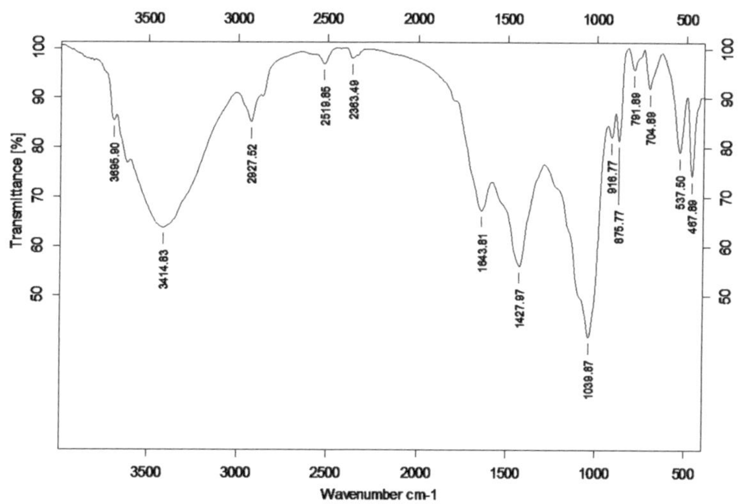

Figure 3.
FTIR spectra of BGMA biomass before adsorption of Ni(II).

the form of the isotherm [32, 64]. According to Giles et al. [65], the isotherm of Ni(II) onto BGMA is detected as an L curve pattern. As a result, it is determined that there is no considerable rivalry between the solvent and the adsorbate for the active sites of BGMA. Also According to Limousin et al., [66], BGMA has a restricted sorption capability for Ni(II) adsorption at the circumstances used in this study.

4.2.1 One parameter model

The experimental data for the adsorption of Ni(II) onto BGMA is fitted to Henry's law (one parameter) model. This model's parameter values and regression coefficient R^2 are shown in **Table 2**. The model fails to match the experimental data under equilibrium conditions due to the small R^2 value.

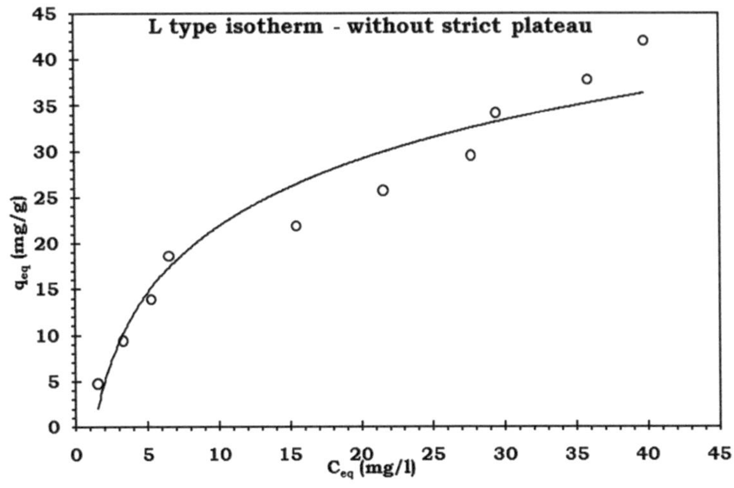

Figure 4.
Experimental results of adsorption of Ni(II) onto BGMA.

Model	Parameter	Value	SSE	R^2	RMSE
Henry's law model	K	1.132	274.4	0.8001	5.522

Table 2.
Parameter values of one parameter (Henry's law) model for adsorption of Ni(II) onto BGMA.

4.2.2 Two parameter models

Table 3 shows the regression coefficients and parameter values of two parameter adsorption isotherm models for Ni(II) adsorption onto BGMA. The R^2 values for the Dubinin-Radushkevich, Hill-de Boer, Fowler-Guggenheim, Halsey, Harkin-Jura, Elovich, and Kiselev models are weak and negative when compared to the experimental data. Though the R^2 values of the Temkin and Flory-Huggins models suggest their relevance, the parameter values found in both models (b_T and n_{FH}) appear to be too high and negative, which are not physically realisable.

Only four models, namely Freundlich, Henry's law with intercept, Jovanovic, and Langmuir, are considered to carry out the following discussion. **Figure 5** depicts a plot of C_{eq} (mg/L) vs. q_{eq} (mg/g) for the two models, as well as experience data.

With experimental data, the Freundlich isotherm model performs better. Its R^2 score indicates its relevance to a large extent. Adsorption sites are stimulated via the surface exchange process, resulting in enhanced adsorption. Because the value of n_F is in the 1–10 range, It means that Ni(II) adsorption from its synthesised solution onto BGMA is favourable. The value of $1/n_F$ is calculated as 0.5608, which is closer to zero, assuring that the active sites of BGMA for Ni(II) adsorption on its surface are more heterogeneous.

Henry's law with intercept model has a high R^2 value, implying its importance. The incorporation of the intercept term considerably improves the linear connection between q_{eq} and C_{eq}.

Langmuir gives improved agreement (R^2 = 0.9432) with experimental adsorption data when followed by the Freundlich and Henry's law with intercept model. It denotes monolayer coverage of the Ni(II) at the BGMA's outer surface. The value of b_L is

S. no.	Model	Parameter	Value	SSE	R^2	RMSE
1	Henry's law with intercept model	K	0.8521	58.67	0.9573	2.708
		m	7.926			
2	Langmuir isotherm model	b_L	0.05284	83.52	0.9432	3.231
		q_{max}	55.75			
		R_L	0.2534			
3	Freundlich isotherm model	a_F (mg/g)	5.029	45.44	0.9669	2.383
		n_F	1.783			
4	Dubinin-Radushkevich model	B_{DR}	0.5688	7027	−4.118	29.64
		K_{DR}	0.1927			
5	Temkin model	A_T	0.7986	97.87	0.9287	3.498
		b_T	240			
6	Hill-de Boer model	K_1	2.024×10^4	1.294×10^4	−1.401	40.22
		K_2	1.535×10^5			
7	Fowler-Guggenheim model	K_{FG}	4.692×10^{-8}	559.2	0.8963	8.36
		W	−1.067			
8	Flory-Huggins isotherm	K_{FH}	0.0008136	5092	0.9012	25.23
		n_{FH}	−1.067			
9	Halsey isotherm model	K_{Ha}	1.91×10^4	3907	−1.845	22.1
		n_{Ha}	2.855			
10	Harkin-Jura isotherm model	A_{HJ}	1.379	23	−5.9015	23
		B_{HJ}	2.559			
11	Jovanovic isotherm model	K_J	0.05745	16.9	0.9359	3.655
		q_{max}	42.04			
12	Elovich isotherm model	K_E	-4.5×10^{-18}	1.66×10^6	−1210	455.8
		q_{max}	0.9139			
13	Kiselev isotherm model	K_{eqK}	85.68	7027	−4.118	29.64
		K_{nK}	85.64			

Table 3.
Parameter values of two parameter adsorption isotherm models for adsorption of Ni(II) onto BGMA.

0.05284 mL/g, which quantifies the affinity of Ni(II) and BGMA. The computed value of R_L is 0.2534, indicating that the adsorption of Ni(II) onto BGMA is favourable.

However, the q_{max} of BGMA calculated by this model (55.75 mg/g) differs from the observed q_{max} value (42.056 mg/g). The variation cannot be significant. The concordance between experimental adsorption data and the Jovanovic isotherm model is quite substantial. It is demonstrated by its R^2 value (0.9359) and q_{max} value (42.04 mg/g).

Because the R^2 values of the four models are high and provide strong mathematical agreement with the experimental results, it cannot be stated that the four isotherms or processes are suitable for adsorption of Ni(II) onto BGMA across the whole concentration range studied. **Figure 6** confirms this by comparing the four models to experimental equilibrium metal uptake and demonstrating the amount of concordance.

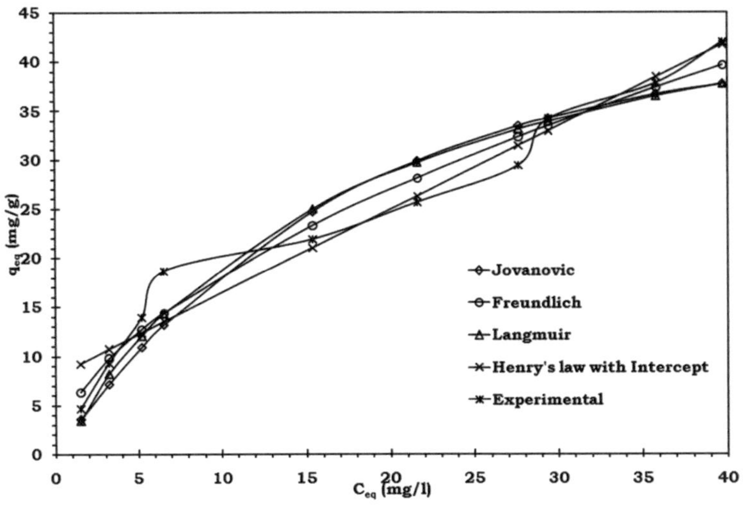

Figure 5.
Comparison of experimental values of equilibrium uptake of Ni(II) with two parameter model values.

The Freundlich isotherm mechanism clearly indicates maximum satisfaction with the equilibrium experimental data based on the R^2, SSE, and RMSE values.

4.2.3 Three parameter models

Table 4 shows the parameter values for three parameter adsorption isotherm models for the adsorption of Ni(II) onto BGMA. Hill, Redlich-Peterson, Langmuir-

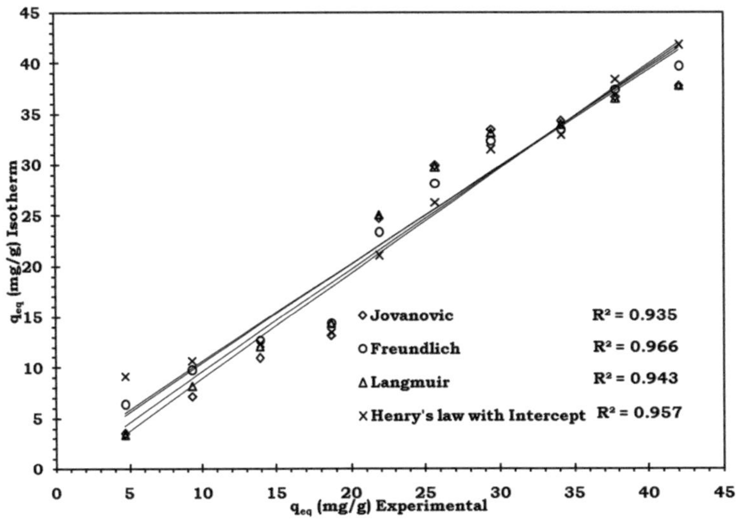

Figure 6.
Concurrence of two parameter model values of equilibrium uptake of Ni(II) with experimental values.

S. no.	Model	Parameter	Value	SSE	R^2	RMSE
1	Hill isotherm model	K_H	423.4	350.2	0.745	7.073
		n_H	0.2637			
		q_{max}	5630			
2	Redlich-Peterson isotherm model	A_{RP}	1.333×10^4	188.3	0.8628	5.187
		B_{RP}	7939			
		β	0.1181			
3	Sips isotherm model	K_S	0.002087	49.55	0.9639	2.661
		q_{max}	2749			
		β	0.5235			
4	Langmuir-Freundlich model	K_{LF}	7.23	1613	0.07953	15.18
		m_{LF}	0.657			
		q_{max}	26.28			
5	Fritz-Schlunder-III isotherm model	K_{FS3}	0.8637	274.7	0.7999	6.265
		M_{FS3}	−8.958			
		q_{max}	1.31			
6	Radke-Prausnits isotherm model-I	K_{RaP1}	115.2	50.34	0.9633	2.682
		M_{RaP1}	0.39			
		q_{max}	0.2358			
	Radke-Prausnits isotherm model-II	K_{RaP2}	5300	158.3	0.8847	4.756
		M_{RaP2}	0.1612			
		q_{max}	1.953			
	Radke-Prausnits isotherm model-III	K_{RaP3}	0.0248	45.5	0.9669	2.549
		M_{RaP3}	0.5569			
		q_{max}	206.7			
7	Toth isotherm model	K_T	0.9765	70.13	0.9489	3.165
		n_T	0.06976			
		q_{max}	1.614×10^5			
8	Khan isotherm model	a_K	0.4462	45.39	0.9669	2.546
		b_K	6.611			
		q_{max}	1.814			
9	Koble-Corrigan isotherm model	A_{KC}	5.277	39.14	0.9715	2.364
		B_{KC}	−0.169			
		n_{KC}	0.3292			
10	Jossens isotherm model	b_J	−0.2596	483	0.6482	8.307
		J	10,710			
		K_J	4789			

S. no.	Model	Parameter	Value	SSE	R^2	RMSE
11	Jovanovic-Freundlich isotherm model	K_{JF}	0.00255	46.92	0.9658	2.589
		n_{JF}	0.5391			
		q_{max}	2133			
12	Brouers-Sotolongo isotherm model	K_{BS}	0.003327	49.44	0.964	2.658
		α_{BS}	0.523			
		q_{max}	1724			
13	Vieth-Sladek isotherm model	K_{VS}	0.6782	33.05	0.9759	2.173
		β_{VS}	0.3985			
		q_{max}	14.37			
14	Unilan isotherm model	K_U	10.59	49.87	0.9637	2.669
		β_U	−3.1			
		q_{max}	−0.08062			
15	Holl-Krich isotherm model	K_{HK}	0.002149	49.6	0.9639	2.662
		n_{HK}	0.5234			
		q_{max}	2671			
16	Langmuir-Jovanovic isotherm model	K_{LJ}	−0.009672	52.96	0.9614	2.751
		n_{LJ}	0.4409			
		q_{max}	820.8			

Table 4.
Parameter values of three parameter isotherm models for adsorption of Ni(II) onto BGMA.

Freundlich, Fritz-Schlunder-III, Radke-Prausnits, and Jossens isotherm models have low R^2 values. However, the models Sips, Radke-Prausnits -I, Radke-Prausnits -III, Toth, Khan, Koble-Corrigan, Jovanovic-Freundlich, Brouers-Sotolongo, Vieth-Sladek, Unilan, Holl-Krich, and Langmuir-Jovanovic isotherm models demonstrate their relevance by strong R^2 values, however the parameter and q_{max} (mg/g) values produced are either negative or excessively high, implying that they are not physically realisable. As a result, all 16 models are dropped from the ongoing debate.

4.2.4 Four parameter models

Table 5 shows the parameter and R^2 values for four parameter isotherm models. To understand the adsorption mechanism of Ni(II) onto BGMA, the Fritz-Schlunder-IV isotherm model, Baudu isotherm model, Weber-van Vliet isotherm model, and Marczewski-Jaroniec isotherm model are investigated.

The Baudu and Fritz-Schlunder-IV models, which have higher R^2 values than the Weber-van Vliet and Marczewski-Jaroniec models, are the most significant of the four models. Unfortunately, the exponents and parameters of all four models are either zero, very low, or excessively high, making them physically impossible to realise. Like a result, just as in the case of the three parameter models, all four parameter models fail to describe the process of adsorption, and the discussion is unnecessary.

Model	Parameter	Value	SSE	R^2	RMSE
Fritz-Schlunder-IV isotherm model	A_{FS5}	5.188×10^{-5}	110.7	0.9194	4.296
	B_{FS5}	−1			
	α_{FS5}	1.092			
	β_{FS5}	-1.58×10^{-5}			
Baudu isotherm model	x	3.491	42.03	0.9694	2.647
	y	0.5412			
	b_o	0.3431			
	q_{max}	5.373			
Weber-van Vliet isotherm model	P1	29.91	1373	-1.861×10^{-10}	1.471
	P2	−6.569			
	P3	8.425			
	P4	0.795			
Marczewski-Jaroniec isotherm model	K_{MJ}	16.93	652	0.5252	10.42
	m_{MJ}	11.11			
	n_{MJ}	0.7926			
	q_{max}	27.26			

Table 5.
Parameter values of four parameter isotherm models for adsorption of Ni(II) onto BGMA.

S. no.	Model	Parameter	Value	SSE	R^2	RMSE
1	Fritz-Schlunder-5 isotherm model	K_{1FS5}	0.3518	45.45	0.9669	3.015
		K_{2FS5}	1.927			
		α_{FS5}	0.5605			
		β_{FS5}	0.0001938			
		q_{max}	41.89			

Table 6.
Parameter values of five parameter isotherm model for adsorption of Ni(II) onto BGMA.

4.2.5 Five parameter model

Table 6 shows the Fritz-Schlunder-V model's parameter values. The R^2 value denotes its relevance. **Figure 7** depicts a comparison of experimental Ni(II) metal uptake with the Fritz-Schlunder-V model under equilibrium conditions. **Figure 8** depicts the agreement of Fritz-Schlunder-V parameter model results for equilibrium Ni(II) uptake with experimental data.

The q_{max} value of 41.89 mg/g for this isotherm model is extremely similar to the experimental q_{max} value of 42.056 mg/g. As a consequence, the Fritz-Schlunder-V isotherm model is firmly established for the adsorption of Ni(II) metal ions from synthetic aqueous solution onto BGMA.

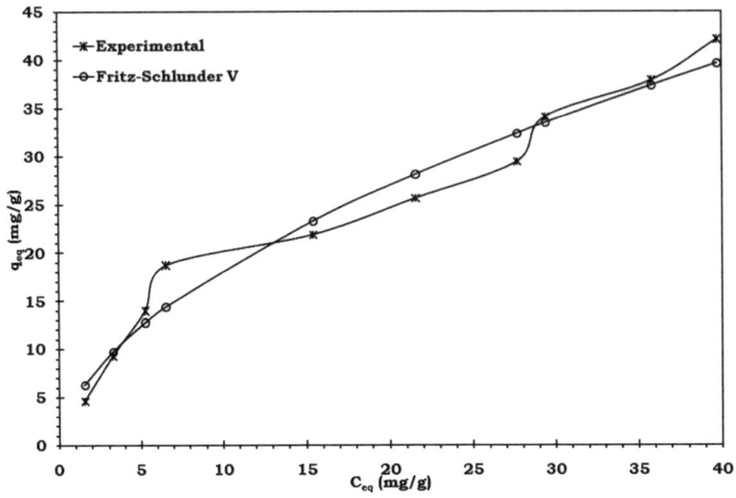

Figure 7.
Comparison of experimental values of equilibrium uptake of Ni(II) with five parameter model values.

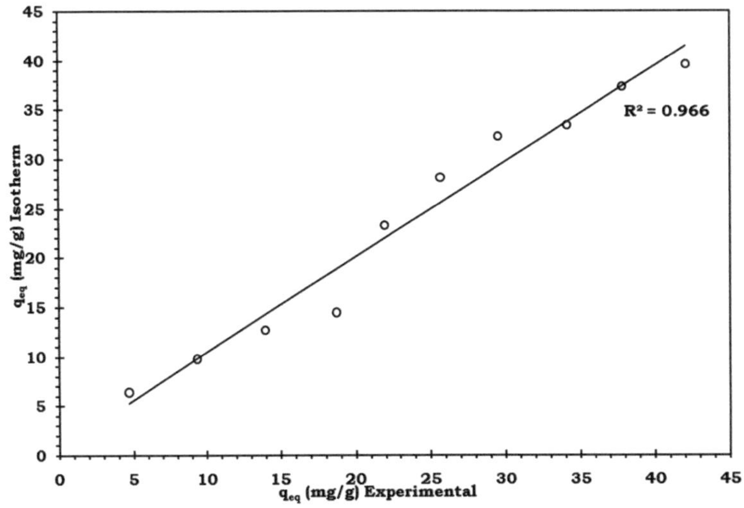

Figure 8.
Concurrence of five parameter model values of equilibrium uptake of Ni(II) with experimental values.

5. Conclusion

The adsorption of Ni(II) metal ions from synthetic aqueous solutions is investigated using BGMA as a low-cost adsorbent. At a pH of 6, 2 g of biomass input, and an agitation speed of 120 rpm, the greatest adsorption capacity of BGMA was determined to be 42.056 mg/g. Because the ionic strength decreases with increasing initial Ni(II) metal ion concentration, the percentage elimination decreases and the equilibrium metal absorption (q_{eq}) increases. The equilibrium experimental data suggests that the isotherm has a L shape, indicating that solvent and Ni(II) are competing for the active sites of BGMA. Furthermore, it suggests that the BGMA has a restricted capacity for

Ni adsorption(II). Furthermore, the efficacy of various isotherms for modelling is investigated using a 1-parameter isotherm, a 13-parameter isotherm, a 16-3-parameter isotherm, a 4-4-parameter isotherm, and a 1-5-parameter isotherm. The experiences are graphically depicted. The Fritz-Schlunder-V isotherm model is obviously relevant in characterising the mechanism of Ni(II) adsorption under the conditions utilised in this work, which was followed by Freundlich. The q_{max} of 41.89 mg/g for this model reveals its significance even more clearly.

Nomenclature

A	Fritz-Schlunder parameter
a_F	Freundlich adsorption capacity (L/mg)
A_{HJ}	Harkin-Jura isotherm constant
a_K	Kahn isotherm model exponent
A_{KC}	Koble-Carrigan's isotherm constant
A_{RP}	Redlich-Peterson isotherm constant (L/g)
A_T	Temkin equilibrium binding constant corresponding to the maximum binding energy
B	Fritz-Schlunder parameter
BGMA	blue green marine algae
b	Langmuir constant related to adsorption capacity (L/mg)
b_0	Langmuir isotherm equilibrium constant
B_{DR}	Dubinin-Radushkevich model constant
B_{HJ}	Harkin-Jura isotherm constant
b_J	Jossens isotherm model parameter
b_K	Khan isotherm model constant
B_{KC}	Koble-Carrigan's isotherm constant
b_L	Langmuir constant related to adsorption capacity (mg/g)
B_{RP}	Redlich-Peterson isotherm constant (L/mg)
b_T	Temkin constant which is related to the heat of sorption (J/mol)
C	Henry's law model intercept
C_{eq}	concentration of adsorbate in bulk solution at equilibrium (mg/L)
C_{in}	initial adsorbate concentration (mg/L)
J	Jossens isotherm model parameter
K	Henry's constant
K_1	Hill-de Boer constant (L/mg)
K_{1FS5}	Fritz-Schlunder-V parameter
K_2	energetic constant of the interaction between adsorbed molecules (kJ/mol)
K_{2FS5}	Fritz-Schlunder-V parameter
K_{BS}	Brouers-Sotolongo model isotherm parameter
K_{DR}	Dubinin-Radushkevich model uptake capacity
K_E	Elovich constant (L/mg)
K_{FG}	Fowler-Guggenheim equilibrium constant (L/mg)
K_{FH}	Flory-Huggins equilibrium constant (L/mol)
K_{FS3}	Fritz-Schlunder III equilibrium constant (L/mg)
K_H	Hill isotherm constant
K_{Ha}	Halsey isotherm constant
K_{He}	Henry's constant
K_{HK}	Holl-Krich isotherm model parameter

K_J	Jossens isotherm model parameter
K_J	Jovanovic constant
K_{JF}	Jovanovic-Freundlich isotherm equilibrium constant
K_K	Kiselev equilibrium constant (L/mg)
K_{LF}	Langmuir-Freundlich equilibrium constant for heterogeneous solid
K_{LJ}	Langmuir-Jovanovic model parameter
K_{MJ}	Marczewski-Jaroniec isotherm model parameter that characterise the heterogeneity of the adsorbent surface.
K_{nK}	equilibrium constant of the formation of complex between adsorbed molecules
K_{RaP}	Radke-Prausnits equilibrium constant
K_S	Sips isotherm model constant (L/mg)
K_T	Toth isotherm constant (mg/g)
K_U	Unilan isotherm model parameter
K_{VS}	Vieth-Sladek isotherm model parameter related to Henry's law
m_{FS3}	Fritz-Schlunder III model exponent
m_{LF}	Langmuir-Freundlich heterogeneity parameter
m_{RaP}	Radke-Prausnits model exponent
n_F	Freundlich adsorption intensity
n_{FH}	number of adsorbates occupying adsorption sites
n_H	exponent of Hill adsorption model
n_{Ha}	Halsey isotherm exponent
n_{HK}	Holl-Krich isotherm model exponent
n_{JF}	Jovanovic-Freundlich isotherm exponent
n_{KC}	Koble-Carrigan's isotherm constant
n_{LJ}	Langmuir-Jovanovic model exponent
n_{MJ}	Marczewski-Jaroniec isotherm model parameter that characterise the heterogeneity of the adsorbent surface
n_T	Toth isotherm exponent
P_1	Weber and van Vliet isotherm model parameter
P_2	Weber and van Vliet isotherm model parameter
P_3	Weber and van Vliet isotherm model parameter
P_4	Weber and van Vliet isotherm model parameter
q_{eq}	amount of adsorbate in adsorbent at equilibrium (mg/g)
q_{max}	maximum quantity of solute adsorbed by the adsorbent (mg/g)
R	gas constant (8.314 J/mol K)
R_L	Langmuir separation factor
T	absolute temperature (K)
W	interaction energy between adsorbed molecules (kJ/mol)
x	Baudu isotherm model parameter
y	Baudu isotherm model parameter

Greek letters

θ	fractional surface coverage
β_{RP}	Redlich-Peterson isotherm exponent
β_S	Sips isotherm exponent
α_{BS}	Brouers-Sotolongo model isotherm parameter is related to adsorption energy
β_{VS}	Vieth-Sladek isotherm model parameter related to Langmuir

β_U Unilan isotherm model exponent
α_{FS5} Fritz-Schlunder-V parameter
β_{2FS5} Fritz-Schlunder-V parameter

Author details

Ramsenthil Ramadoss[1*], Durai Gunasekaran[1] and Dhanasekaran Subramanian[2]

1 Bioprocess Laboratory, Faculty of Engineering and Technology, Department of Chemical Engineering, Annamalai University, Annamalai Nagar, Tamil Nadu, India

2 Mass Transfer Laboratory, Faculty of Engineering and Technology, Department of Chemical Engineering, Annamalai University, Annamalai Nagar, Tamil Nadu, India

*Address all correspondence to: drramsenthil@gmail.com

References

[1] Sočo E, Kalembkiewicz J. Adsorption of nickel(II) and copper(II) ions from aqueous solution by coal fly ash. Journal of Environmental Chemical Engineering. 2013;**1**(3):581-588

[2] Shyam R, Puri JK, Kaur H, Amutha R, Kapila A. Single and binary adsorption of heavy metals on fly ash samples from aqueous solution. Journal of Molecular Liquids. 2013;**178**:31-36

[3] Wu X-W, Ma H-W, Zhang L-T, Wang F-J. Adsorption properties and mechanism of mesoporous adsorbents prepared with fly ash for removal of Cu (II) in aqueous solution. Applied Surface Science. 2012;**261**:902-907

[4] Lu X, Yi-Xu H, Bo-Han Z. Kinetics and equilibrium adsorption of copper(II) and nickel(II) ions from aqueous solution using sawdust xanthate modified with ethane di amine. Transactions of Nonferrous Metals Society of China. 2014;**24**(3):868-875

[5] Cui L, Wu G, Jeong T-S. Adsorption performance of nickel and cadmium ions onto brewer's yeast. The Canadian Journal of Chemical Engineering. 2010; **88**(1):109-115

[6] Malkoc E, Nuhoglu Y. Nickel(II) adsorption mechanism from aqueous solution by a new adsorbent—Waste acorn of *Quercus ithaburensis*. AIChE Environmental Progress and Sustainable Energy. 2010;**29**:297-306

[7] Bulgariu L, Bulgariu D, Macoveanu M. Kinetics and equilibrium study of Nickel(II) removal using peat moss. Environmental Engineering and Management Journal. 2010;**9**(5):667-674

[8] Shen Z, Zhang Y, McMillan O, Jin F, Al-Tabbaa A. Characteristics and mechanisms of nickel adsorption on biochars produced from wheat straw pellets and rice husk. Environmental Science and Pollution Research. 2017; **24**(14):12809-12819

[9] Naskar A, Guha AK, Mukherjee M, Ray L. Adsorption of nickel onto *Bacillus cereus* M^1_{16}: A mechanistic approach. Separation Science and Technology. 2016;**51**(3):427-438

[10] Bulgariu D, Bulgariu L. Sorption of Pb(II) onto a mixture of algae waste biomass and anion exchanger resin in a packed-bed column. Bioresource Technology. 2013;**129**:374-380

[11] Davis TA, Volesky B, Mucci A. A review of the biochemistry of heavy metal biosorption by brown algae. Water Research. 2003;**37**:4311-4330

[12] Volesky B. Advances in biosorption of metals: Selection of biomass types. FEMS Microbiology Reviews. 1994;**14**: 291-302

[13] Demiral H, Güngör C. Adsorption of copper(II) from aqueous solutions on activated carbon prepared from grape bagasse. Journal of Cleaner Production. 2016;**124**:103-113

[14] Ozcan AS, Erdem B, Ozcan A. Adsorption of Acid Blue 193 from aqueous solutions onto BTMN-bentonite. Colloids and Surfaces A: Physicochemical and Engineering Aspects. 2005;**266**:73-81

[15] Temkin MI, Pyzhev V. Kinetics of ammonia synthesis on promoted iron catalyst. Acta Physiochimica USSR. 1940;**12**:327-356

[16] Ruthven DM. Principle of Adsorption and Adsorption Processes.

New Jersey, NJ, USA: John Willey and Sons; 1984

[17] Langmuir I. The constitution and fundamental properties of solids and liquids. Journal of the American Chemical Society. 1916;**38**:2221-2295

[18] Langmuir I. The adsorption of gases on plane surfaces of glass, mica, and platinum. Journal of the American Chemical Society. 1918;**40**:1361-1403

[19] Baroni P, Veira RS, Meneghetti E, Da Silva MGC, Beppu MM. Evaluation of batch adsorption of chromium ions on natural and crosslinked chitosan membranes. Journal of Hazardous Materials. 2008;**152**:1155-1163

[20] Freundlich HMF. Über die adsorption in lösungen. Zeitschrift für Physikalische Chemie. 1906;**57**:385-470

[21] Treybal RE. Mass–Transfer Operations. 3rd ed. McGraw–Hill: Tokyo; 1981

[22] Dubinin MM, Radushkevich LV. The equation of the characteristic curve of activated charcoal. Doklady Akademii Nauk SSSR. 1947;**55**:327-329

[23] Radushkevich LV. Potential theory of sorption and structure of carbons. Zhurnal Fizicheskoi Khimii. 1949;**23**: 1410-1420

[24] Dubinin MM. The potential theory of adsorption of gases and vapors for adsorbents with energetically non-nniform surface. Chemical Reviews. 1960;**60**:235-266

[25] Dubinin MM. Modern state of the theory of volume filling of micropore adsorbents during adsorption of gases and steams on carbon adsorbents. Zhurnal Fizicheskoi Khimii. 1965;**39**: 1305-1317

[26] Temkin MI. Adsorption equilibrium and process kinetics on homogeneous surfaces and with interaction between adsorbed molecules. Zhurnal Fizicheskoi Khimii. 1941;**15**(3):296-332

[27] Hill TL. Statistical mechanics of multi molecular adsorption II. Localized and mobile adsorption and absorption. The Journal of Chemical Physics. 1946; **14**(7):441-453

[28] Hill TL. Theory of physical adsorption. Advances in Catalysis. 1952;**4**:211-258

[29] de Boer JH. The Dynamical Character of Adsorption. Oxford: Oxford University Press; 1953

[30] Kumara PS, Ramalingam S, Kiruphac SD, Murugesanc A, Vidhyarevicsivanesam S. Adsorption behaviour of nickel(II) onto cashew nut shell: Equilibrium, thermodynamics, kinetics, mechanism and process design. Chemical Engineering Journal. 2010; **1169**:122-131

[31] Sampranpiboon P, Charnkeitkong P, Feng X. Equilibrium isotherm models for adsorption of zinc (II) ion from aqueous solution on pulp waste. WSEAS Transactions on Environment and Development. 2014;**10**:35-47

[32] Hamdaoui O, Naffrechoux E. Modeling of adsorption isotherms of phenol and chlorophenols onto granular activated carbon. Part I. Two-parameter models and equations allowing determination of thermodynamic parameters. Journal of Hazardous Materials. 2007;**147**(1):381-394

[33] Hamdaouia O, Naffrechoux E. Modeling of adsorption isotherms of phenol and chlorophenols onto granular activated carbon Part II. Models withore than two parameters. Journal of Hazardous Materials. 2007;**147**:401-411

[34] Amin MT, Alazba AA, Shafiq M. Adsorptive removal of reactive black 5 from wastewater using bentonite clay: Isotherms, kinetics and thermodynamics. Sustainability. 2015;7(11):15302-15318

[35] Ebelegi NA, Angaye SS, Ayawei N, Wankasi D. Removal of congo red from aqueous solutions using fly ash modified with hydrochloric acid. British Journal of Applied Science and Technology. 2017; **20**(4):1-7

[36] Fowler RH, Guggenheim EA. Statistical Thermodynamics. London, England: Cambridge University Press; 1939

[37] Foo KY, Hameed BH. Review: Insights into the modeling of adsorption isotherm systems. Chemical Engineering Journal. 2010;**156**:2-10

[38] Jovanovic DS. Physical sorption of gases. I. Isotherms for monolayer and multilayer sorption. Colloid & Polymer Science. 1969;**235**:1203-1214

[39] Elovich SY, Larinov OG. Theory of adsorption from solutions of non electrolytes on solid (I) equation adsorption from solutions and the analysis of its simplest form, (II) verification of the equation of adsorption isotherm from solutions. Izvestiya Akademii Nauk. SSSR, Otdelenie Khimicheskikh Nauk. 1962;**2**:209-216

[40] Kiselev AV. Vapor adsorption in the formation of adsorbate molecule complexes on the surface. Kolloid Zhur. 1958;**20**:338-348

[41] Hill AV. The possible effects of the aggregation of the molecules of haemoglobin on its dissociation curves. The Journal of Physiology. 1910;**40**:4-7

[42] Redlich O, Peterson DL. A useful adsorption isotherm. The Journal of Physical Chemistry. 1959;**63**:1024-1026

[43] Sips R. On the structure of a catalyst surface. The Journal of Chemical Physics. 1948;**16**:490-495

[44] Valenzuela DP, Myers AL. Adsorption Equilibria Data Handbook. Englewood Cliffs, NJ: Prentice–Hall; 1989

[45] Fritz W, Schlunder EU. Simultaneous adsorption equilibria of organic solutes in dilute aqueous solution on activated carbon. Chemical Engineering Science. 1974;**29**:1279-1282

[46] Radke CJ, Prausnitz JM. Sorption of organic solutes from dilute aqueous solutions on activated carbon. Industrial & Engineering Chemistry Fundamentals. 1972;**11**:445-451

[47] Toth J. State equations of the solid gas interface layer. Acta chimica Hungarica. 1971;**69**:311-317

[48] Khan AR, Ataullah R, Al-Haddad A. Equilibrium adsorption studies of some aromatic pollutants from dilute aqueous solutions on activated carbon at different temperatures. Journal of Colloid and Interface Science. 1997;**194**:154-165

[49] Koble RA, Corrigan TE. Adsorption isotherms for pure hydrocarbons. Industrial and Engineering Chemistry. 1952;**44**:383-387

[50] Jossens L, Prausnitz JM, Fritz W, Schlünder EU, Myers AL. Thermodynamics of multi–solute adsorption from dilute aqueous solutions. Chemical Engineering Science. 1978;**33**:1097-1106

[51] Quiñones I, Guiochon G. Derivation and application of a Jovanovic–Freundlich isotherm model for single–component adsorption on heterogeneous surfaces. Journal of Colloid and Interface Science. 1996;**183**:57-67

[52] Brouers F, Sotolongo O, Marquez F, Pirard JP. Microporous and heterogeneous surface adsorption isotherms arising from Levy distributions. Physica A: Statistical Mechanics and its Applications. 2005; **349**:271-282

[53] Ncibi MC, Altenor S, Seffen M, Brouers F, Gaspa S. Modelling single compound adsorption onto porous and non-porous sorbents using a deformed Weibull exponential isotherm. Chemical Engineering Journal. 2008;**145**:196-202

[54] Vieth WR, Sladek KJ. A model for diffusion in a glassy polymer. Journal of Colloid Science. 1965;**20**:1014-1033

[55] Chern JM, Wu CY. Desorption of dye from activated carbon beds: Effects of temperature, pH, and alcohol. Water Research. 2001;**35**:4159-4165

[56] Hadi M, McKay G, Samarghandi MR, Maleki A, Aminabad MS. Prediction of optimum adsorption isotherm: Comparison of chi–square and log-likelihood statistics. Desalination and Water Treatment. 2012;**49**:81-94

[57] Parker GR. Optimum isotherm equation and thermodynamic interpretation for aqueous 1,1,2–trichloroethene adsorption isotherms on three dsorbents. Adsorption. 1995;**1**: 113-132

[58] Shahbeig H, Bagheri N, Ghorbanian SA, Hallajisani A, Poorkarimi S. A new adsorption isotherm model of aqueous solutions on granular activated carbon. WJMS. 2013; **9**:243-254

[59] M. Baudu. Etude des interactions solute–fibres de charbon actif. Application et regeneration [Ph.D. diss.]. Universite de Rennes I; 1990

[60] McKay G, Mesdaghinia A, Nasseri S, Hadi M, Aminabad MS. Optimum isotherms of dyes sorption by activated carbon: Fractional theoretical capacity and error analysis. Chemical Engineering Journal. 2014;**251**:236-247

[61] van Vliet BM, Weber WJ Jr, Hozumi H. Modeling and prediction of specific compound adsorption by activated carbon and synthetic adsorbents. Water Research. 1980; **14**(12):1719-1728

[62] Parker GR Jr. Optimum isotherm equation and thermodynamic interpretation for aqueous 1,1,2-trichloroethene adsorption isotherms on three adsorbents. Adsorption. 1995;**1**(2): 113-132

[63] Sivarajasekar N, Baskar R. Adsorption of basic red 9 onto activated carbon derived from immature cotton seeds: Isotherm studies and error analysis. Desalination and Water Treatment. 2014;**52**:1-23

[64] Schay G. On the definition of interfacial excesses in a system consisting of an insoluble solid adsorbent and a binary liquid mixture. Colloid and Polymer Science. 1982;**26**:888-891

[65] Giles CH, MacEwan TH, Nakhwa SN, Smith D. Studies in adsorption. Part XI. A system of classification of solution adsorption isotherms, and its use in diagnosis of adsorption mechanisms and in measurements of specific surface areas of solids. Journal of the Chemical Society. 1960;**10**:3973-3993

[66] Limousin G, Gaudet J-P, Charlet L, Szenknect S, Barthes V, Krimissa M. Review: Sorption isotherms: A review on physical bases, modeling and measurement. Applied Geochemistry. 2007;**22**:249-275

Chapter 13

Microalgae: An Exquisite Oil Producer

Ishita Bhattacharya

Abstract

With the influx in population and shortage of conventional energy-sources, an exponential-rise of the microalgal oil-production has been observed in the past two decades. The algal bio-oil is used in various industries viz. food, pharmaceutical, cosmetic and biodiesel plants. The present study is focused towards the production of oil from oleaginous microalgae in photo-bioreactors and open water systems. Moreover, microalgae can thrive in non-cultivable waters like seawater, salt water and even wastewater which make the algal technology more attractive in terms of soil and water preservation. Using sunlight and nutrients like salts of magnesium, potassium, sodium etc. the autotrophic microalgae can grow in large quantities in indoor photo-bioreactors and in open ponds. Microalgae are able to produce approximately 10,000 gallons of oil per acre as compared to the higher plants that produces only 50 gallons per acre (soy), 110 to 145 gallons per acre (rapeseed), 175 gallons per acre (Jatropha), 650 gallons per acre (palm). The biomass productivity is 10 times higher than that of the phytoplanktons and 20–30% higher than that of the terrestrial biomass. In terms of the fatty acid composition, the microalgal oil can well match with the plant-derived oil, mainly C16 and C18 fatty acids. Some microalgae are also rich in valuable polyunsaturated-fatty-acids, which have multiple health benefits.

Keywords: microalgae, bio-oil, photo-bioreactor

1. Introduction

In this twentieth century due to the shortage in the conventional energy sources as well as exponentially rising trend of environmentally harmful products, microalgae have been chosen as an alternative source for a wide variety of metabolic products, viz. dietary supplements, pharmacological compounds, lipids, enzymes, biomass, polymers, toxins, pigments, wastewater treatment, and "renewable energy".

The microalgal cultivation chiefly follows autotrophic growth. Almost all microalgae are photosynthetic in nature having chlorophyll a, chlorophyll b and bacterio-chlorophyll in some blue-green algae (cyanobacteria) [1], and thus are significant solar energy convertors, so, they are cultivated in illuminated environments either in open or closed cultivation systems [2].

For the last two decade microalgae have been recognized as a prominent alternative source for oil production. Several oleo genic species of microalgae can be manipulated to overproduce specific lipids and fatty acids through alteration of their physical and chemical properties of the culture medium. Microalgae can accumulate substantial amounts of lipids – up to 50% of dry cell weight in certain species [3]. Many microalgal species can thrive well in water with high salt concentration, viz. brackish water or seawater, thereby avoiding the demand for fresh water which has been designated as a limited resource in many parts of the world [4].

In microalgae, lipids play an important role in the synthesis of plasma membranes lipid protein lipid structure, in maintaining buoyancy and as an energy reserve during adverse growth conditions [5]. Accumulation of lipids in the microalgae can be attributed to the consumption of sugars at a rate higher than that of the rate of cell doubling, which promote conversion of excess sugar into lipids which favor the algae in its stationary phase of growth to fight the nutrient depletion [6].

Different saturated Fatty acids (SFA), monounsaturated Fatty acids (MUFA), and polyunsaturated Fatty Acids (PUFA) have been reported in microalgae. Hexadecanoic acid (C16:0) and oleic acid (C18:1) are common fatty acids in the microalgae. Omega-3 (ω-3) FA) can be traced in some [7]. Significantly, the microalgae can produce essential Fatty Acids, viz. alphalinolenic acid, which can be converted to eicosapentaenoic acid (EPA) and docosahexaenoic acid (DHA) in mammals by metabolic pathways [8]. The concentration of EPA is insignificant in the terrestrial plants. Seaweeds, such as Palmaria palmata [9] can produce EPA, but in lesser concentration in comparison to the microalgae [10].

Microalgae have been designated as a producer of different types of renewable biofuels viz. methane produced by anaerobic digestion of the biomass [11], biodiesel derived from oil [12–14]; and biohydrogen can be produced photo-biologically [15, 16]. The concept of using microalgae as a source of fuel is not new [17], but for the past few decades it has being taken seriously because of the escalating price of petroleum and, most importantly, the emerging concern about global warming that is associated with the burning of fossil fuels [18].

2. Attributes of microalgae

2.1 Cultivation of microalgae

Cultivation of microalgae for biodiesel production aims at maximizing the lipid productivity along with the growth rate of the microalgae. In the batch cultivation system, microalgae are exponentially grown in the log phase to increase their biomass and then they subjected to a starvation phase by omitting or limiting the nutrient supply towards the end of the stationary phase of growth. As, algal oil is a secondary metabolite, so nutrient deprivation can lead to a higher yield.

The two most common methods of microalgae cultivation are open cultivation systems and controlled closed cultivation systems.

Open-air cultivation systems comprise natural or artificial ponds, raceway ponds, and the inclined surface systems driven by paddle wheels, usually operating at water depths of 15–30 cm [19]. They represent the classical processes used for production of algal biomass. Although different types of open reactors have been studied since last few decades by different research groups, but the most commonly used systems are shallow big ponds, tanks, circular ponds, and raceway ponds. Some of the major advantages of an open cultivation system are minimal capital, operating costs, and lower energy requirement for culture mixing. The disadvantages are open systems require large areas to scale up, susceptibility to contamination (by birds, small insects and rotifers), adverse weather conditions, difficulty to regulate growth parameters viz. evaporation rate, culture temperature, etc. A scientific investigation reported the damaging effects of the occurrence of rotifers to the cultures of *Tetraselmis, Chlorella, Nannochloropsis*, *Scenedesmus*, and the damage caused by amoeba to diatoms.

On the other hand, closed and controlled cultivation systems employ photobioreactors to attain axenic single-species culture of microalgae. Photobioreactors are successfully used for producing large quantities of microalgal biomass [20]. The different types of photobioreactor (PBR) include horizontal or serpentine tube, flat-plate, bubble column, airlift column and stirred tank. PBRs can be designed and calibrated according to the research need and the experimental organism. This closed system utilizes relatively little space, while increasing the light availability and minimum contamination issues. However, the limitations of PBRs include bio-fouling, overheating, benthic algal growth, cleaning issues, growth limitation due to high build-up of dissolved oxygen and costlier operation [21, 22].

2.2 Oil productivity

Oleaginous microalgae (**Table 1**) are a promising source for the production of renewable biofuels because of their efficient photosynthetic capabilities. Moreover, microalgal growth requires less area in comparison to the terrestrial plants, and they are capable to channel the majority of the acquired energy into cell division, which increases the biomass yield [23]. Microalgae can be subdivided into four different groups depending on the carbon source (inorganic and organic), namely, autotrophic, mixotrophic, heterotrophic, and photoheterotrophic [24].

The synthesis of triacylglycerol in microalgae takes place mostly in the chloroplast and endoplasmic reticulum through multiple enzymatic reactions [25]. Fatty acid synthesis in the chloroplast, assembly of glycerolipids in endoplasmic reticulum, and accumulation of TAGs into the oil bodies are the three major steps involved in the accumulation of lipids in the microalgae [26]. It has been proven that facilitate the synthesis of high amounts of lipids is influenced by different stress conditions

Microalgae	Oil yield (%)
Scenedesmus sp	38
Chlorella protothecoides	49
Botryococcus braunii	28
Tetraselmis elliptica	14
C. vulgaris	89
Auxenochlorella protothecoides	66
Chlamydomonas reinhardtii	59

Table 1.
Some prominent oleaginous microalgae.

such as physical, chemical, or environmental, individually or in combination [27]. Under the aforementioned stress conditions, microalgae can switch their metabolism towards the synthesis of neutral lipids in the form of TAGs, which serves as a form of carbon and energy storage [28–30]. Microalgae employ the de novo pathway to synthesize lipids. It starts in the chloroplast by CO2 fixation into sugars, which are further metabolized to acetyl-CoA, which acts as a precursor of fatty acid synthesis [31].

Marine microalgae have a higher content of PUFA in comparison to the freshwater species because they need to produce more unsaturated fatty acids to survive in the salty marine environment [32]. Thus, cultivation of marine microalgae can render higher economic interest to the cultivars. According to reported literature, the marine oleaginous diatom Fistulifera solaris when cultivated in photoautotrophic conditions can produce 135.7 mg/(L·day) EPA. On the otherhand, the heterotrophic growth of the marine diatom Nitzschia laevis, when supplemented with glucose, resulted in EPA production of 174.6 g/(L·day) [33].

2.3 Extraction of oil

During lipid extraction, the microalgal biomass is exposed to an organic eluting solvent which extracts the lipids out of the cell cytoplasm. A lipid extraction technology for microalgal oil production needs to be highly specific towards the lipids in order to avoid the co-extraction of non-lipid contaminants, viz. protein and carbohydrates. The lipid extraction technology should be more selective towards acylglycerols than other lipid fractions as they are not readily convertible to biodiesel, such as polar lipids and non-acylglycerol neutral lipids (free fatty acids, hydrocarbons, sterols, ketones, carotenes, and chlorophylls) [34]. Moreover, the technology should be efficient (both time and energy saving), non-reactive with the lipids, relatively cheap (capital cost and operating cost), and safe (environmentally and mechanically) [35]. Dewatering of the microalgal biomass beyond a paste consistency (200 g dried microalgal biomass/L culture) is energy consuming, so, it will be economically friendly if the selected lipid extraction technology is effective for the wet feedstock, i.e. concentrate or disrupted concentrate with concentrations between 10 and 200 g dried microalgal biomass/L culture [36].

2.3.1 Solvent extraction

The principles underlying solvent extraction of microalgal lipids are based on the concept of chemistry 'like dissolving like'. The long hydrophobic fatty acid chains interact with neutral lipids through weak van der Waals forces, thus forms globules in the cytoplasm [37]. The mechanism for organic solvent extraction is depicted in **Figure 1**. When a microalgal cell is exposed to a non-polar organic solvent, such as hexane or chloroform, the organic solvent penetrates through the cell membrane into the cytoplasm and interacts with the neutral lipids trough van der Waals forces to form an organic solvent-lipids complex. This organic solvent–lipids complex, driven by a concentration gradient, diffuses across the cell membrane. The neutral lipids are thus extracted out of the cells and remain dissolved in the non-polar organic solvent. However, some neutral lipids remain as a complex with polar lipids in the cytoplasm. The complex is strongly linked via hydrogen bonds to the proteins in the cell membrane. The van der Waals interactions formed between non-polar organic solvent and

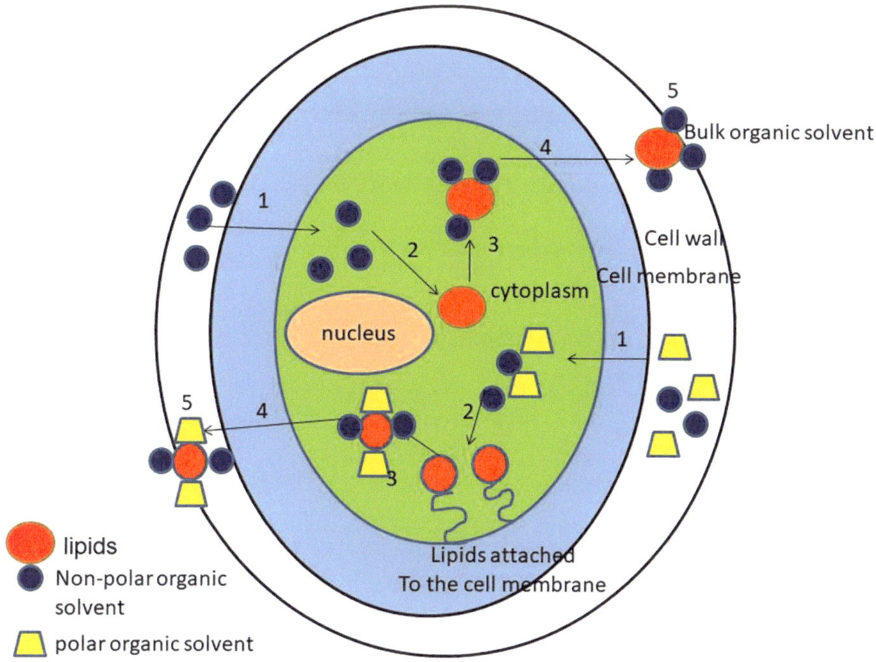

Figure 1.
Schematic representation depicting the solvent extraction of oil from oleaginous microalgae. 1. Penetration of the solvent inside the cell cytoplasm. 2.Solvent interaction with the lipids. 3. Formation of lipid-solvent complex. 4. Diffusion of the complex through the cell membrane. 5. Diffusion of the complex in the surrounding bulk organic solvent.

neutral lipids are inadequate to disrupt these membrane-based lipid–protein associations. On the other hand, polar organic solvent (viz. methanol or isopropanol) is able to disrupt the lipid–protein associations by forming hydrogen bonds with the polar lipids in the complex [38].

Figure 2 depicts the extraction steps generally, undertaken for laboratory-scale production of microalgal oil and finally trans-esterified to biodiesel using an organic solvent mixture.

2.4 Compositional analysis of algal oil

Algal oil is very high in unsaturated fatty acids [39]. The mentioned fatty acids are present in abundance in algal oil [40].

- Arachidonic acid (AA)

It is an unsaturated, essential long chain (C20) fatty acid and acts as a precursor in the biosynthesis of prostaglandins, thromboxanes, and leukotrienes.

- Eicospentaenoic acid (EPA)

EPA is an omega-3 fatty acid. Omega-3 fatty acid helps lower risk of heart disease, lower triglycerides in the blood, blood pressure, and inflammation.

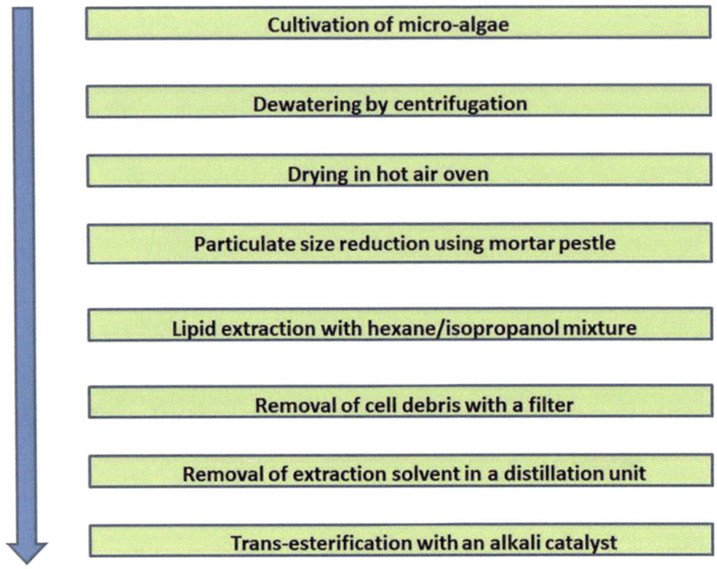

Figure 2.
Microalgal oil and further trans-esterification to bio-diesel extraction procedure.

- Docasahexaenoic acid (DHA)

DHA is a long-chain, highly unsaturated omega-3 fatty acid. It affects cell and tissue physiology and function by altering the membrane structure. It also has prominent roles in membrane protein function, cellular signaling and production of lipid mediator.

- Gamma-linolenic acid (GLA)

GLA is an essential fatty acid belonging to omega-6 fatty acid. Omega-6 fatty acids play a crucial role in brain function, growth and development, stimulate skin and hair growth, maintain bone health, regulate metabolism, and maintain the reproductive system.

- Linoleic acid (LA)

It is an essential fatty acid belonging to the omega-3 group. It has been reported to inhibit the synthesis of prostaglandin resulting in reduced inflammation and prevention of certain chronic diseases.

2.4.1 Elemental (CHNO) analysis of the algal oil

CHNO analysis of the algal oil depicts the elemental constitution in order to assess its economic and environmental significance. The Carbon (C), Hydrogen (H), Nitrogen (N) and Oxygen (O) percentage (wt/wt) are reported as 64.2, 16.11, 0.87, and 21.8% respectively in the oil obtained from *C.vulgaris* [41]. The higher heating value (HHV) of the algal oil has been reported as high as 40.43 MJ/kg [42]. The high fraction of carbon and HHV ensure large energy content of the oil that

Microalgae	Growth phase (batch cultivation)	FAME lipids	Ash	Carbohydrates	Proteins
N. granulata	Early	12.30	14.2	8.92	32.7
	Mid	25.52	13.6	11.12	23.1
	Late	57.33	5.1	10.89	9.4
S.acutus	Early	9.15	4.5	16.88	46.3
	Mid	17.03	1.8	49.71	17.4
	Late	38.55	2.2	39.42	7.85
C.vulgaris	Early	12.07	6.7	11.12	43.27
	Mid	15.02	4.4	35.69	24.00
	Late	23.14	5.3	38.00	15.2

Table 2.
Compositional analysis of the microalgae.

can be further trans-esterified to bio-diesel or any bio-oil. A negligible fraction of nitrogen (0.89%) and no sulfur content advocate that there is probably no or minimal chance of NO_x and SO_x emission during the utilization of the algal oil for energy generation.

2.4.2 Analysis of lipids as total fatty acid methyl ester (FAME), protein, carbohydrate, and ash content of the microalgae

According to **Table 2** the composition of different microalgal species viz. *S. acutus*, *C. vulgaris*, and *N. granulata* varies significantly depending on the growth-stage and the growth environment. As per a scientific investigation [43], longer cultivation time and nutrient depletion, the lipid content increases and protein content decreases. The carbohydrate content in *N. granulata* and *S. acutus* increased from early to mid growth stage, while decreased from mid to late stage. On the other hand, an increase in carbohydrate content has been noted in *C. vulgaris* in the starvation phase.

3. Conclusion

The present study advocated the different attributes of oil production through micro-algal route. Conventional algal cultivation systems include raceway ponds with lower investment cost but also renders lower yields while high value products (micro-algae and oil) are produced with closed photobioreactors with high cost investments. However, fresh water demand can be reduced employing the photobioreactors, using waste water, brackish water or seawater as the culture broth.

The presence of high amount of polyunsaturated fatty acids gives micro-algal biofuel good flow properties under low temperatures, thus, reducing the risk of cold filter plugging and making it suitable as aviation fuel. Most importantly, the oil composition can be manipulated by selecting the desired strain and the cultivation parameters. This significant feature of the algal oil probably can be a solution to overcome the falling economy. Micro-algal biofuel production can possibly be solutions to reduce conventional energy demand. But increase in yields by enhanced reactor designs, optimized process control, adapted separation technologies as well as integral concepts considering biology, process design for cultivation, and downstream processing to meet the different requirements for the specific target markets should be addressed in biorefinery concepts to be integrated with coal-based power plants to ensure a cleaner and greener tomorrow.

Conflict of interest

The authors declare no conflict of interest.

Thanks

A heartiest thank to my daughter, Vivita Mukherjee for her awesome co-operation in my work.

Author details

Ishita Bhattacharya
Department of Life Science, Binod Bihari Mahto Koyalanchal University, Dhanbad, India

*Address all correspondence to: ibhattacharya222@gmail.com

References

[1] Pradhan L, Bhattacharjee V, Mitra R, Bhattacharya I, Chowdhury R. Biosequestration of CO2 using power plant algae (Rhizoclonium hieroglyphicum JUCHE2) in a flat plate Photobio-bubble-reactor-experimental and modelling. Chemical Engineering Journal. 2015;**275**:381-390. DOI: 10.1016/j.cej.2015.04.037

[2] Kiran Kumar P, Vijaya Krishna S, Verma K, Pooja K, Bhagawan D, Srilatha K, et al. Bio oil production from microalgae via hydrothermal liquefaction technology under subcritical water conditions. Journal of Microbiological Methods. 2018;**153**:108-117. DOI: 10.1016/j.mimet.2018.09.014

[3] Singh SK, Bansal A, Jha MK, Dey A. Comparative studies on uptake of wastewater nutrients by immobilized cells of *Chlorella minutissima* and dairy waste isolated algae. Indian Chemical Engineer. 2011;**53**(4):211-219. DOI: 10.1080/00194506.2011.706441

[4] Arif M, Bai Y, Usman M, Jalalah M, Harraz FA, Al-Assiri MS. Highest accumulated microalgal lipids (polar and non-polar) for biodiesel production with advanced wastewater treatment: Role of lipidomics. Bioresource Technology. 2020;**298**:122299. DOI: 10.1016/j.biortech.2019.122299

[5] Araujo GS, Matos LJBL, Goncalves LRB, Fernandes FAN, Farias WRL. Bioprospecting for oil producing microalgal strains: Evaluation of oil and biomass production for ten microalgal strains. Bioresource Technology. 2011;**102**(8):5248-5250. DOI: 10.1016/j.biortech.2011.01.089

[6] Chen F, Johns MR. Effect of C/N ratio and aeration on the fatty acid composition of heterotrophic *Chlorella sorokiniana*. Journal of Applied Phycology. 1991;**3**:203-209. DOI: 10.1007/BF00003578

[7] Francisco EC, Neves DB, Jacob-Lopes TT, Franco E. Research review paper: Biodiesel from microalgae. Biotechnology Advances. 2007;**25**:294-306. DOI: 10.1016/j.biotechadv.2007.02.001

[8] Guillard RRL, Ryther JH. Microalgae as feedstock for biodiesel production: Carbon dioxide sequestration, lipid production and biofuel quality. Journal of Chemical Biotechnology. 2010;**85**:395-403. DOI: 10.1002/jctb.2338

[9] Lee RF, Valkirs AO, Seligman PF. Studies on marine planktonic diatoms I. *Cyclotella nana* Hustedt and *Detonula confervacea* (Cleve) gran. Canadian Journal of Microbiology. 1962;**8**:229-239. DOI: 10.1139/m62-029

[10] Lee RF, Valkirs AO, Seligman PF. Importance of microalgae in the biodegradation of tributyltin in estuarine waters. Environmental Science and Technology. 1989;**23**:1515-1518. DOI: 10.1021/es00070a010

[11] Balke TD, Diosady LL. Rapid aqueous extraction of mucilage from whole white mustard seed. Food Research International. 2000;**33**:347-356. DOI: 10.1016/S0963-9969(00)00055-7

[12] Banu RJ, Kavitha SP, Gunasekaran M, Kumar G. Microalgae based biorefinery promoting circular bioeconomy-techno economic and life-cycle analysis. Bioresource Technology. 2020;**302**:122-822. DOI: 10.1016/j.biortech.2020.122822

[13] Bargut TCL, Santos LP, Machado DGL, Aguila MB, Mandarim-de-Lacerda CA. Eicosapentaenoic acid (EPA) vs. docosahexaenoic acid (DHA): Effects in epididymal white adipose tissue of mice fed a high-fructose diet. Prostaglandins Leukotrienes and Essential Fatty Acids. 2017;**123**:14-24. DOI: 10.1016/j.plefa.2017.07.004

[14] Baskar G, Kalavathy G, Aiswarya R, Selvakumari IA. Advances in bio-oil extraction from nonedible oil seeds and algal biomass. In: Azad AK, editor. Advances in Eco-Fuels for a Sustainable Environment. 7th ed. Sawston: Woodhead Publishing; 2019. pp. 187-210. DOI: 10.1016/B978-0-08-102728-8.00007-3

[15] Chen J, Li J, Dong W, Zhang X, Tyagi DR, Drogui P. The potential of microalgae in biodiesel production. Renewable and Sustainable Energy Reveiws. 2018;**90**:336-346. DOI: 10.1016/j.rser.2018.03.073

[16] Chen YH, Walker HT. Fed-batch fermentation and supercritical fluid extraction of heterotrophic microalgal *Chlorella protothecoides* lipids. Bioresource Technology. 2012;**114**:512-517. DOI: 10.1016/j.biortech.2012.03.026

[17] Chisti Y. Biodiesel from microalgae. Biotechnology Advances. 2007;**25**:294-306. DOI: 10.1016/j.biotechadv.2007.02.001

[18] Chua TE, Schenk MP. A biorefinery for *Nannochloropsis*: Induction, harvesting, and extraction of EPA-rich oil and high-value protein. Bioresource Technology. 2017;**244**:1416-1424. DOI: 10.1016/j.biortech.2017.05.124

[19] Chung YS, Lee JW, Chung CH. Molecular challenges in microalgae towards cost-effective production of quality biodiesel. Renewable and Sustainable Energy Reviews. 2017;**74**:139-144. DOI: 10.1016/j.rser.2017.02.048

[20] Gaeda PV, Vinsente VA, Fernandes B. Microalgal biomass cultivation. In: Rastogi RP, Madamwar D, Pandey A, editors. Algal Green Chemistry. 1st ed. Amsterdam: Elsevier; 2017. pp. 257-284. DOI: 10.1016/B978-0-444-63784-0.00013-8

[21] Molina Grima E, Fernández FGA, García Camacho F, Chisti Y. Photobioreactors: Light regime, mass transfer, and scale-up. Journal of Biotechnology. 1999;**70**:231-247. DOI: 10.1016/S0168-1656(99)00078-4

[22] Chisti Y. Biodiesel from microalgae beats bioethanol. Trends in Biotechnology. 2008;**26**:126-131. DOI: 10.1016/j.tibtech.2007.12.002

[23] Bailleul B, Cardol P, Breyton C, Finazzi G. Electrochromism: A useful probe to study algal photosynthesis. Photosynthesis Research. 2010;**106**:179-189. DOI: 10.1007/s11120-010-9579-z

[24] Simão FA, Waterhouse RM, Loannidis P, Kriventseva EV, Zdobnov EM. BUSCO: Assessing genome assembly and annotation completeness with single-copy orthologs. Bioinformatics. 2015;**31**:3210-3212. DOI: 10.1093/bioinformatics/btv351

[25] Mansfeldt C, Richter L, Ahner B, Cochlan W, Richardson R. Use of *De novo* transcriptome libraries to characterize a novel oleaginous marine chlorella species during the accumulation of triacylglycerols. PLoS One. 2016;**11**:e0147527. DOI: 10.1371/journal.pone.0147527

[26] Alric J, Redox ATP. Control of photosynthetic cyclic electron flow in Chlamydomonas reinhardtii: (II) involvement of the PGR5-PGRL1

pathway under anaerobic conditions. Biochimica et Biophysica Acta. 2014;**1837**:825-834. DOI: 10.1016/j.bbabio.2014.01.024

[27] Pollock SV, Colombo SL, Prout DL, Godfrey AC, Moroney JV. Rubisco activase is required for optimal photosynthesis in the green alga Chlamydomonas reinhardtii in a low-CO2 atmosphere. Plant Physiology. 2003;**133**:1854-1861. DOI: 10.1104/pp.103.032078

[28] Li T. Assessment of photosynthesis regulation in mixotrophically cultured microalga Chlorella sorokiniana. Algal Research-Biomass Biofuels and Bioproducts. 2016;**19**:30-38. DOI: 10.1016/j.algal.2016.07.012

[29] Alves A, Euverink G, Santos H, Dijkhuizen L. Different physiological roles of ATP- and PPi-dependent phosphofructokinase isoenzymes in the methylotrophic actinomycete Amycolatopsis methanolica. Journal of Bacteriology. 2001;**183**:7231-7240. DOI: 10.1128/JB.183.24.7231-7240.2001

[30] Ettema T, Ahmed H, Geerling A, van der Oost J, Siebers B. The non-phosphorylating glyceraldehyde-3-phosphate dehydrogenase (GAPN) of Sulfolobus solfataricus: A key-enzyme of the semi-phosphorylative branch of the Entner-Doudoroff pathway. Extremophiles. 2008;**12**:75-88. DOI: 10.1007/s00792-007-0082-1

[31] Roche TE, Baker JC, Yan X, Hiromasa Y, Gong X, Peng T. Distinct regulatory properties of pyruvate dehydrogenase kinase and phosphatase isoforms. Progress in Nucleic Acid Research and Molecular Biology. 2001;**55**:33-54. DOI: 10.1016/s0079-6603(01)70013-x

[32] Alboresi A. Light remodels lipid biosynthesis in *Nannochloropsis gaditana* by modulating carbon partitioning between organelles. Plant Physiology. 2016;**171**:2468-2482. DOI: 10.1104/pp.16.00599

[33] Gérin S, Mathy G, Franck F. Modeling the dependence of respiration and photosynthesis upon light, acetate, carbon dioxide, nitrate and ammonium in *Chlamydomonas reinhardtii* using design of experiments and multiple regression. BMC Systemic Biology. 2014;**8**:96. DOI: 10.1186/s12918-014-0096-0

[34] Banerjee A, Sharma R, Chisti Y, Banerjee UC. *Botryococcus braunii*: A renewable source of hydrocarbons and other chemicals. Critical Reviews on Biotechnology. 2002;**22**:245-279. DOI: 10.1080/07388550290789513

[35] Bligh EG, Dyer WJ. A rapid method of total lipid extraction and purification. Canadian Journal of Biochemistry and Physiology. 1959;**37**:911-917. DOI: 10.1139/o59-099

[36] Boldor D, Kanitkar A, Terigar BG, Leonardi C, Lima M, Breitenbeck GA. Microwave assisted extraction of biodiesel feedstock from the seeds of invasive Chinese tallow tree. Environmental Science and Technology. 2010;**44**:4019-4025. DOI: 10.1021/es100143z

[37] Brown TM, Duan P, Savage PE. Hydrothermal liquefaction and gasification of *Nannochloropsis* sp. Energy and Fuels. 2010;**24**:3639-3646. DOI: 10.1021/ef100203u

[38] Brujan EA, Nahen K, Schmidt P, Vogel A. Dynamics of laser induced cavitation bubbles near an elastic boundary. Journal of Fluid Mechanics. 2001;**433**:251-281. DOI: 10.1017/S0022112000003335

[39] Chemat F, Zill EH, Khan MK. Applications of ultrasound in food technology: Processing, preservation and extraction. Ultrasonic and Sonochemistry. 2011;**18**:813-835. DOI: 10.1016/j.ultsonch.2010.11.023

[40] Bhattacharya I, Gupta VK, Perveen H, Aiman U, Banerjee P, Gupta N, et al. Production of algal oil from power plant algae. Indian Chemical Engineer. 2020;**63**(2):184-192. DOI: 10.1080/00194506.2020.180674

[41] Ghosh S, Das S, Chowdhury R. Effect of pre-pyrolysis biotreatment of banana pseudo-stem (BPS) using synergistic microbial consortium: Role in deoxygenation and enhancement of yield of pyro-oil. Energy Conversion and Management. 2019;**195**:114-124. DOI: 10.1016/j.enconman.2019.04.094

[42] Chowdhury R, Ghosh S, Debnath B, Manna D. Indian agro-wastes for 2G biorefineries: Strategic decision on conversion processes. In: De S, Bandyopadhyay A, Assadi M, Mukherjee DA, editors. Sustainable Energy Technology and Policies. 1st ed. Vol. 1. Singapore: Springer; 2018. pp. 353-373. DOI: 10.1007/978-981-10-7188-1_16

[43] Dong T, Van Wychen S, Nagle N, Pienkos PT, Laurens LML. Impact of biochemical composition on susceptibility of algal biomass to acid-catalyzed pretreatment for sugar and lipid recovery. Algal Research. 2016;**18**:69-77. DOI: 10.1016/j.algal.2016.06.004

Chapter 14

Revisiting Microalgae as an Additive for Nutraceuticals: A Review

Kausthubh Sumanth, Sanjana Subramanya, Sourav Umashankar, Supriya Gummalam, Rajeswari Mallikarjunaiah, Ashwani Sharma and Nagashree Nagaraj Rao

Abstract

In order to meet the ever-growing global demands for food, healthcare, and energy, among other sources, the twenty-first century has seen a significant surge in the use of microalgae. They have seen applications in varied industries ranging from pharmaceuticals to energy to even the food industry, where its role as a source of proteins shines the most among other bioactive compounds. The microalgal biomass has the innate ability to grow in varied ecological conditions and has diverse compositions. While not economically competitive with fossil fuels or other renewable energy sources such as solar and wind, microalgal sources are technically viable, and a multitude of resources and time have been poured into the research of microalgal renewable fuels (biodiesel, ethanol, hydrogen, etc.). The rich diversity of microalgae, which is still underutilized, provides a variety of physiologically active metabolites of economic importance. These bioactive metabolites have antioxidant, antibacterial, antifungal, antiviral, anti-inflammatory, and anticancer properties. The microalgal biomass is a rich source of various compounds such as fatty acids, carotenoids, polysterols, and phenolics that can be utilized to synthesize pharmaceutical compounds and other nutraceuticals. Considering microalgae as a superfood, space food, functional food, strong agent for detoxification with high content of micro and macronutrients has found potential application in occupational, systematic, and life style disorders subsequently enhancing immunity. The path from algal research to the launching of new food products or dietary supplements is strongly affected by industrial, regulatory, and nutritional considerations. Our purpose is to review and assess what is known about different food components (i.e., proteins, polysaccharides, lipids, vitamins, minerals, and antioxidants, potential toxicants) in the context of improving knowledge about the efficacy of algal foods as nutraceuticals. This review will add be an asset for food, pharma, nutra, and cosmetic sector.

Keywords: microalgae, nutraceutical, bioactive, food, immunity

IntechOpen

1. Introduction

1.1 Microalgae

Microalgae have a vast biodiversity and are an almost unexplored resource. Microalgae, also known as microphytes, are minute, microscopic algae that cannot be seen with the human eye. They are phytoplankton that live in both the water column and the sediment and can be found in both freshwater and marine systems [1]. Microalgae and bacteria form the foundation of the food web, providing energy to all trophic levels above them. Chlorophyll a concentrations are frequently used to quantify microalgae biomass, and they can be a good indicator of prospective production [2]. They are unicellular organisms that live individually, in chains, or in groups. Their diameters can range from a few micrometers to a few hundred micrometers depending on the species. Microalgae, unlike higher plants, lack roots, stems, and leaves. They've evolved to thrive in an environment dominated by viscous forces.

It is estimated that there are between 200,000 and 800,000 species in various genera, with roughly 50,000 species described [3]. Chemically, over 15,000 new chemicals derived from algal biomass have been identified [4]. Carotenoids, antioxidants, fatty acids, enzymes, polymers, peptides, toxins, and sterols are some examples [5]. Microalgae, capable of performing photosynthesis, are important for life on earth; they produce approximately half of the atmospheric oxygen [6] and develop photoautotrophically while using carbon dioxide as a greenhouse gas. Microalgae and cyanobacteria, collectively known as phytoplankton, dominate photosynthesis in the ocean [7]. Microalgae chemical composition is not a continuous factor; it fluctuates depending on a variety of circumstances, including species and growth conditions. Some microalgae have the ability to adapt to changes in environmental conditions by changing their chemical composition in response to variation in the environment. Their ability to substitute phospholipids with non-phosphorus membrane lipids in phosphorus-depleted settings is a particularly striking example [8]. Changing environmental parameters such as temperature, illumination, pH, CO_2 supply, and nutrients can help microalgae collect desired products to a great amount [9–20].

1.2 Major classification of microalgae

Chlorophyta: The phylum's members can be found in freshwater, marine, or even terrestrial habitats. It includes unicellular and multicellular organisms with chlorophylls a and b in a single chloroplast surrounded by two envelope membranes. Unicellular representatives of the phylum Chlorophyta, such as *Chlorella vulgaris*, *Dunaliella salina*, and *Haematococcus pluvialis*, are used in commercial manufacturing today and *Parietochloris incisa* and *Botryococcus braunii* have the ability to produce lipids and hydrocarbons respectively.

Rhodophyta: The phylum Rhodophyta is mostly made up of marine multicellular species, with a few freshwater or unicellular species thrown in for good measure. A huge single chloroplast is encircled by two envelope membranes with a single central pyrenoid, and cells are spherical with an eccentric nucleus. The brown-to-olive-colored unicells are coccoid, nonmotile, and have a single pyrenoid-containing chloroplast, and the mucilaginous sheath can be thickened unilaterally.

Haptophyta: Haptophyta algae are mostly marine and unicellular or colonial, while several freshwater species have been discovered. The two most well-known Haptophyta species utilised as feed microalgae in aquaculture are *Isochrysis* aff.

galbana (T-ISO) and *Pavlova salina*. All haptophytes have one or more pyrenoid-containing chloroplasts and an antapical nucleus, with the nuclear envelope connected to the chloroplast ER and a peripheral ER beneath the plasma membrane.

Dinophyta: Members of the phylum Dinophyta are unicellular and mostly marine, with a few freshwater species. Only around half of the Dinophyta are photosynthetic, with the remaining 50% being heterotrophs lacking chloroplasts. It is a marine, heterotrophic, colourless dinoflagellate with dinokont flagellation, in which the transverse flagellum is encircled by a medial encircling cingulum that is displaced and drops downhill [21].

1.3 Nutraceuticals

A nutraceutical, often known as a "bioceutical," is a pharmaceutical substitute that claims to have physiological benefits [22, 23]. A product that provides nutritional values as well as pharmaceutical values was coined as *nutraceutical* by Dr. Stephen DeFelic in 1989. There have been many definitions for this uprising term. The American

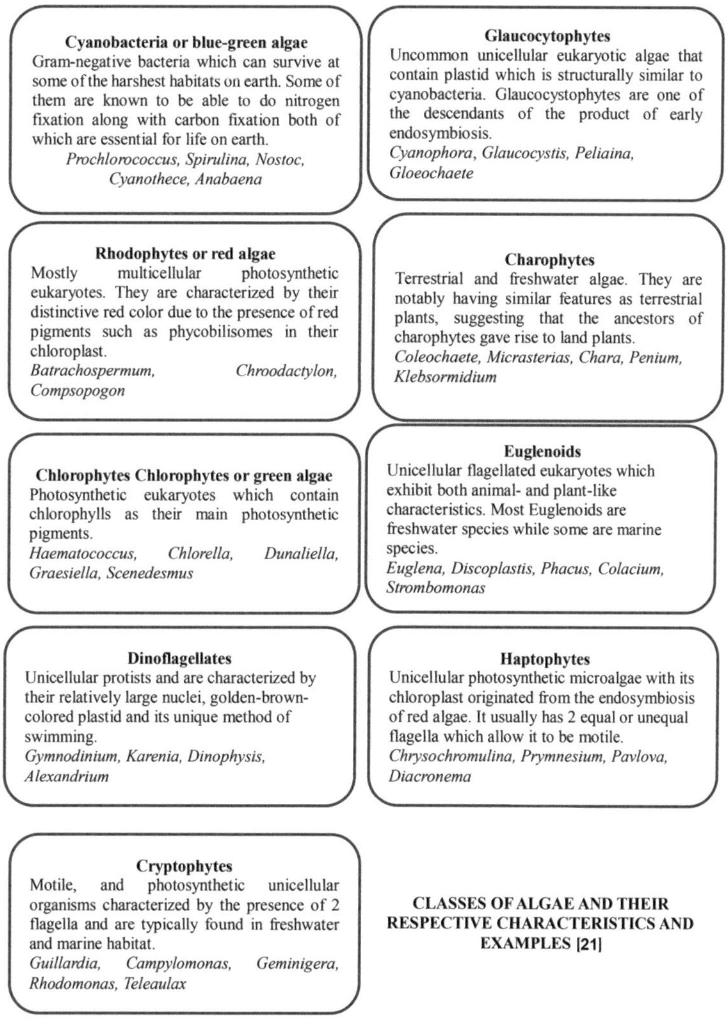

CLASSES OF ALGAE AND THEIR RESPECTIVE CHARACTERISTICS AND EXAMPLES [21]

Nutraceutical Association has defined the term as follows: "A nutraceutical is any substance conceived as a food, or part of a food which provides medical or health benefits, as well as the prevention and treatment of a disease" while according to the US Institute of Medicine, nutraceuticals include "any substance that is a food or part of a food which provides medicinal or health benefits including the prevention and treatment of disease, beyond the traditional nutrients it contains." Nutraceuticals, in general, have a favorable impact in improving customers' health and wellness. Therefore, health-promoting substances generated from food or food items to aid in the prevention or treatment of disease and/or dysfunction have also been considered as nutraceuticals.

2. Nutraceuticals from microalgae

2.1 Role of microalgae

Microalgae have been making its waves in the scientific community over the last few decades for the multitude of derivatives that can be obtained from them and the potential they hold in a vast number of fields. Microalgae have a massive biodiversity and are an essentially unexplored resource. It is believed that there are between 200,000 and 800,000 species in various genera, with approximately 50,000 species described [4]. Over 15,000 new chemicals derived from algal biomass have been chemically identified [5]. The various types of products include carotenoids, antioxidants, fatty acids, enzymes, polymers, peptides, toxins, sterols, etc. [6].

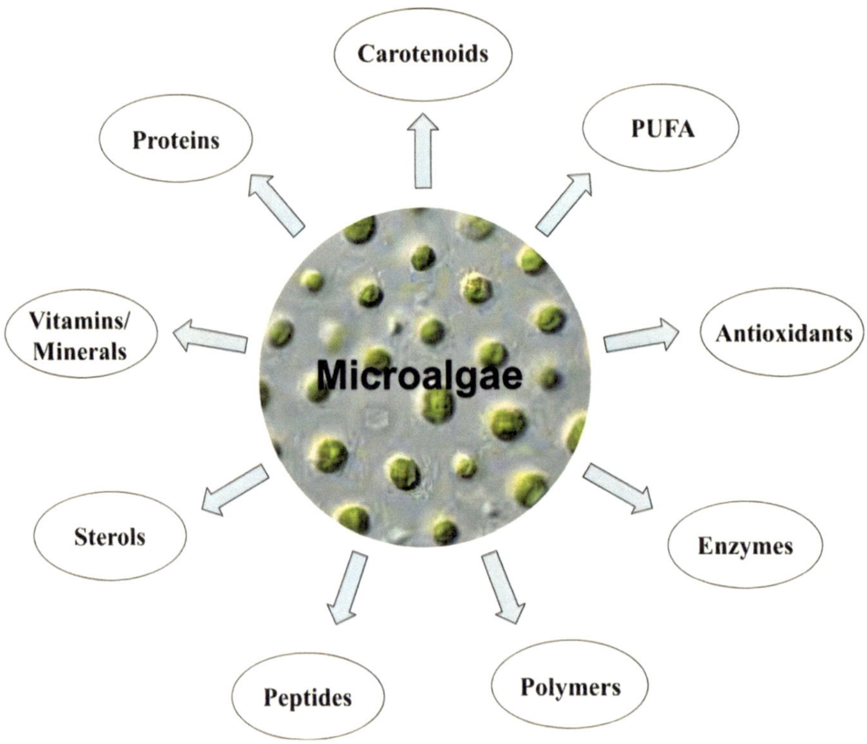

Figure 1.
An overview of nutraceuticals used from microalgae.

2.2 Microalgae and nutraceuticals

The market for nutraceuticals extracted from microalgae is dominated, if not composed majorly of products from around four species of microalgae—*Spirulina, Chlorella, Haematococcus,* and *Dunaliella*. Other species that are involved in nutraceutical production include *Odontella aurita, Schizochytrium sp, Phaeodactylum,* etc.

There are numerous essential compounds of interest (as show in **Figure 1**) that are and can be produced by means of these microalgae. These compounds are then further used for production of their respective functional and nutraceutical feed. Some of the important nutrients include

- Long-Chain Polyunsaturated Fatty Acids (PUFAs) – Arachidonic acid (AA), Eicosapentaenoic acid (EPA), etc.
    - These are defined as fatty acids with more than 18 carbon atoms and at least one double bond in their chemical structures. These can be synthesized as food supplements and are essential nutrients for humans and animals that are not produced internally. They have been recorded to have antimicrobial, anti-inflammatory, antioxidant properties.
- Pigments/Carotenoids – β-Carotene, Astaxanthin, etc.
    - Carotenoids and other pigment groups such as phycobilins are the most industrially produced algal pigments. Carotenoids are known to be strong antioxidants and to provide photoprotection to cells. Lately these algal carotenoids have been shown to have anticancer properties as well.
- Phenolic Compounds – Caffeic acid, *p*-coumaric acid, etc.
    - Phenolic compounds are one of the most significant types of natural antioxidants that may be used as dietary supplements. These molecules are primarily engaged in the defense against biotic stimuli and stress, such as grazing and UV radiation, bacteria colonization, or other fouling organisms, or metal contamination.
- Proteins/Amino Acids/Peptides – Lysine, Isoleucine, Tryptophan, etc.
    - Proteins being one of the major constituents of microalgae are now being studied as viable alternative protein sources. These compounds have been found to have hepatoprotective, anti-inflammatory, immunomodulating, anticancer, and antioxidant properties.
- Vitamins – Vitamin C, E, Riboflavin (B2), Pyridoxine (B6), etc.
    - Microalgae have shown good capability to produce important vitamins. *Spirulina* and *Chlorella* have been observed to produce quality bioavailable vitamin B12.
- Minerals – Zinc, Phosphorus, Potassium, etc.
    - Microalgae have the ability to accumulate trace elements and can be consumed as a daily nutritional supplement for minerals. Minerals provide significant

functions and are either incorporated into compounds or stay in their elemental state.

- Polysaccharides – Extracellular polysaccharides, sulfated polysaccharides, etc.
    - Microalgal polysaccharides have been reported to be rather complex polysaccharides such as immulina and other compounds that contain sugars such as galactose, xylose, fucose, etc.
- Sterols – Phytosterol, Poriferasterol, clionasterol, etc.
    - While sterols play a fundamental role in microalgae physiology, particularly with respect to their membrane integrity, they have gained popularity for their potential to lower LDL cholesterol and boost cardiovascular health.

Furthermore, sterols have been linked to anti-inflammatory and anti-atherogenicity, anticancer, and anti-oxidation actions, as well as protection against nervous system illnesses such as autoimmune encephalomyelitis, amyotrophic lateral sclerosis, and Alzheimer's disease.

2.3 Important microalgal species

2.3.1 Spirulina

Spirulina (classified as *Arthrospira sp*) is a prokaryotic cyanobacterium that has been commercially produced for over 30 years for a variety of applications such as fish food, vitamin supplements, aquaculture, medicines, and nutraceuticals. *Spirulina*, by means of photosynthesis, converts sunlight into a lot of life essential nutrients (Fatty Acids, Carbohydrates, proteins, etc.). It is often regarded as a superfood and is widely farmed to fulfill current demand, particularly in specially built raceway ponds and photobioreactors.

Spirulina is one of the algae being researched for large-scale commercial cultivation. *Spirulina* has 60–70% protein by weight (containing several amino acids) and up to 10 times more beta-carotene per unit mass than carrots. *Spirulina* is high in B vitamins, phycocyanin, chlorophyll, vitamin E, omega-6 fatty acids, and minerals.

2.3.2 Chlorella

Chlorella is a photoautotrophic, single-cell, spherical (2–10 m in diameter) green microalga with no flagella. It is simple to cultivate and produces massive amounts of biomass in a short period of time. It grows quickly and requires just CO2, water, sunshine, and a minimal quantity of minerals to thrive. They contain about 11–58% protein, 12–28% carbohydrate, and 2–46% lipids of its dry weight. Other composites include β-carotene, inositol, vitamin B6, vitamin B12, etc. *Chlorella* has been observed to lower cholesterol levels, decrease blood pressure, and even enhance the immune system.

2.3.3 Haematococcus

Haematococcus pluvialis (*H. pluvialis*) is a unicellular freshwater green microalgae that is most known for being the major producer of Astaxanthin, a red pigment that has shown innovative anti-inflammatory and antioxidant applications in human nutrition. It has also been recently observed to have preventive powers for diabetes and certain neurodegenerative diseases. The Astaxanthin levels of *H. pluvialis* are around 1.5–3% of its dry weight making it the largest known natural source of the same.

2.3.4 Dunaliella

Dunaliella is a green unicellular algae that has been observed to contain large quantities of β-carotene, protein, and glycerol. *Dunaliella* has the ability to grow in a relatively vast array of conditions and does not need specific waters for its cultivation. The extraction of the nutritional compounds can take place through its thin cell wall. It is known for the production of many carotenoid pigments, and these pigments are potent free-radical scavengers that have shown to reduce levels of enzyme inactivation and lipid peroxidation.

2.4 Therapeutic applications of microalgal nutraceutical substances

Microalgae are known for being a fantastic source of protein, which enables it to meet the ever-growing demands of a growing population. Microalgae-derived protein products exhibit a high concentration of protein with complete essential amino acid profiles. Other than proteins, microalgae are known to contain crucial vitamins such as A, B12, C, D, E and minerals such as iron, calcium, potassium, and more. Foods derived from common microalgae species *Chlorella sp.* and *Spirulina platensis* are known for their high protein content and nutritional value and are available in form of capsules, powders, and liquids. These have great abilities to combat various types of diseases due to their antimicrobial, anti-inflammatory, anticancer, and immunosuppressive properties [24].

Generation of reactive oxygen species induced by prooxidants leads to oxidative stress and is a cause of swelling and lysis of mitochondria and mutagenic actions. They accelerate the aging process and contribute to various threatening chronic and degenerative diseases such as diabetes mellitus, rheumatoid arthritis, and cancer. Enzymatic and nonenzymatic antioxidants from microalgae have the capability to quench free radicals and have high radical scavenging activities [25]. The antimicrobial properties of microalgae-based food products are mainly due to its lipid composition. Fatty acids affect the membranes of microbes resulting in the damage of internal constituents. This has a direct impact on the cells' metabolism [26].

Monounsaturated and polyunsaturated fatty acids (PUFAs) in microalgae derived food products also contribute to their antibacterial properties. Microalgal fatty acids intrusively interfere with bacterial mechanisms and can result in fatal effects such as cell leakage, reduced uptake of nutrients and inhibition of cellular respiration. Antiviral and antifungal properties of microalgae food products are attributed to polysaccharides and lipid fractions respectively [27]. High concentrations of polyphenols, polysaccharides, and phycobiliproteins in microalgae-derived nutraceuticals are the reason for their exhibition of anticancer properties due to their abilities to induce apoptosis in tumor cells [28]. The biological response due to diverse factors such as pathogens, allergens, irritants is termed as inflammation, and this is mainly observed in damaged or infected cells and tissues [29]. Chronic inflammation can directly contribute to chronic illnesses such as diabetes, cardiovascular and neurodegenerative disorders, and several inflammatory responses are a direct consequence of oxidative stress. Anti-inflammatory compounds such as carotenoids produced by microalgae reinvigorate the body's immunity and accelerate healing [27].

Despite the obvious benefits, due to lack of awareness, microalgae are underexploited as a potential food source. There are a plethora of microalgal species that have the ability to produce therapeutic bioactive compounds with their respective functional properties, which are impacted by the varying biotic and abiotic stresses [27].

Carotenoids produced by microalgae have excellent therapeutic effects on humans due to their antioxidant properties, thus yielding protection against oxidative and free radical stresses [24].

They play a significant role in the inhibition of oxidative injury to cells and tissues. Two of the industrially important carotenoids include β-carotene and astaxanthin. The 9-cis-isomer of β-carotene has been observed to possibly prevent the advancement of artherosclerosis in humans. Studies have also evaluated the positive effects of β-carotene in cardiovascular health. Other than its addition in multivitamin tablets, β-carotene can also be used in food products such as cheese and butter. Astaxanthin is another important carotenoid that is known for its exceptional antioxidant properties. It has been proved to reduce oxidative stress and inflammation and aids in improving the immune system to combat cardiovascular diseases [27].

Another valuable bioactive compound extracted from the microalgal biomass is chlorophyll, which possesses powerful antioxidant and antimutagenic properties. Studies have demonstrated its wound healing properties and stimulation of tissue growth. Due to its facilitation of rapid oxygen and carbon dioxide interchange, it is used in the treatment of oral sepsis and ulcers. By trapping mutagens in the gastrointestinal tract, chlorophyll contributes to prevention of cancer [30]. Chlorophyll also consists of a component called chlorophyllin, which has the ability to target multiple carcinogen pathways and invade their cell cycle [24]. Despite the various benefits, utilization of chlorophyll for dietary needs has disadvantages of its own. Apart from being an expensive natural food dye, it is also unstable to changes in pH conditions, and to ensure its stability, it has to undergo chemical modification [30].

Carbohydrates from microalgae are in the form of reducing sugars such as sucrose, lactose, fructose, and polysaccharides. Despite limited food applications, microalgal polysaccharides have garnered attention due to their nontoxicity and biocompatibility [31]. They boost the immune system functionality and have the ability to block tumorigenesis. Polysaccharides such as immulina and immurella demonstrate great anticancer properties while β-1,3 glucan from *Chlorella* species alleviates gastric ulcers and atherosclerosis [27].

Most microalgae consist of polar lipids such as phospholipids and galactolipids in the exponential growth phase and then tend to accumulate triacylglycerols under stressful conditions [31]. In general, microalgae are rich in poly monounsaturated fatty acids, which play crucial roles in cellular metabolism, balancing of membrane fluidity and electron-oxygen transport. External administration of lipids is important for humans due to the lack of ability of synthesizing lipids for the maintenance of homoeostasis. Poly monounsaturated fatty acids have also been proven to decrease the prevalence of various chronic diseases and have demonstrated health benefits with respect to the nervous system [27]. Microalgal fatty acids have also shown therapeutic effects against inflammation and numerous cardiovascular diseases such as myocardial infarction and cardiac arrhythmia. Docosahexaenoic acid (DHA) shows excellent cardiovascular and nervous system benefits and is therefore used as a nutritional supplement in infant formula [32]. Species of microalgae such as *Cryptothecodininum cohnii* are cultivated in fermenters for the mass production of DHA [31].

Microalgal protein extraction has a lot of benefits regarding enhanced nutritional value, productivity, and efficiency [33]. Phycobiliproteins are hydrophilic complexes used in popsicles, gum, soft drinks, and dairy products. These pigments are found

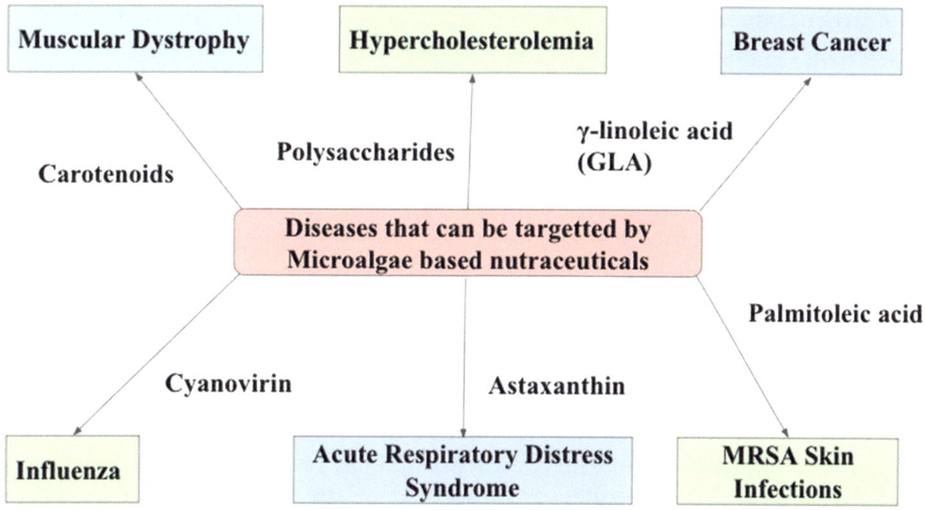

Figure 2.
This figure summarizes some of the various diseases that can be targeted by bioactive compounds derived from microalgae based food products [27, 35].

majorly in Cyanobacteria and have innate wound healing, antiviral, antioxidative properties and can act as neuroprotective agents. It also has the ability to function as an immunotoxin against B-cell lymphoma and also has an enhanced resistance to UV-induced stress [24, 27].

Microalgal species contain a plethora of vitamins (e.g., Pro-Vitamin A, Vitamin B12, Vitamin C, Vitamin D, Vitamin E), which are used as food supplements and are proven to help detoxify, revitalize cells and are involved in the activation of the immune system. The precursor of Vitamin A produced by microalgae has been seen to hinder the development of tumors in various types of cancers such as bladder, lung, and skin cancers. Cobalamin (Vitamin B12) has been demonstrated to aid in DNA repair, whereas ascorbic acid acts as an immunomodulatory agent for the prevention of severe diseases. Through the blockage of cell cycle progression, Vitamin D has demonstrated anticancer properties. Tocopherols and tocotrienols (Vitamin E) meanwhile have shown beneficial effects against atherosclerosis and pancreatic cancer [27, 34].

Microalgae consist of high levels of sterols and are components of the cellular membrane. The sterol composition is directly impacted by algal strain and the external conditions they are subjected to. Microalgae-derived sterols have been reported to demonstrate good anticancer, antioxidant, and anti-inflammatory properties. These have also been utilized as supplements for lowering cholesterol levels leading to the decrease in cardiovascular disease risks [27]. **Figure 2** summarizes some of the various diseases that can be targeted by bioactive compounds derived from microalgae based food products.

2.5 Challenges and future prospects

Despite the discovery of several compounds of high biological value and health benefits, microalgae continue to remain one of the most sparsely explored groups of organisms with an overwhelming majority of microalgal compounds yet to be isolated and identified. Although several thousand species of microalgae are believed to exist, only a small handful of these are currently cultivated and used for industrial applications.

This dearth of information calls for intensive research in the field of microalgal bioprospecting. These developments, however, give rise to a new set of challenges.

While the nutritional content of a few microalgal species has been thoroughly researched upon, their palatability, digestibility, and nutritional value ultimately depend upon the genetic makeup of the individual strains used, as well as the processes involved in large-scale biomass production. A major constraint in the use of large amounts of microalgae for human nutrition is the presence of excessive quantities of nucleic acids, which when metabolized in the human body to form urea may prove deleterious to human health. Microalgal cultivation strategies in the past include the use of natural lagoons or artificial ponds. While these open air systems are relatively economical, they suffer from a number of challenges including frequently varying climatic conditions, the prevalence of non-axenic cultures, etc. To combat these issues, the recent years have seen a rise in the use of closed systems. However, major constraints encountered with the use of closed culturing systems are the lack of inexpensive equipment and the inefficient use of light, both of which make scale-up problematic. Despite these challenges, closed reactors are promising candidates for the production of high-value nutraceuticals by microalgae including their production by recombinant technologies. Food safety is a highly pertinent matter in microalgal technology, particularly when open air systems are involved, and these products are thoroughly scrutinized by regulatory bodies in developed countries to determine their safety and efficacy prior to human consumption.

Despite being rich sources of proteins, microalgal foods and nutraceuticals have failed to carve a market niche for themselves due to the unpleasant odors and tastes associated with them, and better marketing strategies need to be developed to attract a larger consumer base There are also several considerations involving the specific doses to be prescribed to consumers. Proper labeling and marketing must be ensured to specify the quantities of all the ingredients, including details on allergenics and side effects.

Another key area of future research is the whole-genome sequencing of microalgal species producing high quantities of nutraceuticals to better understand the genes, enzymes, and metabolic pathways involved in nutraceutical production as well as the

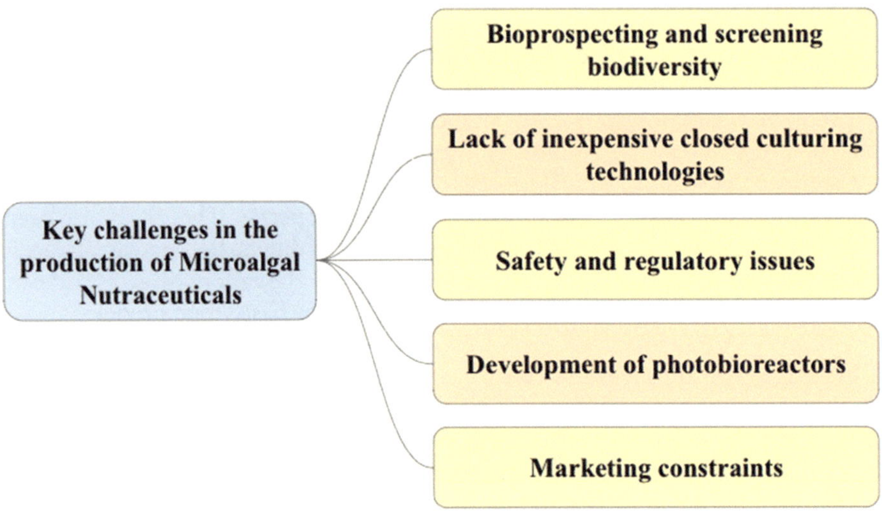

Figure 3.
Depicts challenges faced during microalgal nutraceuticals production.

mechanisms involved in the upregulation/downregulation of these genes and pathways. The development of engineered microalgae to boost protein expression, metabolism, photosynthesis, etc., is integral to large-scale production in the future. **Figure 3** depicts challenges faced during microalgal nutraceuticals production.

3. Conclusion

Microalgae have been used since time immemorial to provide food to humans and animals, but their nutritional largess has only been exploited very recently to provide high-value products on an industrial scale. Microalgae are currently used for pharmaceuticals, cosmetics, food and food additives, therapeutics, etc. While the effects of microalgae and their nutraceutical products have been tested worldwide to positive results, further health benefits are likely to be discovered with more intensive research. Advances in bioprospecting are required to further isolate and characterize microalgal compounds as they remain among the least explored groups of organisms on the planet. Other key areas of research include the development of efficient and economical microalgal cultivation technologies, metabolic engineering, genomics, and synthetic biology. Increased research and development will no doubt enable greater production as well as the identification and characterization of more microalgal species and continued investigation into their benefits for human health and nutrition.

Author details

Kausthubh Sumanth, Sanjana Subramanya, Sourav Umashankar,
Supriya Gummalam, Rajeswari Mallikarjunaiah, Ashwani Sharma
and Nagashree Nagaraj Rao*
Department of Biotechnology, R.V. College of Engineering, Bengaluru, Karnataka,
India

*Address all correspondence to: nagashreenao@rvce.edu.in

References

[1] Thurman HV. Introductory Oceanography. New Jersey, USA: Prentice Hall College; 1997

[2] Thrush S, Hewitt J, Gibbs M, Lundquist C, Norkko A. Functional role of large organisms in intertidal communities: Community effects and ecosystem function. Ecosystems. 2006;**9**(6):1029-1040. DOI: 10.1007/s10021-005-0068-8. S2CID 23502276

[3] Starckx S. A place in the sun—Algae is the crop of the future, according to researchers in Geel Flanders Today. 2012. Available from: http://www.flanderstoday.eu/current-affairs/place-sun. [Accessed: February 15, 2021]

[4] Cardozo KH-M, Thais G, Barros MP, Falcão VR, Tonon AP, Lopes NP, et al. Metabolites from algae with economical impact. Comparative Biochemistry and Physiology C. 2006;**146**:60-78

[5] Ratha SK, Prasanna R. Bioprospecting microalgae as potential sources of "Green Energy"—challenges and perspectives. Applied Biochemistry and Microbiology. 2012;**48**(2):109-125

[6] Williams R. Microscopic algae produce half the oxygen webreathe. The Science Show. ABC, 2013. Retrieved 11 November 2020.

[7] Parker MS, Mock T, Armbrust EV. Genomic insights into marine microalgae. Annual Review of Genetics. 2008;**42**:619-645

[8] Bonachela J, Raghib M, Levin S. Dynamic model of flexible phytoplankton nutrient uptake. PNAS. 2012;**108**(51):20633-20638

[9] Cardozo KHM et al. Metabolites from algae with economical impact. Comparative Biochemistry and Physiology C: Toxicology and Pharmacology. 2007;**146**(1-2):60-78

[10] Ratha SK, Prasanna R. Bioprospecting microalgae as potential sources of 'Green Energy'—challenges and perspectives (Review). Applied Biochemical Microbiology. 2012;**48**(2):109-125

[11] Nilesh Hemantkumar J, Ilza Rahimbhai M. Microalgae and its use in nutraceuticals and food supplements. In: Microalgae—From Physiology to Application. IntechOpen; 2020

[12] de Raposo MF, de Morais RMSC, de Morais AMMB. Bioactivity and applications of sulphated polysaccharides from marine microalgae. Mar. Drugs. 2013;**11**(1):233-252

[13] Alsenani F, Ahmed F, Schenk P. Nutraceuticals from Microalgae. In: Nutraceuticals and Functional Foods in Human Health and Disease Prevention. CRC Press; 2015. pp. 673-684

[14] Nicoletti M. Microalgae nutraceuticals. Foods. 2016;**5**(3):54

[15] Kaur P. Microalgae as nutraceutical for achieving sustainable food solution in future. In: Environmental and Microbial Biotechnology. Singapore: Springer Singapore; 2020. pp. 91-125

[16] Andrade LM. Chlorella and Spirulina microalgae as sources of functional foods, nutraceuticals, and food supplements; An overview. MOJ Food Processing and Technology. 2018;**6**(1):62-65

[17] West MB, Zubeck HM. Evaluation of microalgae for use as nutraceuticals and nutritional supplements. Journal of Nutrition Food Science. 2012;**2**(5)

[18] García JL, de Vicente M, Galán B. Microalgae, old sustainable food and fashion nutraceuticals. Microbial Biotechnology. 2017;**10**(5)

[19] Paniagua-Michel J. Handbook of Marine Microalgae—Biotechnology Advances. pp. 255-267

[20] Singh Joginder, Vyas Ashish, Wang Shanquan, Prasad Ram. Microbial Biotechnology: Basic Research and Applications. 2020

[21] Heimann K, Huerlimann R. Microalgal classification. In: Handbook of Marine Microalgae. 2015. pp. 25-41

[22] Sarris J, Murphy J, Mischoulon D, Papakostas GI, Fava M, Berk M, et al. Adjunctive nutraceuticals for depression: A systematic review and meta-analyses. American Journal of Psychiatry. 2016;**173**(6):575-587. DOI: 10.1176%2Fappi.ajp.2016.15091228

[23] Maciej B, Maria PA, Vincenza GR, Cicero Arrigo FG, Atanasov Atanas G, Gani B, et al. The role of nutraceuticals in statin intolerant patients. Journal of the American College of Cardiology. 2018;**72**(1):96-118

[24] Koyande AK, Chew KW, Rambabu K, Tao Y, Chu D-T, Show P-L. Microalgae: A potential alternative to health supplementation for humans. Food Science and Human Wellness. 2019;**8**(1):16-24. DOI: 10.1016/j.fshw.2019.03.001

[25] Liu Z, Ren Z, Zhang J, Chuang C-C, Kandaswamy E, Zhou T, et al. Role of ROS and nutritional antioxidants in human diseases. Frontiers in Physiology. 2018;**9**:477. DOI: 10.3389/fphys.2018.00477

[26] de Morais MG, da Vaz B, de Morais EG, Costa JAV. Biologically active metabolites synthesised by microalgae. BioMed Research International. 2015;**2015**:835761

[27] Kiran BR, Venkata Mohan S. Microalgal cell biofactory-therapeutic, nutraceutical and functional food applications. Plants. 2021;**10**(5):836. DOI: 10.3390/plants10050836

[28] Martínez Andrade KA, Lauritano C, Romano G, Ianora A. Marine microalgae with anti-cancer properties. Marine Drugs. 2018;**16**(5). DOI: 10.3390/md Starckx 16050165

[29] Chen L, Deng H, Cui H, Fang J, Zuo Z, Deng J, et al. Inflammatory responses and inflammation-associated diseases in organs. Oncotarget. 2018;**9**(6):7204-7218. DOI: 10.18632/oncotarget.23208

[30] Hosikian A, Lim S, Halim R, Danquah MK. Chlorophyll extraction from microalgae: A review on the process engineering aspects. International Journal of Chemical Engineering. 2010;**2010**:1-11. DOI: 10.1155/2010/391632

[31] Barkia I, Saari N, Manning SR. Microalgae for high-value products towards human health and nutrition. Marine Drugs. 2019;**17**(5):304. DOI: 10.3390/md17050304

[32] Koletzko B, Bergmann K, Brenna JT, Calder PC, Campoy C, Clandinin MT, et al. Should formula for infants provide arachidonic acid along with DHA – A position paper of the European Academy of Paediatrics and the Child Health Foundation. The American Journal of Clinical Nutrition. 2020;**111**(1):10-16. DOI: 10.1093/ajcn/nqz252

[33] Wang Y, Tibbetts SM, McGinn PJ Bonachela. Microalgae as sources of high-quality protein for human

food and protein supplements. Foods (Basel, Switzerland). 2021;**10**(12):3002. DOI: 10.3390/foods10123002

[34] Galasso C, Gentile A, Orefice I, Ianora A, Bruno A, Noonan DM, et al. Microalgal derivatives as potential nutraceut Comparative Biochemistry and Physiology ical and food supplements for human health: A focus on cancer prevention and interception. Nutrients. 2019;**11**(6):1226. DOI: 10.3390/nu11061226

[35] Debasis B, Harry P, Anand S. Nutraceuticals and functional foods in human health and disease prevention. Nutraceuticals from Microalgae. 2015;**9**:673-684

Chapter 15

Microalgae and Fish Nutrition

Nasreen Mohi Alddin Abdulrahman

Abstract

Fish has long been a source of "rich food for poor people" and has played an important role in increasing food security and nutrition in developing countries. Because various chemicals in algae can have confusing effects, the results of experimental research can be difficult to understand. Algae has been associated with strengthening immune systems, lipid metabolism, antiviral and antibacterial action, improved gut function, stress resistance besides providing a source of protein, amino acids, fatty acids, vitamins and minerals, and other biologically active phytochemicals in cattle and aquaculture feeds, even when used in modest amounts. The addition of algae to the fish diet modified the growth performance of the fish, causing it to improve. Its use resulted in a decrease in feed conversion ratio expenses, which plays an important part in determining aquaculture costs, an increase in feed efficiency ratio, and a decrease in feed conversion ratio. In accordance with the findings of chemical composition, various statements were acquired wherein the high proportion of algae significantly affects the protein and fat ratio. The outcomes demonstrated that algae could be a decent option as an additive for fish feed.

Keywords: microalgae, freshwater fish, health, growth

1. Introduction

Diet supplementation is an important aspect of aquaculture management, especially in intensive or semi-intensive fish culture, and is promising for increasing fish production [1]. Protein, on the other hand, is required for normal tissue function, fish body protein maintenance and replenishment, and growth. Due to the high cost of protein, it is more cost-effective to use all of the protein for tissue repair and growth rather than catabolizing it for energy [2].

This cost is determined by a variety of factors, including the amount of protein in the product, the source and kind of ingredients sourced from plant or animal sources, and manufacturing techniques. Different feeding management solutions, such as on-demand feeding regimes, are in addition to generating low-cost diets and/or good husbandry and pond management [3] possibly aid in the growth of fish. One issue that fish farmers face is finding a balance between rapid fish growth and the most efficient use of available feed. Because managing their feed intake in accordance with their energy needs is expected to improve when fish are fed with a proper feeding frequency, growth and feed conversion ratio are expected to improve.

The concept of a feeding schedule was created to account for fluctuations in the protein requirements and digestibility of farmed fish. In terms of practicality, the ideal condition would maximize the utilization of dietary protein for growth while reducing the usage of proteins for functional protein synthesis, gluconeogenesis, lipogenesis, and energy production [4].

Algae are photosynthetic creatures that provide the ultimate source of cellular carbon as well as chemical energy to other organisms. As a result, they were frequently referred to as primary producers. Microalgae (seaweed) and macroalgae (seaweed) are the two types of algae (unicellular). Microalgae require light, carbon dioxide, and nutrients to grow. Microalgae are grown for food, to produce valuable compounds, as biofilters to remove nutrients and other contaminants from wastewater, in the cosmetic and pharmaceutical industries, and for aquaculture. Also, due to their high oil content and quick biomass production, microalgae could be a viable source of biofuel [5].

Because the fish expends energy to collect prey but gains no energy from ingesting it, some plankton pass through the gut of planktivorous fishes undigested. In this circumstance, the fish may detect and reject such undesirable creatures. In the early stages of a fish's life, natural food is essential [6].

Cyanobacteria have been used in photosynthesis and its genetic control, photoregulation of genetic expression, cell differentiation and N2fixation, nitrogen, carbon, and hydrogen metabolism, resistance to environmental stress, and molecular evolution due to their benefits for humans, animals, and the environment (photosynthesis) [7]. When carp are removed, algal output decreases due to nutrient depletion, macrophytes grow due to reduced turbidity, and zooplankton increases due to increased macrophyte cover.

Spirulina is a cyanobacterium that has been commercially produced for over 10 years because of its high nutritional content, which includes protein, amino acids, vitamins, minerals, vital fatty acids, and carotene [8]. *Spirulina* can be used as a nutritional supplement for people, as well as a feed supplement for animals with economic benefits. For example, when fed to trout, sea bass, fancy carp, red tilapia, shrimp, and mollusk, it can be a good food supplement. The alga has been discovered to be a good source of protein as well as a way to improve the color, flavor, and quality of meat [9].

Spirulina has been shown to have therapeutic effects in animals, including fish, as a growth promoter, probiotic, and immune system booster [10]. *Spirulina* is used to help livestock, poultry, prawns, carp, canaries, and exotic birds develop faster [11]. Preclinical research indicates *Spirulina* possesses antimutagenic, hypocholesterolemic, immunological, and antiviral effects.

Spirulina's chlorophyll functions as a purifying and cleansing agent against harmful chemicals. It is also utilized as a probiotic agent and as a food supplement to increase color in ornamental fish. *Spirulina* contains protein (60–70%), necessary amino acids and fatty acids, phycocyanin (14%), chlorophyll (1%) and carotenoid colors (0.37%), vitamin B-12, and minerals that play key functions in animals in a variety of ways [10, 12].

Spirulina is a source of linolenic acid (GLA), an important fatty acid with therapeutic effects. Iron, calcium, chromium, copper, magnesium, manganese, phosphorus, potassium, sodium, and zinc are among the minerals found in *Spirulina*. By breaking down indigestible feed components, *Spirulina* promotes the intestinal flora of fish, according to Bargey's Manual of Determinative Bacteriology, *Spirulina* is an oxygenic photosynthetic bacterium that belongs to the Cyanobacteria and Prochlorales families. In this classification, the sequence of the rRNA subunit 16S is considered. In 1989, these microorganisms were classified into two genera, according to the suggestion by Gomont in 1892 [12].

Spirulina is becoming a popular health food all over the world. It is a filamentous Cyanobacterium that belongs to the Cyanophyte class of algae. Furthermore, *Spirulina* is a natural resource that is high in protein, amino acids, vitamins, minerals, essential fatty acids, B-complex, and -carotene [13].

Spirulina has been shown to be capable of breaking down indigestible feed components and improving the intestinal flora in fish in previous studies [12]. In fish, the creation and release of enzymes that transfer lipids for growth rather than storage. Furthermore, the -carotene in *Spirulina* helps to keep the mucous membrane in place, preventing hazardous materials from entering the body. *Spirulina*'s chlorophyll functions as a purifying and cleansing agent against harmful chemicals [14].

Phosphorus and nitrogen from agricultural and industrial effluents, as well as home wastewater, can produce major eutrophication in any aquatic body. These nutrients, on the other hand, can be used to boost plant growth, such as phytoplankton, which can be used as natural fish food or in pharmaceuticals. Due to its great nutritional content, *Spirulina* is one of the most promising microalgae for culture [15].

2. Importance of fish and aquaculture to alleviate poverty and malnutrition

The nutritional benefits of fish and fish oil consumption on human health, such as cancer, diabetes, and heart disease prevention, have long been known. The global demand for aquatic foods is predicted to continue to climb as public knowledge of the health advantages of fish intake grows [16].

Furthermore, by 2050, the world's population is predicted to increase by more than 30%, resulting in an additional 2.3 billion mouths to feed, with the majority of this expansion occurring in developing countries where fish is the primary source of protein [17].

The progressive intensification of production systems has resulted in the aquaculture sector's exponential rise during the last two decades. The use of manufactured feeds intended to fulfill the nutritional requirements of the targeted fish species is a major contributor to this intensive production system. For many fish species, feeds account for up to 70% of the variable cost of a commercial aquaculture operation [18].

The cost of fishmeal, an important protein source in fish diets, drives feed production prices. In recent years, the price of fishmeal has climbed more than twofold. It increased from around US$600 per metric ton in 2005 to around US$2000 in the first quarter of 2010 [19].

3. Using of algae as a supplement to enhance the nutritional value of fish

In place of artificial vitamin and mineral pre-mixes, 15% of mineral-rich seaweed has been included in commercial salmon meals [20]. Final testing revealed that salmon fed the "seaweed" diets were healthier and more energetic, with superior flavor and texture, possibly due to bromophenolic chemicals contained in seaweeds. In other studies, adding *Enteromorpha prolifera* and *Cladophora* sp. to laying hens' diets improved egg weight and eggshell thickness.

The vitamin content of algal biomass varies a lot depending on the species. According to Brown and Miller [21], ascorbic acid has the most variability, which could be related to changes in algal processing, drying, and storage, as ascorbic acid is particularly heat sensitive. This demonstrates the disadvantage of obtaining essential micronutrients from natural sources: there is too much variability due to the combined effects of different algal species, growing seasons, culture conditions, and processing methods to reliably supply the required micronutrients in a pre-determined manner. As a result, algal biomass in animal diets is primarily used as a supplement rather than a complete replacement for produced minerals or vitamins.

Carotenoids are a group of pigments that exist naturally in the living world and are yellow, orange, or red in color. Only bacteria, fungus, algae, and higher plants can synthesize carotenoids from scratch; therefore, animals must rely on the pigment or a similarly comparable precursor being provided in their diets, which would otherwise have gone down the food chain.

Due to the inclusion of fishmeal and fish oil in formulated aquafeeds, farmed fish and shellfish are rich sources of long chain, highly unsaturated fatty acids (HUFA). HUFA are essential for human health since they aid in the prevention and treatment of coronary heart disease, hypertension, diabetes, arthritis, and other inflammatory and autoimmune diseases. Due to a global lack of fish oil and fishmeal, researchers are increasingly looking at other lipid sources, such as algal biomass [22].

Unlike terrestrial crops, algae can directly produce HUFA such as arachidonic acid (AA, 20:4n-6) (*Porphyridium*), eicosapentaenoic acid (EPA, 20:5n-3) (*Nannochloropsis*, *Phaeodactylum*, *Nitzschia*, *Isochrysis*, *Diacronema*) and docosahexaenoic acid (DHA, 22:6n-3) (*Crypthecodinium*, *Schizochytrium*). While most of these algae are not acceptable for direct human consumption, adding them to animal feeds could increase their nutritional value for people indirectly. However, only a few studies have been conducted to date to assess microalgal lipids in farmed fish meals [23].

	% Crude Protein	**% Crude Lipid**	**% Crude Carbohydrate***	**% Ash**	**Gross Energy MJ/kg**
Fishmeal	63.0	11.0	—	15.8	20.1
Poultry meal	58.0	11.3	—	18.9	19.1
Corn-gluten	62.0	5.0	18.5	4.8	21.3
Soybean	44.0	2.2	39.0	6.1	18.2
Wheat meal	12.2	2.9	69.0	1.6	16.8
Spirulina	58.0	11.6	10.8	13.4	20.1
Chlorella	52.0	7.5	24.3	8.2	19.3
Tetraselmis	27.2	14.0	45.4	11.5	18.0
Gracilaria sp[1]	34.0	1.5	37.1	26.9	13.4
Gracilaria sp[2]	10.0	0.9	50.1	34.0	11.2
Ulva lactuca[1]	37.4	2.8	42.2	17.4	15.7
U. lactuca[2]	12.5	1.0	57.0	24.5	11.2
Schizochytrium[3]	12.5	40.2	38.9	8.4	25.6

**Carbohydrates calculated as the difference % DM – (% protein + % lipid + % ash).*
[1]*Cultured in the effluent of fish tanks.*
[2]*Collected from natural habitat.*
[3]*Commercial product, Martek Biosciences.*

Table 1.
Typical composition of commercially available feed ingredients and algae species (per dry matter).

Despite the low lipid content of seaweeds, Dantagnan et al. [24] found that including *Macrocystis pyrifera* meal in the diet at a rate of 6% increase the level of PUFAs in trout flesh. Micro- and macroalgae have also been investigated as potential alternatives to fish oil and flaxseed for increasing the HUFA content in hens' eggs [25].

The **Table 1** compares the usual nutritional profiles of commercially available animal feed ingredients with some selected micro- and macroalgae to aid in evaluating algae as a potential source of protein and energy in the form of carbs and fats.

4. Algae

Cyanobacteria (blue–green algae) are Gram-negative oxygenic photosynthetic autotrophs that are among the most successful and oldest living organisms on the planet [26, 27]. The majority of oxygen in the early atmosphere originated from cyanobacteria's oxygenic photosynthesis [27]. They are important primary producers on a global scale and play important roles in nitrogen, carbon, and oxygen biogeochemical cycles (30% of the annual oxygen production on earth) [28, 29].

They are the organisms that deliver oxygen to the earth and hence played an important part in the evolution of life. Some cyanobacteria have a unique biological mechanism (which combines N2-fixation and oxygenic photosynthesis) and can be used as a model to research significant biological activities or capabilities. Unicellular, colonial, filamentous, and branched filamentous forms are all included [30]. They are broken down into five pieces [31].

Cyanobacteria are also responsible for the origin of eukaryotic plant life on the planet, as the chloroplast of eukaryotic cells is descended from a cyanobacterial predecessor. It is a filamentous Cyan bacterium that belongs to the Cyanophyta class of algae. *Spirulina* was deemed "the best for tomorrow" by the United Nations (UN) world food conference, and it has gained appeal as a nutritional supplement in recent years [32].

The utilization of cyanobacteria as a nontraditional food and protein source appears to be promising [33–35]. Extremophyles are cyanobacteria that live in severe settings, such as *Spirulina* (alkalophilic), Extremophyle mass cultures are expected to be free of microbial contamination due to their high needs, avoiding a serious problem in outdoor cultures [34].

Pigments, such as chlorophyll a, carotenoids, and phycobiliproteins, are abundant in cyanobacteria. *Spirulina* phycocyanin buffer extract is utilized in eye shadow, eyeliner, and lipsticks. Because the product is water-insoluble, it does not fade or irritate the skin when exposed to water or sweat [36]. Cyanobacteria manufacture carbohydrates, particularly the compatible solutes glucosyl glycerol, trehalose, and sucrose, under various osmotic conditions.

In both animals and people, cyanobacteria can help lower cholesterol levels. When a high cholesterol meal was supplemented with cyanobacteria, the levels of total cholesterol, low-density lipoprotein, and very low-density lipoprotein cholesterol in rat serum were lowered. Mollusks, fish, and crabs feed on cyanobacteria. It has been found that the cyanobacterium *Spirulina* not only increases protein content but also improves the color of fish flesh. Cyanobacteria, in conjunction with bacteria, perform a crucial function in regulating the water body's O2 and CO2 balance, supporting aquaculture. Cyanobacteria assist in the removal of phosphate and nitrogen from polluted water while also producing biomass.

Cyanobacteria are vulnerable to unexpected physical and chemical changes in environmental factors, such as light, salinity, temperature, and nutrient constraint, in their native habitat. *Spirulina* is prokaryotic cyanobacteria that are spirally coiled or filamentous and have a lot of similarities morphologically (as shown in **Figures 1** and **2**) [37]. The loosely coiled trichomes of varied width with cross-walls, visible in light microscopy, are the most distinctive feature of *Spirulina* [31]. The morphology of these related strains has traditionally been used to distinguish them: helix type, distribution of pores in the cell wall, appearance of septa under light microscopy, and trichome diameter and fragmentation type [38].

The cyanobacterium *Spirulina* platensis is cultivated commercially as a possible source of proteins and medicines. Diatoms, dinoflagellates, green and yellow–brown flagellates, and blue–green algae are all examples of phytoplankton.

These groups, as photosynthetic organisms, play a critical role in ocean productivity and form the foundation of the marine food chain. *Spirulina* and *Chlorella* are two alga genera that require special attention because of their value as human food and *in vitro* and/or *in vivo* antioxidant capacity. These algae can be widely farmed to produce

Figure 1.
Spirulina in natural shape.

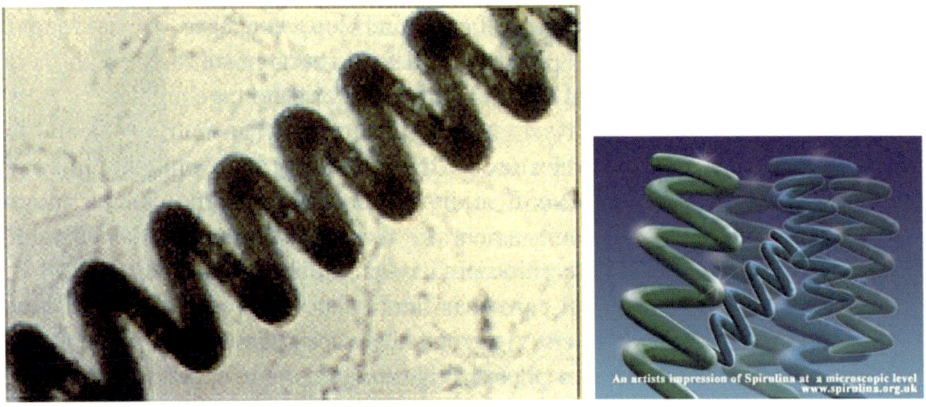

Figure 2.
Spiral shaped of Spirulina.

a protein-rich material for alimentary (diet supplementation) or industrial usage (blue pigments, emulsifiers, thickening, and gelling agent) [39, 40].

Due to a wide spectrum of vital elements, such as vitamins, minerals, and proteins, *Spirulina*'s chemical makeup suggests that it has a high nutritional value [41]. Microalgae play a vital function in aquaculture as a source of zooplankton for fish and larvae to eat [42]. *Spirulina* may boost carotenoid and pigment levels, according to a study by Lu et al. [43].

Furthermore, the usage of *Spirulina* meal in the animal feed business is growing [44, 45]. Aquaculture of macro- and microalgae is a lucrative global business. Macroalgae are grown for both their hydrocolloids and their nourishment.

In the commercial rearing of many aquatic species, microalgae are an essential food source and feed supplement. Algae are the natural food source for these creatures, therefore their relevance in aquaculture is unsurprising, only a few algae species contain components that have antioxidant properties. It has been reported that including *Spirulina* in ayu's feed produces in better flavor, firmer flesh, and brighter skin color. Other research has found that a 5% dietary *Spirulina* supplement reduces muscle lipids and improves the flavor and texture of striped bass jack.

Spirulina has been identified as a potential protein source for fish feed. Earlier research looked into how adding dry *Spirulina* powder to a diet changes the taste and quality of fish. *Spirulina* supplementation in freshwater fish feed has been shown to improve growth and promote gonad development and maturation, according to other studies.

Antioxidants from marine organisms, including alga extracts from several species, were studied. Many algae species have been shown to be powerful antioxidants. Due to a wide spectrum of vital elements, such as vitamins, minerals, and proteins, *Spirulina*'s chemical makeup suggests that it has a high nutritional value. Aztecs have been collecting and using *Spirulina* (now known as Arthrospira) [34].

Externally, *Spirulina* is used as a poultice to treat certain disorders. The International Association of Applied Microbiology designated *Spirulina* as a "great future food source" in 1967. While no microbe ever delivered on its promise of inexpensive protein, *Spirulina* continues to spur research and production, owing to its claimed nutritional benefits [46].

The ability of this microbe to use ammonia as a nitrogen source at high alkaline pH values may be due to a comparatively high cytoplasmic pH (4.2–8.5). *Spirulina* contains a higher percentage of high-quality protein (59–65%) than other regularly used plant sources such as dried soybeans (35%), peanuts (25%), or cereals (8–10%). Due to the absence of cellulose in its cell walls (as is the case for eukaryotic green microalgae, such as *Chlorella*, Ankistrodesmus, Selenastrum, Scenedesmus), *Spirulina* has a unique value: after 18 hours, more than 85% of its protein has been digested and assimilated.

Spirulina is also a common ingredient in ornamental fish feed, such as carp, because it improves coloration. Algal carotenoids may also operate as a growth factor, which could lead to yet another use for algae in aquaculture diets. Because of its incredible ability to generate high-quality concentrated food, cyanobacteria, particularly *Spirulina*, is being developed as the "Food of the Future." *Spirulina* is said to offer a full protein content of 65–70%, with all essential amino acids in perfect balance.

When the necessary circumstances for producing *Spirulina* can be attained, culturing this organism is not difficult. *Spirulina*, on the other hand, has a high protein and vitamin content despite its low protein output by an order of magnitude. 20 g dried *Spirulina* offers 100% of the recommended daily allowance of vitamin B12,

70% of the recommended daily allowance of thiamine, 50% of the recommended daily allowance of riboflavin, and 12% of the recommended daily allowance of niacin. *Spirulina* also has a high level of p-carotene (provitamin A) and important unsaturated fatty acids, which are both beneficial nutritionally. For the feeding of artificially grown clams, a semi-commercial concept on Cape Cod, USA, uses three different and relatively pure algae cultures in unheated water. The three species employed enable seasonal changes in growing conditions to be compensated for. After that, the algae is diluted with seawater and circulated through the hatchery beds, where the clams filter feed to get the protein source. Algae has also been discovered to give a growth factor to the larvae's culture media, improving their survival and growth.

The following are some of the primary advantages of using *Spirulina* in aquaculture, according to their promotional literature:

1. Because of *Spirulina*'s intrinsic palatability, better growth rates are achieved and less feed is lost. Fish fed this cyanobacterium have reduced belly fat, indicating that the energy has been transferred to growth. In feeding trials with Cherry salmon, this theory was tested and confirmed.

2. In terms of meat flavor, consistency, and color, fish-fed *Spirulina* has a higher quality. Henson [47] cites research in which *Spirulina* supplements improved the coloration of Sea Bream, Mackerel, Yellowtail, and ornamental koi carp.

3. Henson [47] also mentions research in which yellowtail was shown to have improved survival rates after being reared with *Spirulina*, with mortality rates lowering by 14%.

4. The blue pigment phycocyanin has been blamed for the reduced toxicity and greater effectiveness of fish medicines in *Spirulina*-fed fish. *Spirulina* reduced the hazardous effects of heavy metal poisoning in particular fish, according to Henson [47].

Due to the unpopularity of artificial dyes, one of the key areas of research into the aquacu1tural relevance of *Spirulina* has been the color improvement potential. *Spirulina* is used to improve the color of ornamental koi carp, trout, salmon, and shrimp, sweet smelt, red tilapia, and the striped jack [48].

The high production costs of pure-culture-produced biomass have hampered the use of *Spirulina* as a protein and pigment source in aquaculture. As a result, the algae are either utilized as a beginning feed for larvae or as a specialty diet for adults (e.g., for color enhancement in ornamental fish). Given *Spirulina*'s nutritionally complete nature, it appears that if production costs could be kept to a low, this cyanobacterium might provide a novel feed source for aquaculture creatures.

Spirulina was named "The Best for Tomorrow" by the United Nations World Food Conference, and it has gained appeal as a food supplement in recent years. *Spirulina*, planktonic blue–green microalgae, has been proposed as a future food source that is both acceptable and safe. Due to its antioxidant, anti-inflammatory, antimetastasis, and blood cholesterol-lowering properties, it has recently been considered a source for possible medicines. *Spirulina*, for example, increased interferon production and natural killer cell activity when given orally [49].

Aquacultural systems based on microalgae and their animal consumers, which can be considered an indirect use of microalgae in human food, have been far more successful thus far, however, the uptake of microalgal biomass by commercially important filter-feeders is very promising from an energetic standpoint. Microalgae are the biological starting point for energy transfer in most aquatic ecosystems and are, hence, the foundation of many aquaculture operations' food chains.

Spirulina is also used in fish farming, primarily for colored fishes [50], as a good source of antioxidant pigments, such as carotenoids, lutein, astaxanthin, zeaxanthin, and others, for intracellular protection of fish larvae against various diseases as well as the bright coloration of fishes [50]. *Spirulina* supplementation has been shown to prevent ischemic brain damage [51].

Despite its widespread distribution and economic importance, little is known about the feeding ecology of the common carp in natural settings. The influence of this cyprinid species on macrophytes has been well described, as has the functional anatomy of its feeding mechanism.

However, the majority of diet research has been conducted in fish culture ponds. The risk of consuming *Spirulina* was evaluated, and after a subchronic therapy, mice showed no harmful effects. *Spirulina* maxima oil extract or defatted fraction feeding reduced carbon tetrachloride-induced fatty liver growth in rats, showing a hepato-protective activity. This lower plant group contains a large range of vitamins, colors, and practically all-important nutrients, including PUFA (polyunsaturated fatty acid), and is also a good source of proteins and carbs. A lot of algae have been validated over time due to their remarkable impact on fish development and vitality, but only about 40 genera have gained widespread use in aquaculture.

Furthermore, some *Spirulina* species lack a cell wall, resulting in enhanced digestion and absorption. A number of studies have previously reported that dietary inclusion of *Spirulina* improves fish growth [13]. The ability of *Spirulina* to act as an antiviral, anticancer, hypercholesterolemia, and health improvement agent is receiving interest as a nutraceutical and a possible pharmaceutical source. When *Spirulina* alga is fed to young prawns and fingerlings, the fish have good coloring, a low death rate, and a high growth rate. These studies have also discovered an increase in the amounts of linoleic acid, GLA, protein, and an improved color in the meat of the fish when compared to fish fed on standard instant feeds [52].

The use of plant products as protein sources in fish meals has a lot of potential for aquaculture around the world. *Spirulina* is multicellular, filamentous blue–green algae that have grown in popularity in the health food sector and is increasingly being included in people's diets. Because muscle protein deposition is the primary cause of growth in fish, the flow of amino acids (A.A) from diet to developing biomass must be maintained. Fish require a variety of essential elements, including protein, fat, carbohydrate, vitamins, and minerals, although these requirements differ depending on the species. In comparison to the basal diet, 1–10% *Spirulina* supplementation boosted growth rate (up to 1.5 times), survival rate, and feed efficiency. There was also evidence of illness resistance to bacterial infection.

Bermejo et al., [53] found that the biliproteins found in *Spirulina*, such as phycocyanin, are responsible for the majority of the antioxidant capacities of this microalga's protean extract, and they suggested that *Spirulina* could be used to make a natural dietary antioxidant supplement or added to healthy food products like cereals, fruit bars, or drinks to prevent some chronic diseases involving free radicals.

Furthermore, *Spirulina* is gaining popularity due to its bioactive components, which have antioxidant properties [51]. Supplementing with live *Spirulina* enhanced fish growth and feed utilization, which could be related to improved feed intake and nutrient digestibility. *Spirulina*, on the other hand, contains a number of nutrients, including vitamins and minerals, that may aid in the development of growth.

Increased fish appetite may have resulted in higher feed intake and improved growth in *Spirulina*-enriched diets, leading to better feed intake and growth. Changes in protein and lipid content in the fish body, on the other hand, could be linked to changes in their synthesis, muscle deposition rate, and/or different growth rates. Additionally, it works as an immunomodulator [54]. *Spirulina* platensis is more extensively dispersed and found primarily in Africa, Asia, and South America. Several studies have been undertaken using dried *Spirulina* as a feed supplement [55]. *Spirulina* has been shown to have an accessible energy content of 2.50–3.29 kcal/gram and a phosphorous availability of 41%.

Spirulina typically contains only trace amounts of zinc (21–40 g/g), although it is easily enhanced [56] (Azina®: 6000 g Zn/g). There are simple methods for obtaining zinc-rich *Spirulina* [46]. Magnesium is abundant in *Spirulina*, and its bioavailability is excellent [57]. *Spirulina* has been designated a national food in China [58].

It is said to have been consumed as food in Mexico during the Aztec civilization 400 years ago. It is still eaten by the Kanembu tribe in the Republic of Chad's Lake Chad region, where it is sold as dried bread known as "dihe." [59].

Chlorella is a freshwater single-celled microalga with a grassy odor. Its distinctive emerald-green hue and lovely grass odor are attributed to its high chlorophyll concentration, which is the highest of any known plant. The name "*Chlorella*" comes from the Latin words "chlor" which means "green" and "ella" which means "little." Its size ranges from 2 to 8 microns, making it only visible through a microscope. It is about the same size as a human red blood cell, but the shape is different: *Chlorella* is spherical, whereas human red blood cells are disc-shaped. *Chlorella* reproduces quickly, dividing into four new cells every 17–24 hours. This exceptional ability to reproduce indicates a high level of "qi," or life energy [60].

Chlorella spp. is being investigated as a potential source of a wide range of nutrients (carotenoids, vitamins, minerals) that are widely used in the healthy food industry, as well as in animal feed and aquaculture Gastric ulcers, wounds, constipation, anemia, hypertension, diabetes, newborn malnutrition, and neurosis are all problems that *Chlorella* spp. can help. Glycolipids and phospholipids are also thought to have antiatherogenic and antihypercholesterolemic properties, whereas glycoproteins, peptides, and nucleotides have antitumor properties. However, a beta-1,3-glucan, which is a strong immunostimulator, a free-radical scavenger, and a blood lipid reducer, appears to be the most important component in *Chlorella* spp. [61]. These groups, as photosynthetic organisms, play a critical role in ocean productivity and form the foundation of the marine food chain. Spirulina and *Chlorella* are two alga genera that require special attention because of their importance as human meals and *in vitro* and *in vivo* antioxidant capacity. These algae can be widely grown to produce a protein-rich material for alimentary (diet supplementation) or industrial usage (blue pigments, emulsifiers, thickening, and gelling agent) [39, 40].

At technical medium, *C. vulgaris* grew satisfactorily. Up to 10% phyto-s, 57.63% crude protein, 5.84% fat, 6.44 mg/gram beta-carotene, 4.12 mg/gram vitamin C, and 1.32 mg/gram vitamin E *Chlorella vulgaris* has the potential to be a natural and ASUH feed additive, and Phyto-s can be employed for mass production nutrition [62]. *C. vulgaris* is a spherical, unicellular microalga that grows in fresh water and has a diameter of 2–10 M. It grows quickly under ideal conditions and is resistant to invaders and the harsh climate. In the aqueous medium, light and CO2 are the bare minimum conditions for algae formation. Their development is expedited and targets the synthesis of a specific set of compounds by changing the medium and changing the circumstances [63].

5. *Chlorella* as a feed supplement for humans

Microalgae are effective producers of high-protein biomass due to their quick growth rates and use of renewable resources. Microalgae are photosynthetic heterotrophic organisms that include vital amino acids, protein, minerals, vitamins, chlorophylls, antioxidants, and bioactive compounds [64]. Microalgae have been used in food and medicine because of their qualities. Researchers have recently become interested in the immunostimulating effects of microalgae.

The use of algae as animal feed is more frequent than the use of algae in human diets. A vast number of nutritional and toxicological studies revealed that algal biomass can be employed as a beneficial feed supplement that can effectively replace conventional protein sources (soy, fish meal, rice bran, etc.) [65]. Seaweeds are also high in minerals, including salt, potassium, and iodine, as well as fiber. Supplementation of seaweeds to improve the texture of foods is another potential area where their use becomes crucial [66]. Both national governments and intergovernmental organizations have a role to play in re-evaluating the potential of *Spirulina* to meet both their own food security demands as well as a tool for their international development and emergency response initiatives [39]. Algae are high in vitamins, minerals, proteins, polyunsaturated fatty acids, antioxidants, and other nutrients [67]. The enormous potential of microalgae arises from the fact that they are less thoroughly studied than agricultural crops, that they may be cultivated in conditions that are inappropriate for plants (requiring less or no seasonality), and that some species produce several times more than plants. Their potential for producing useful molecules or biomass is generally recognized, and they can be employed to improve the nutritional value of food and feed since they use sunlight energy more efficiently.

6. Feeding algae to fish

The utilization of *Spirulina* (*Arthrospir platensis*) as a growth and immunity enhancer for Nile tilapia, *Oreochromis niloticus*, was investigated (L.). Fish have been shown to benefit from *Spirulina*'s growth-promoting properties. Although *Spirulina* supplementation boosted protein deposition in the fish body, especially when fed a 1.25–5.0 g/kg diet, there were no significant differences in fish survival among the three treatments. When fish were provided a *Spirulina* supplement, their physiological indicators improved. With increased *Spirulina* levels in fish diets, total fish mortality 10 days after IP injection with *A. hydrophila* and its count following incubation with fish serum decreased. When fish were given 5.0–10.0 g *Spirulina*/kg, the lowest mortality and bacterial levels were observed. These findings suggest that *Spirulina* supplementation is effective in preventing disease in tilapia culture, with an optimum dose of *Spirulina* in the diet of 5.0 to 10.0 g per kg of food [68].

In the study of Al-Koye [69], replacing fishmeal with 10% *Spirulina* had a positive impact on all growth metrics, including weight gain, daily growth rate, specific growth rate, relative growth rate, and productivity, particularly food efficiency ratio and survival. The protein content of the fish carcass was also affected, as was the lipid content of the fish diet, and had a significant impact on blood parameters. Different *Ulva* levels in the diet of [70] were utilized, and the fish maintained at 10, 15, and 20% nutritional *Ulva* had the most significant (P 0.05) values of protein efficiency ratio (PER), protein production value (PPV percent), and energy retention (ER percent). Green seaweeds (*Ulva* sp.) could thus be added to the diet of red tilapia

(*Oreochromis* sp.) at a rate of 15% to boost growth performance without affecting feed efficiency or survival rate.

The goal of [71]'s study was to see how a diet including *Ulva lactuca*, green macroalgae, affected the growth, feed consumption, and body composition of African catfish *Clarias gariepinus*. Weight gain, specific growth rate, and feed consumption all showed significant differences. Overall, the experiment revealed that African catfish-fed diets containing 20% and 30% *U. lactuca* had poorer growth and feed utilization than the control group and fish-fed diets containing 10% *U. lactuca*. Güroy et al. [72] demonstrated that adding dietary low-level *Ulva* meal to numerous fish species, including rainbow trout *Oncorhynchus mykiss* and tilapia *O. niloticus* [73], improved growth performance and lipid deposition.

The effects of two algal meals (*Cystoseira barbata* or *Ulva rigida*) on feed consumption, development, and nutrient usage in young Nile tilapia, *O. niloticus*, were examined in a 12-week feeding experiment. The fish fed the 5% Cystoseira diet, control diet, and 5% *Ulva* diet gained the most weight (156%, 151%, and 150%, respectively), but the values were not significantly different ($P > 0.05$) from the other treatments, with the exception of the fish fed the 15% *Ulva* diet (P 0.05), which gained the least weight. 15% of the diet is made up of fish. The feed change ratio of *Ulva* meal was poor (FCR). At the maximal supplementation level of 15%, protein and energy utilization contribute to a decline in the groups fed algal meals. Carcass lipid levels fell as *Ulva* meal concentrations increased, but carcass lipid levels increased as Cystoseira meal concentrations increased (P 0.05). According to the findings, *C. barbata* or *U. rigida* meals could be utilized in tilapia diets in small amounts [74].

Rybak et al. [75] investigated the ability of freshwater *Ulva* species (*Ulva*ceae, Chlorophyta) to serve as metal bioindicators in rivers and lakes. From June to August 2010, researchers looked at changes in heavy metal (Ni, Cd, and Pb) and alkaline earth metal (Ca and Mg) concentrations in freshwater *Ulva* thalli. Ni was detected in the highest concentration among the heavy metals studied in thalli, whereas Cd was found in the lowest concentration. Metal concentrations in macroalgae, water, and sediment had statistically significant connections. *Ulva* populations from freshwater habitats were more efficient in bioaccumulating nickel than those from marine ecosystems.

7. Nutritional considerations of algal usage

Algae must compete with similar feed ingredients, namely fishmeal and oilseed meals, to break into the animal feed market and be economically viable (soya, etc.). Green, blue–green, and colored flagellates have all been utilized as animal feeds in the past, with the benefit that artificially farmed algae are very efficient protein producers in terms of land and water utilization. In animal production systems, good nutrition is critical for producing a healthy, high-quality product at a low cost. Nutrition is crucial in fish farming since feed accounts for 40–50% of production costs. With the introduction of new, unbiased commercial foods that support optimal fish development and health, fish nutrition has evolved dramatically in recent years. The creation of new species-specific diet formulations aids the aquaculture (fish farming) business in meeting the rising demand for inexpensive, safe, and high-quality fish and seafood [76].

Abdulrahman [77] evaluated the effect of replacing fishmeal with *Spirulina* spp. at four different levels, 0, 5, 10, 15, and 20%, such as T1, T2, T3, T4, and T5 on carcass mean weight (CMW) with head and without peripheral organs and CMW without head and peripheral organs, where the third and fifth treatments give the higher value in CMW with head and without peripheral organs, and the fifth treatment gives the highest value in C When it comes to chemical composition, the T3 and T5 have a greater significant difference in crude protein than the other treatments, while the T5 has a greater significant difference in crude fat.

The purpose of [78] was to look into the effects of *Chlorella* powder (CHP) as a feed additive on the growth of juvenile Korean rockfish, Sebastes schlegeli (Hilgendorf). *Chlorella* powder (CHP) was added to six experimental diets at 0, 0.5, 1.0, 1.5, 2.0, and 4.0% (CHP0, CHP0.5, CHP1.0, CHP1.5, CHP2.0, and CHP4.0, respectively) of the dry matter basis. These findings imply that the optimum dietary CHP supplementation amount for juvenile Korean rockfish growth and feed utilization is around 0.5% of the diet, with no deleterious impacts on blood parameters or body composition.

Abdulrahman researched the effects of varying quantities of the alga *Spirulina* spp. in the fish laboratory of Sulaimani University's Animal Production Department (2014). T1 was a control treatment with no *Spirulina* spp., T2 was a treatment with 1 gm *Spirulina* /kg diet, T3 was a treatment with 3 gm *Spirulina* /kg diet, and T4 was a treatment with 5 gm *Spirulina* /kg diet. Weight gain of 6.89, Daily growth rate of 0.17, Specific growth rate of 0.147, Relative growth rate of 15.31, and Food change ratio of 2.14 were all significantly higher in the fourth treatment than in the other treatments, while the Food effectiveness ratio was significant in T3 and T4 (62.48 and 62.47), respectively.

According to Abdulrahman et al., [79], adding *Spirulina* platensis to a fish's diet as a feed additive or a partial replacement for expensive fishmeal results in significant improvements in growth, coloration, reproduction, and flesh quality. According to the findings of Sleman et al., [80], *Chlorella* supplementation in the diet may have an effect on blood and biochemical parameters.

According to Abid [81], it can also be added to the diets of common carp in various quantities to have an effect. (T4) algae as a feed additive in combination with a 7.5 gm *Chlorella*/kg diet had a positive impact on weight increase, daily growth rate, and relative growth rate. The findings of chemical studies (proximate analyses) revealed that common carp flesh had a positive impact on protein, lipids, ash, and moisture. Blood parameters, such as monocytes, granulocytes, RBC, HGB, and glucose, were also affected. The addition of (T3) algae as a feed additive to a diet containing 5 gm *Chlorella*/kg food had an effect on the relative growth rate. It also had a positive impact on feed utilization, such as the protein efficiency ratio. The condition factor had a significant impact on fish meat indices, as well as on fish weight without viscera. The effects of lipid and moisture on proximate analysis and some blood picture parameters were positive.

Author details

Nasreen Mohi Alddin Abdulrahman
College of Veterinary Medicine/University of Sulaimani/Sulaimaniyah/, Iraq

*Address all correspondence to: nasreen.abdulrahman@univsul.edu.iq

References

[1] Abdel-Tawwab M, Abdelghany AE, Ahmad MH. Effect of diet supplementation on water quality, phytoplankton community structure, and the growth of Nile tilapia, *Oreochromis niloticus* (L.), common carp, *Cyprinus carpio* L., and silver carp, *Hypophthalmichthys molitrix* V. polycultured in fertilized earthen ponds. Journal of Applied Aquaculture. 2007;**19**(1):1-24

[2] Gauquelina F, Cuzona G, Gaxiolab G, Rosasb C, Arenab L, Bureauc DP, et al. Effect of dietary protein level on growth and 84 energy utilization by *Litopenaeussty lirostris* under laboratory conditions. Journal of Aquaculture. 2007;**271**(1-4):439-448

[3] Kim KD, Kang YJ, Kim KW, Kim KM. Effects of feeding rate on growth and body composition of juvenile flounder, *Paralichthy solivaceus*. Journal of the World Aquaculture Society. 2007;**38**(1):169-173

[4] Jamabo NA, Alfred-Ockiya JF. Effects of dietary protein levels on the growth performance of *Heterobranchus bidorsalis* (GeoffroySaint-Hilaire, 1809) fingerlings from the Niger Delta. African Journal of Biotechnology. 2008;**7**(14):2483-2485

[5] Hattab M, Ghaly A. Effects of Light Exposure and Nitrogen Source on the Production of Oil from Freshwater and Marine Water Microalgae. American Journal of Biochemistry and Biotechnology. 2014;**10**(4):208-230

[6] Al-Harbi AH, Siddiqui AQ. Effects of tilapia stocking densities on fish growth and water quality in tanks. Asian Fisheries Science. 2000;**13**(4):391-396

[7] Choonawala BB. *Spirulina* production in brine effluent from cooling towers. PhD dissertation, Durban University of Technology, Durban; 2007. p. 421

[8] Pandey JP, Tiwari A. 2020. Optimization of Biomass Production of *Spirulina maxima*. Journal of Algal Biomass. 2010;**1**(2):20-32

[9] Chacon-Lee TL, Gonzalez-Marino GE. Microalgae for healthy foods—Possibilities and challenges. Comprehensive Reviews in Food Science and Food Safety. 2010;**2010**. DOI: 10.1111/j.1541-4337.2010.00132. xVol.9

[10] James XH, Sampath K, Thangarathinam R, Vasudevan V. Effect Of Dietary *Spirulina* Level on Growth, Fertility, Coloration and Leucocyte Count in Red Swordtail. The Israeli Journal of Aquaculture – Bamidgeh. 2006;**58**(2):97-104

[11] Nandeesha M, Gangadhara B, Maniseery J, Venkataraman L. Growth performance of two Indian mahor carps, (*Catla catla* and *Labeo rohita*) fed diets containing different levels of *Spirulina platensis*. Bioresource Technology. 2001;**80**(2):117-120

[12] Ramakrishnan CM, Haniffa MA, Manohar M, Dhanaraj M, Arokiaraj AJ, Seetharaman S, et al. Effects of probiotics and *Spirulina* on survival and growth of juvenile common carp (*Cyprinus carpio*). Israeli Journal of Aquaculture – Bamidgeh. 2008;**60**(2):128-133

[13] Tongsiri S, Mang-Amphan K, Peerapornpisal Y. Effect of replacing fishmeal with *Spirulina* on growth, carcass composition and pigment of the mekong giant catfish. Asian Journal of Agricultural Science. 2010;**2**(3):106-110

[14] James R. Effect of dietary supplementation of *Spirulina* on growth and phosphatase activity in copper-exposed carp (*Labeo rohita*). Journal of Aquatic Bamidgeh. 2010;**62**(1):19-27

[15] Toyub MA, Miah MI, Habib MAB. Growth performance of *Hypophthalmichthys molitrix* and *Barbodesgonionotus* fingerlings by feeding microalgae cultured on fertilizer factory effluent. Bangladesh Journal of Science and Industrial Research. 2010;**45**(4):315-322

[16] Gina S. Future aquaculture. Feeds and feed costs: The role of fish meal and fish oil. In: Offshore aquaculture, in the United States: Economic considerations, implications and opportunities. 2009

[17] UN. World Economic Situation and Prospects 2010 (United Nations publication, Sales No. E.10.II.C.2), 2010. Released in January 2010. Available from: http://www.un.org/esa/policy/wess/wesp.html

[18] FAO. Report of the global conference on small-scale Fisheries. Bangkok, Thailand, 13-17 October 2008. Rome: 190 pp; 2009.

[19] IMF. The International Monetary Fu. annual report 2010 Supporting A Balanced Global Recovery I: pp. 84; 2010

[20] Kraan S, Mair C. Seaweeds as ingredients in aquatic feeds. International Aquafeeds. 2010;**13**(6):10-14

[21] Brown MR, Miller KA. The ascorbic acid content of eleven species of microalgae used in mariculture. Journal of Applied Phycology. 1992;**4**:205-215

[22] Atalah E, Hernández Cruz CM, Izquierdo MS. Two microalgae *Crypthecodinium cohnii* and *Phaeodactylum tricornutum* as Alternative source of essential fatty acids in starter feeds for sea bream (*Sparus aurata*). Aquaculture. 2007;**270**(1-4):178-185

[23] Ganuza E, Benítez-Santana T, Atalah E, Vega-Orellana O, Ganga R, Izquierdo MS. *Crypthecodinium cohnii* and *Schizochytrium* sp. as potential substitutes to fisheries-derived oils from seabream (*Sparus aurata*) microdiets. Aquaculture. 2008;**277**(1-2):109-116. DOI: 10.1016/J.AQUACULTURE.2008.02.005

[24] Dantagnan P, Hernandez A, Borquez A. Inclusion of macroalgae meal (*Macrocystispyrifera*) as Feed ingredient for rainbow trout (*Oncorhynchus mykiss*): Effect on flesh fatty acid composition. Aquaculture Research. 2009;**41**(1):87-94

[25] Kassis N, Drake SR, Beamer SK, Matak KE, Jaczynski J. Development of nutraceutical egg products with omega-3-rich oils. LWT—Food Science and Technology. 2010;**43**:777-783

[26] Gademan K, Portman C. Secondary metabolites from cyanobacteria: Complex structures and powerful bioactivities. Current Organic Chemistry. 2008;**12**:326-341

[27] Schopf JW. The fossil records: Tracing the roots of the cyanobacterial lineage. In: Whitton BA, Potts M, editors. The Ecology of Cyanobacteria. Dordrecht: Kluwer; 2000. pp. 13-35

[28] DeRuyter YS, Fromme P. Molecular structure of the photosynthetic apparatus. In: Herrero A, Flores E, editors. The Cyanobacteria: Molecular Biology, Genomics and Evolution. Norfolk: Caister Academic Press; 2008. pp. 217-270

[29] Karl D, Michaels A, Bergman B, Capone D, Carpenter E, Letelier R, et al. Dinitrogen fixation in the world's oceans. Biogeochemistry. 2002;**57**(58):47-98

[30] Thajuddin N, Subramanian G. Cyanobacterial biodiversity and potential applications in biotechnology. Current Science. 2005;**89**:47-57

[31] Castenholz RW. General characteristics of the cyanobacteria. In: Boone DR, Castenholz RW, editors. Bergey's Manual of Systematic Bacteriology. 2nd ed. New York: Springer; 2001. pp. 474-487

[32] Tomitani A, Knoll AH, Cavanaugh C, Terufumi O. The evolutionary diversification of cyanobacteria: Molecular-phylogenetic and paleontological perspectives. In: Proceedings of the National Academy of Sciences. 2006;**103**(14):5442-5447. DOI: 10.1073/pnas.0600999103

[33] Gantar M, Svircev Z. Microalgae and cyanobacteria: Food for thought. Journal of Phycology. 2008;**44**:260-268

[34] Pulz O, Gross W. Valuable products from biotechnology of microalgae. Applied Mi-crobiology and Biotechnology. 2004;**65**(6):635-648

[35] Rosenberg JN, Oyler GA, Wilkinson L, Betenbaugh MJ. A green light for engineered algae: Redirecting metabolism to fuel a biotechnology revolution. Current Opinion in Biotechnology. 2008;**19**:430-436

[36] Reddy CM, Bhat VB, Kiranmai G, Reddy MN, Reddanna P, Madyastha K. Selective inhibition of cyclooxygenase-2 by Cphycocyanin, a biliprotein from *Spirulina platensis*. Biochemical Biophysical and Research Communication. 2000;**3**:599-603

[37] Turpin. 1829. Available from: https://diatombase.org/aphia.php?p=taxdetails&id=962857

[38] Vonshak A, Tomaselli L. *Arthrospira* (*Spirulina*): Systematic and ecophysiology. In: Whitton A, Potts M, editors. The Ecology of Cyanobacteria. the Netherlands: Kluwer Academic Publishers; 2000. pp. 505-522

[39] Habib MAB, Parvin M, Huntington TC, Hasan MR. A Review on culture, production and use of *Spirulina* as food for human and feeds for domestic animals and fish. FAO Fisheries and Aquac. Circular, Roma, No. 1034: 41pp; 2008

[40] Hasan MR, Chakrabarti R. Use of algae and aquatic macrophytes as feed in small-scale Aquac: A review. FAO Fisheries and Aquac. Technical Paper. No. 531. Rome, FAO: 135p; 2009

[41] Dillon JC, Phan PA. Spirulina as source of protein in human nutrition. Bulletin de l'Institut Oceanographique de Monaco. 1993;**12**:103-107

[42] Chakraborty RD, Chakraborty K, Radhakrishnan EV. Variation in fatty acids composition of Artemia salina nauplii enriched with microalgae and baker's yeast for use in larviculture. Journal of Agricultural and Food Chemistry. 2007;**55**:4043-4051

[43] Lu J, Yoshizaki G, Sakai K, Takeuchi T. Acceptability of raw Spirulina to larval tilapia *Oreochromis niloticus*. Fisheries Science. 2002;**68**:51-58

[44] Gebhardt SE, Robin G. Nutritive Value of Foods. U.S. department Agriculture, Agricultural Research Service, Home and Garden Bulletin 72; 2002

[45] Morais MG, Radmann EM, Andrade MR, Teixeira GG, Brusch LRF, Costa JAV. Pilot scale semi continuous production of Spirulina biomass in southern Brazil. Aquaculture. 2009;**294**:60-64

[46] Falquet J. A Sustainable Response to Malnutrition in Hot Regions: The Local Production of Spirulina. Geneva: Antenna Technologies; 2006

[47] Henson R. Spirulina improves Japanese fish feeds. Aquaculture Magazine. 1990;**6**:38-43

[48] Hardy T, Herman E, Hu G, Krogdahl Å, Nelson R, Overturf K, et al. Expanding the utilization of sustainable plant products in aquafeeds: A review. Aquaculture Research. 2007;**38**:551-579

[49] Akao Y, Ebihara T, Masuda H, Saeki Y, Akazawa T, Hazeki K, et al. Enhancement of antitumor natural killer cell activation by orally administered *Spirulina* extract in mice. Cancer Science. 2009;**100**(8):1494-1501. DOI: 10.1111/j.1349-7006.2009.01188.x

[50] Chiu L, Huei M, Emily S, Chang Y. Techniques in finfish larviculture inTaiwan. Aquaculture. 2001;**200**:1-31

[51] Wang L, Pan B, Sheng J, Xu J, Hu Q. Antioxidant activity of *Spirulina platensis* extracts by supercritical carbon dioxide extraction. Food Chemistry. 2007;**105**:36-41

[52] Promya J, Chitmanat C. The effects of *Spirulina platensis* and *Cladophora algae* on the growth performance, meat quality and immunity stimulating capacity of the African sharptooth catfish (*Clarias gariepinus*). International Journal of Agriculture and Biology. 2011;**13**:77-82

[53] Bermejo-Pareja F, Benito-León J, Vega S, Medrano MJ, Román GC. Neurological Disorders in Central Spain (NEDICES) Study Group. Incidence and subtypes of dementia in three elderly populations of central Spain. Journal of the Neurological Sciences. 2007;**264**(1-2): 63-72. DOI: 10.1016/j.jns.2007.07.021. PMID: 17727890

[54] Takeuchi T, Lu J, Yoshizaki G, Satoh S. Effect on the growth and bodycomposition of juvenile tilapia *Oreochromis niloticus* fed raw *Spirulina*. Fisheries Science. 2002;**68**:34-40

[55] Torres P, Chow F. Growth and photosynthetic pigments of *Gracilariopsis tenuifrons* (Rhodophyta, Gracilariaceae) under high light in vitro cuture. Journal of Applied Phycology. 2015;**27**:11243-11251

[56] Cogne G, Lehmann B, Dussap CG, Gros JB. Uptake of macrominerals and trace elements by the cyanobacterium *Spirulina platensis* under photoautotrophic conditions: Culture medium optimization. Biotechnology and Bioengineering. 2003;**81**:588-593

[57] Planes P, Rouanet J-M, Laurent C, Baccou J-C, Besancon P, Caporiccio B. Magnesium bioavailability from magnesium-fortified *Spirulina* in cultured human intestinal Caco-2 cells. Food Chemistry. 2002;**77**:213-218

[58] Henrikson R. Earth Food Spirulina. Hawaii: Ronore Enterprises, Inc; 2009. p. 188

[59] Abdulqader G, Barsanti L, Tredici M. Harvest of *Arthrospira platensis* from Lake Kossorom (Chad) and its household usage among the Kanembu. Journal of Applied Phycology. 2000;**12**:493-498

[60] DOE (U.S. Department of Energy). National Algal Biofuels Technology Review. U.S. Department of Energy, Office of Energy Efficiency and Renewable Energy, Bioenergy Technologies Office, pp. 212, 2016.

[61] Spolaore P, Joannis-Cassan C, Duran E, Isambert A. Commercial applications of microalgae. Journal of Bioscience and Bioengineering. 2006;**101**:87-96

[62] Marlida Y, Purwati E. The Optimizing of Growth and Quality of *Chlorella vulgaris* as ASUH feed supplement for Broiler. International Journal on Advanced Science, Engineering and Information Technology. 2014;**4**(4):90-93

[63] Safi C, Zebib B, Merah O, Pontalier PY, Vaca-Garcia C. Morphology, composition, production, processing and applications of *Chlorella vulgaris*: A review. Renewable and Sustainable Energy Reviews. 2014;**35**(C):265-278

[64] Kwak JH, Baek SH, Woo Y, Han JK, Kim BG, Kim OY, et al. Beneficial immunostimulatory effect of short-term *Chlorella* supplementation: Enhancement of natural killer cell activity and early inflammatory response (randomized, double-blinded placebo-controlled trial). Nutrition Journal. 2012;**11**:53

[65] Becker EW. Microalgae as a source of protein. Biotechnology Advances. 2007;**25**:207-210

[66] Chojnacka K, Saeid A, Michalak I. The possibilities of the application of algal biomass in the agriculture. Science. 2012;**11**:1242-1248

[67] Gouveia L, Batista AP, Sousa I, Ray-mundo A, Bandarra NM. Micro-algae in novel food products. In: Papa-doupoulos K, editor. Food Chemistry Research Develop-ments. New York: Nova Science Publishers; 2008. pp. 75-112

[68] Abdel-Tawwab M, Abdel-Rahman AM, Ismael NEM. Evaluation of commercial live bakers' yeast, *Saccharomyces cerevisiae* as a growth and immunity promoter for fry Nile tilapia, *Oreochromis niloticus* (L.) challenged in situ with *Aeromonas hydrophila*. Aquaculture. 2008;**280**:185-189

[69] Al-Koye H. Effect of Using *Spirulina* spp. Instead of Fishmeal on Growth, Blood Picture and Microbial Load of Common carp *Cyprinus carpio* (MSc Thesis). Iraq, College of Agriculture. Erbil: University of Salahaddin; 2013. p. 131

[70] El-tawil NE. Effects of Green Seaweeds (*Ulva* sp.) as Feed Supplements in Red Tilapia (*Oreochromis* Sp.) Diet on Growth Performance, Feed Utilization and Body Composition. Journal of the arabian aquaculture society. 2010;**5**(2):179-194

[71] Abdel-warith AA, Younis EMI, Al-asgah NA. Potential use of green macroalgae *Ulva lactuca* as a feed supplement in diets on growth performance, feed utilization and body composition of the African catfish, *Clarias gariepinus*. Saudi Journal of Biological Sciences. 2016;**23**(3):404-409

[72] Güroy D, Güroy B, Merrifield DL, Ergun S, Tekinay AA, Yigit M. Effect of dietary *Ulva* and *Spirulina* on weight loss and body composition of rainbow trout, *Oncorhynchus mykiss* (Walbaum), during a starvation period. Journal of Animal Physiology and Animal Nutrition. 2011;**95**:320-327

[73] Ergun S, Soyuturk M, Guroy B, Guroy D, Merrifield D. Influence of *Ulva* meal on growth, feed utilization, and body composition of juvenile Nile tilapia (*Oreochromis niloticus*) at two levels of dietary lipid. Aquaculture International. 2009;**17**:355-361

[74] Güroy B, Cirik S, Güroy D, Sanver F, Tekinay AA. Effects of *Ulva rigida* or *Cystoseira barbata* meals as a feed additive on growth performance, feed utilization, and body composition in Nile tilapia, *Oreochromis niloticus*. Turkish Journal of Veterinary and Animal Sciences. 2007;**31**:91-97

[75] Rybak A, Messyasz B, Łe B. Chemosphere Freshwater *Ulva* (Chlorophyta) as a bioaccumulator of selected heavy metals (Cd, Ni and Pb) and alkaline earth metals (Ca and Mg). 2012;**89**:1066-1076

[76] Craig S, Specialist E, College V. Medicine V, Tech V. Understanding Fish Nutrition, Feeds and Feeding. Virginia Cooperative Extension programs and employment. 1890 Extension Program. Petersburg: Virginia State University; 2009. p. 6. Available from: www.ext.vt.edu

[77] Abdulrahman NM. Effect of replacing fishmeal with *Spirulina* spp. on carcass chemical composition of common carp *Cyprinus carpio* L. Iraqi. Journal of Veterinary Science. 2014;**28**(2):67-70

[78] Sungchul C. Effects of *Chlorella* powder as a feed additive on growth performance in juvenile Korean rockfish, Sebastes schlegeli (Hilgendorf). 2016

[79] Abdulrahman NM, Hama HJ, Hama SR, Hassan BR. Effect of microalgae *Spirulina* spp. as food additive on some biological and blood parameters of common carp *Cyprinus carpio* L. Iraqi Journal of Veterinary Sciences. 2018;**33**(1):27-31

[80] Sleman H, Abdulrahman NM, Hassan N, HamaSalih H. Evaluation of blood, biochemical and biological effects of microalgae *Chlorella* and germinated barley powder as a source of prebiotic on common carp *Cyprinus carpio* L. Iraqi Journal of Veterinary Sciences. 2021;**35**(2):271-277

[81] Abid SH. Physiological Effect of Microalgae *Chlorella* Sp. On The Common Carp (*Cyprinus carpio* L.) Msc thesis, Bingöl University Institute of Science. 2018. p. 100

Chapter 16

Exploring the Anti-cancer Potential of Microalgae

Abhishek Saxena, Aditi Raj and Archana Tiwari

Abstract

Cancer, the deadliest disease in the world, is taking away the lives of millions of people. The disease and its property of metastasis are still understudied. Various therapies have been discovered to cure this malignancy, but nearly all of them introduce a lot of side effects. Therapies such as radiation, chemotherapy, surgery, etc., are in vogue but are not so economical and approachable for many needy people. Since the nature of cancerous cells is very complex among different individuals, it becomes even more complex to treat them. In modern times, biologically active compounds extracted from plants, weeds, and, most importantly, algae (marine drugs) found in the sea have proved to possess excellent anti-cancer potential. However, the major bottlenecks are the extraction of active substances in ample quantity with high quality. This chapter describes the role of microalgae as anticancer agents. Several aspects of bioactive compounds and challenges linked to microalgae will be discussed. A brief account of nanotechnology and its role in the treatment of cancer in the context of microalgae will be highlighted. The level of algal properties that affect cell proliferation, cell arrest, and apoptosis is elaborated. The current scenario of this investigation is extensively discussed in the study, along with the chemical structure, pros and cons.

Keywords: anti-cancer, cancer, microalgae, nanotechnology, bioactive compounds

1. Introduction

Cancer, the second leading cause of death worldwide, is severely affecting the health of our society and the circumstances do not appear to improve [1]. The continuous and exponential cell division amplifies the tumor and leads to metastatic cancer [2]. Several drugs are designed to treat the disease. Therefore, the cost touches the sky, it requires that the cell lines which are isolated from the biopsy are found to be different varieties that require personalized medication [3]. The standard chemotherapy comprises radiation, medication, and drugs, that after some time show the side effects [3, 4]. Besides this, the ideology used is that the target is to destroy the cell line which is killing the natural defense cells as well as the other cells which are captured by these tumor cells and leading to proliferation [5]. Sometimes, the cancer is cured completely but after years, it comes back, and this stage is secondary metastasis [6].

Nature might have the solution to this problem as well [7]. The microalgae, which can be easily cultivated and can tolerate extreme environmental conditions, prove to be suitable for various purposes [4, 8]. The extensive variety of microalgae can open the doors for various chemical and biological interactions, which could lead to treatment pathways [4]. The unique compounds found in the microalgae are derived and experimented with cancer cell lines [9]. This treatment can be effective and has a high biological activity for the purpose [10]. The microalgae can be effortlessly cultured and nurtured in bioreactors. From this, enormous biomass is obtained, which signifies it is renewable for various purposes [11]. It is still inadequately investigated for drug innovations [12]. They utilize the energy from the sun and fix carbon dioxide, which amplifies the greenhouse gas outcomes and eliminates nitrogen and phosphorus [13]. The activity depends on the variant of replicas' nurturing conditions. It is a maker of stable isotopes 13C, 15N, and 2H [13]. The entity has many roles to play in pharmaceuticals, cosmetics, nutraceuticals, aquaculture, etc. [14].

The components of great importance found in them are lipids, proteins, carbohydrates, and nucleic acids, supplemented by pigments, vitamins, minerals, and polysaccharides [13]. The biologically active compounds found in the variants are for example acetylic acids, β-carotene, agar, agarose, alginate, PUFAs, vitamin B, lutein, etc. [15]. For instance, sterols are utilized to cure cardiovascular diseases. Most face, skin, and sun protection cosmetics incorporate microalgae into their composition [14]. For example, *Tetraselcuis suecica, which belongs to the Chlorophyceae,* gives 74g PUFA per kg. It is effective against lung adenocarcinoma. A549 is an alveolar base epithelial cell and H460 is a lung tumor cell line with an operative quantity of 5-μg μL^{-1} [16, 17]. *Chlorella vulgaris* arrests cell proliferation and metastasis of lines H1299, which are non-small lung cancer cells extracted from lymph nodes, A549 adenocarcinoma alveolar base cells, and H1437 which is the same caused by p53 mutations. It is treated which affects the cell arrest at 200 $\mu g\ mL^{-1}$ [17–19]. The cost of biomass processing is high due to its poorly researched situation. It can be further augmented to obtain maximum benefit for various purposes. The exploitation of resources is yet to be done. Proper protocols and ethical issues must be retained [12]. Nature should not be hampered by treatments, industrial requirements, and research work [6]. It possesses antimicrobial, antifungal, antiviral, anti-HIV, anti-neurological, and anticancer qualities [20]. It also has its part in environmental issues like wastewater management and greenhouse effects. The component here is natural, safe, and renewable which is the biggest aspect to be researched, it invites the various departments of biological sciences, and the motif of fighting the cancer is attracted towards it [17].

2. Microalgae: a potential anticancerous source

Structurally, they are unicellular, photosynthetic microorganisms [13] and are seen as colorful plumes [21]. They have a simple cell structure [22], which is typically a marine or freshwater organism [23]. They transform sunlight, water, and carbon dioxide into algal biomass. Phosphorus, nitrogen, and carbon are the basic nutrients, which are required in high quantities for their growth and productivity [13]. They can form colonies and can also exist as filaments or spheres with the same kind of cells. The ability to carry out food making process, which is photosynthesis, is possible because of photosynthetic pigments [22]. The major chemical constituents are lipids, proteins, and carbohydrates [22]. Their ability to carry out the food making process, which is photosynthesis, is possible because of photosynthetic pigments [22].

The color is due to the pigments found in it, which are categorized as *Chlorophyta* (green), *Rhodophyta* (red), and *Phaeophyta* (brown) [13]. The different modes of nutrition are found in the variants whether it is autotrophs, heterotrophs, or mixotrophs [22]. They are capable of nitrogen fixation. The symbiotic relationship can also be found in fungi in lichens [22, 24]. Situations and nutrition when favorable can lead to algal bloom, which is red tide. The increased production of this releases a toxin that kills the fish. The species can absorb oxygen and hydrogen from water [6].

The microalgae have various aspects of being a potential source for medication [9]. It has several positive outlooks, which make it the best choice for people suffering from cancer [1]. The major priority basis is its production sources, bioactive substances, survival in extreme areas, adaptability, and cost efficiency [5]. The production of these species is very easy in the laboratory as well as in real-life areas. They are easily found, but their unique benefits can be reaped only when they are cultivated in different parts of the world [9]. The second aspect is the bioactive elements that are found in the variants [8]. They are proved to be very much fit to be applicable in the biotechnology sector. They are used in medications for various diseases and work as a drug delivery systems too [9]. It is a very good option for various processes, and in previous sections of this article, it is highlighted that it works effectively against malignancy with lesser or fewer side effects [25]. It is found in different areas, climates, and environmental conditions, so different varieties could be extracted and assessed to determine which element is suitable to fit in which sector and treatments [9]. It is cost-efficient because it is cheap and readily available, but everything which comes from nature expects humanity to pay the price for it [25].

The ethical issues need to be taken care of precisely. Nature should not be hampered, and it should not exceed the limit [9]. The medication for cancer and the different elements extracted from it work for different stages of the disease [26].

Algae name	Algal group	Mechanism of action	Cancer cell line	References
Enteromorpha intestinalis	Green algae	• Invasion • Cease cancer cell growth	HepG2	[31–34]
Gayralia oxysperma	Green algae	• Myotoxicity	U-87 MG	[35–38]
Fucus evanescens	Brown algae	• Anti-tumour activity	RPMI-7951	[39, 40]
Laminaria japonica	Brown algae	• Induce apoptosis	HeLa cells HT29	[41, 42]
Laurencia obtusa	Red algae	• Anti-oxidative • Cytotoxic • Anti-metastatic	THP-1	[43, 44]
Padina tetrastromatica	Brown algae	• Anti-proliferative	HeLa cells	[45, 46]
Sargassum horneri	Brown algae	• Anti-Neuroinflammatory	DLD-1	[36, 38, 47–49]
Laurencia papillosa	Red algae	• Cell cycle arrest • Apoptosis	MCF-7	[49, 50]
Sargassum plagiophyllum	Brown algae	• Anti-inflammatory action	HepG2 AG549	[49, 51]
Sargassum wightii	Brown algae	• Cytotoxic	AG549 MCF7	[36, 38, 49, 51]

Table 1.
Algae variants with mechanism of action and the cancer cell lines.

The concentration of that medicine needs to be altered according to the patient's stage of cancer, his body and health condition, and the characteristics of the medicine that depend on which phase it is made to act. Some medicines and compounds are naturally designed to inhibit or completely cease malignancy [27]. Some could act in an intravasation state, and some could lead to dormancy. The dormancy can lead to secondary metastasis, which then could relapse after some years [28]. The concentration of elements specifically matters, which is different for different patient's according to the patient reports and conditions [27]. *Chemo-prevention:* the quality which either prevents or delays the onset of cancer [9]. *Microalgae are* adaptive, fast proliferation, renewable, and of interest for genetic process and modification for industrial purposes, and can be utilized as a whole [29]. It is autotrophic, which uses light for both positive and negative effects, like photo-inhibition. It requires secondary sources or metabolites for its various mechanisms, like protection, defense, etc. [9]. For instance, amino acids and peptides from *Chlorella vulgaris* protect the genetic evidence, hydroxyl radicals, scavenger effects, and gastrointestinal enzyme resistance, which comprises of 7.5 μm [30]. Algae variants with the mechanism of action and the cancer cell lines are presented in **Table 1**.

3. Types of bioactive compounds

The biological elements found in a different classes of microalgae are beneficial for various diseases, and they acts as a treatment, which further helps in lowering the complexity of the ailment. Many therapeutic and pharmacological benefits can be availed [52]. The compounds act upon the tumors and could suppress their ability to metastasize. High-value chemicals (HVC) are isolated from variants of microalgae that could toxicant the cell at the cellular level, reduce the invasiveness of the tumor, and increase the possibility of apoptosis [17]. In **Table 2** different types of bioactive compounds effective against cancer cell lines are presented in **Table 2**. A detailed overview of the different bioactive from microalgae are summarized below:

3.1 Pigments

Fucoxanthin, which is a carotenoid, can excite the *tumor suppressor genes* and cease the life cycle of the cell, but unfortunately, it cannot be effective for apoptosis. It is purified from microalgae, diatom, and brown seaweeds [67, 68]. When the agent *5-fluorouracil* was blended with *colorectal cancer*, it showed good results. It works by natural medication inhibition and cure of reactive oxygen species (ROS) [69]. The basis is the deoxyribonucleotide acid cleavage of the topoisomerases complex, which is Box-2 [70]. It is the intermediate stability in the working catalytic aspect of *topoisomerase* that stimulates apoptosis [62]. An example includes astaxanthin-KATO-III and SNU-1 cell lines for gastric cancer. There is an escalated expression of p27 in gastric cell lines because of the kinase regulated by the extracellular environment (ERK) [62]. To treat rats with induced colon carcinoma, astaxanthin was utilized, but it is very expensive [64]. The anti-inflammatory action of nuclear factor Kappa (Nf-kB) [55] is responsible for creating disruptions leading to cancer, and one of them is breast cancer [57, 58]. Acetylenic carotenoids like Fucoxanthin HL-60, which homosapiens leukemia cell line [58]. *C-Phycocyanin* is a product of *Spirulina platensis*, which labors for pathological amendment, the formation of fragments of the genetic material, and the *regression of Bcl-2 proteins,* accelerates and excites *caspase 2, 3,4,6,8,9,10 HeLa MCF-7 cell line [9]*.

Effective extract	Potential	Cancer line/Treatment	Reference
Fucoxanthin	Induction of Apoptosis	• Apoptosis of HL-60 cell line for leukaemia. • Procaspase-3 is cleaved by mitochondrial permeabilization.	[53, 54]
Cryptophycin-1	Activity against solid tumour	• Mouse lymphatic cells	[55, 56]
Borophycin	Cytotoxic effect	• Human epidermoid cancer • Colorectal adenocarcinoma	[57–60]
Lyngbya majuscule	Suppress binding capacity of tubulin polymerization.	• A549 Human lung cancer cell line	[9]
Coibamide - A	Cytotoxicity	• H460 Lung cancer cell	[56, 61]
Astaxanthin	DNA-topo cleavage complex	• KATO-III SNU-1 Gastric cancer cell line	[17, 62]
Dicosahexoienoic acid (DHA)	Cytotoxicity Lipid upregulation	• HT29 Colon tumour cell lines	[9, 57, 63]
Yessotoxin	Apoptotic death	• HeLa cells death	[53, 64]
Scytonemin	Antiproliferative Anti-inflammatory	• Human fibroblast • Endothelial cells	[62, 65]
C-phycocyanin	Alteration of DNA and fragmentation Downregulate BCL-2 expression	• HeLa cells Cervical cancer line • MCF-7 Breast tumour cell line	[9]
Fucoxanthin	Induction of Apoptosis	• Apoptosis of HL-60 cell line for leukaemia. • Procaspase-3 is cleaved by mitochondrial permeabilization.	[54, 66]
Cryptophycin-1	Activity against solid tumour	• Mouse lymphatic cells	[55, 56]
Borophycin	Cytotoxic effect	• Human epidermoid cancer • Colorectal adenocarcinoma	[57–60]
Lyngbya majuscule	Suppress binding capacity of tubulin polymerization.	• A549 Human lung cancer cell line	[9]
Coibamide - A	Cytotoxicity	• H460 Lung cancer cell	[56, 61]
Astaxanthin	DNA-topo cleavage complex	• KATO-III SNU-1 Gastric cancer cell line	[17, 62]
Dicosahexoienoic acid (DHA)	Cytotoxicity Lipid upregulation	• HT29 Colon tumour cell lines	[9, 57, 63]
Yessotoxin	Apoptotic death	• HeLa cells death	[30, 53, 64]
Scytonemin	Antiproliferative Anti-inflammatory	• Human fibroblast • Endothelial cells	[62, 65]
C-phycocyanin	Alteration of DNA and fragmentation Downregulate BCL-2 expression	• HeLa cells Cervical cancer line • MCF-7 Breast tumour cell line	[9]

Table 2.
Bioactive compounds effective against cancer lines.

Scytonemin is a natural sunscreen, and it ceases proliferation and inflammation. It is purified from *Stigonema sp.* It is not in favor of human fibroblast and endothelial cells [65].

3.2 Polysaccharide and Polyunsaturated fatty acids (PUFAs)

Polysaccharides are intracellular storage compounds. They also include an extracellular mucilaginous matrix or exopolysaccharides (EPS') and sulfated polysaccharides [63] They provide several health benefits, such as anti-cancer, anti-inflammatory, and immunostimulant [71]. Chrysolaminarin, a food storage polymer, inhibits the proliferation of tumoral colon cells [72]. For example, chrysolaminarin from *Chaetoceros muelleri* showed an immuno-stimulatory effect. Sulfated polysaccharides have radical scavenging activity that prevents oxidative stress, which otherwise cause l chronic degenerative diseases [73]. Fucoidan-sulphated, which has a high amount of fucose and is isolated from kelp, is effective against HeLa cells and human uterine cancer [74]. GA3P (D-galactose+ L-(+) Lactic acid) constrains the DNA topoisomerase I and II duties and is further obtained from the dinoflagellate Gyminidium [70]. The microalgae are rich in docosahexaenoic acid (DHA) which results in toxic accumulation in cells and highly controlled lipid peroxidation [9]. It stimulates programmed cell death of colonic malignant cell HT29 and aggregation of DHA in cardiolipin. In stomach cell lines, the compound DHA escalates in-cell cytochrome c, Bax, and p53 [57].

3.3 Toxins and polyketide

Algal species, *Protoceratium reticulatum*, *Yessotoxins* are isolated, which intensifies the apoptotic death in HeLa cells. It could upturn pathological group of cell alterations, segregation of genetic material, and changes in membrane potential of a powerhouse of the cell [53]. The neurotoxin *Cryptophycin1,* which is obtained from *Nostoc sp. GSV 224,* shows promising results against solid tumors and cell lines [56]. *Cryptophycin-8* is a semi-synthetic alternative of *Cryptophycin*, which exhibits action on proliferation in vivo [9] *Borophycin* is an acetate derivative of polyketide. It is mined from marine *Spongioform* [59]. It shows its activity in kB cell lines, which are supposed to contain *human papillomavirus18 sections*, and *LoVo lines*, which were first found in a *supraclavicular region* with the diagnosis of colon cancer. It works as a cytotoxic agent against *human epidermoid* and *colorectal tumors* [60]. Bio-product by *Lyngbya majuscule, Cucarin-A* is operative for breast and renal carcinoma [75]. It dominates the binding capacity of the process of making polymers of tubulin into the formation of a Colchicine pocket for binding [9].

3.4 Peptides

As basic knowledge or key concept, which delivers the point that peptide, is obtained from enzymatic hydrolysis, which means the cleavage of bonds by water in which enzyme is involved [9, 76]. Dolastatin works for human ovarian and colon malignant tumors. The B16, colon 26 adenocarcinoma, and M50 76 sarcoma and HT29 [77]. The medicine is extracted from *Lyngbya* sp. *and Sympleca sp*. The *Chlorella vulgaris* isolated peptide stops AGS gastric cancer after the stage of the G1 step of the life of a cell. Algal sulfolipids, which are subjected to Nostoc ellipsosporum, are operational against HIV [78]. It can also be extracted from E.coli, which could even deactivate HIV. The glycolipids are effective against HT29 in colorectal cell lines which also use 5-fluorouracil and oxaliplatin for the cure [79].

4. Cancer: a leading cause of death

Cancer is the leading cause of death worldwide [9]. It is a disorder, which has no standard treatment feasible for all. To treat cancer, chemotherapy, radiation, targeted drug delivery are standardized [12]. However, the matter of fact is that there are numerous cancer cell lines, which makes treatment a tedious process. As a basic fact, the structure and type of tumor line found in a person are often different for different people [80]. The cells multiply exponentially [9]. They are very invasive, and the process of extravasation goes hand in hand, which makes it even more miserable [57]. Studies show that the neoangiogenesis in these cancer cells is highly disorganized. It all is detected later [80]. In addition, by that time tumors would have invaded the primary site and even metastasized to the secondary site for invasion and development [3]. The technology and methodology we possess are not able to detect the growth at the earliest. The treatment and detection technology should increase its accuracy and should be able to detect to the earliest. That is why microalgae are combined with different technologies to treat malignancy [14]. The basic objective is to at least cease the process if it cannot be removed completely [80]. The area of marine chemotherapeutics is scarcely researched and if done it has sure that the cure of the disease lies on this planet and this land only [9]. **Figure 1** depicts the hallmarks of cancer.

4.1 Bio-engineered microalgae: systematic approach

Microalgae are produced naturally but not in so much abundance that they could help in research, medical, and other fields [66]. Therefore, it is cultured and grown artificially to meet the requirements. To protect environmental and

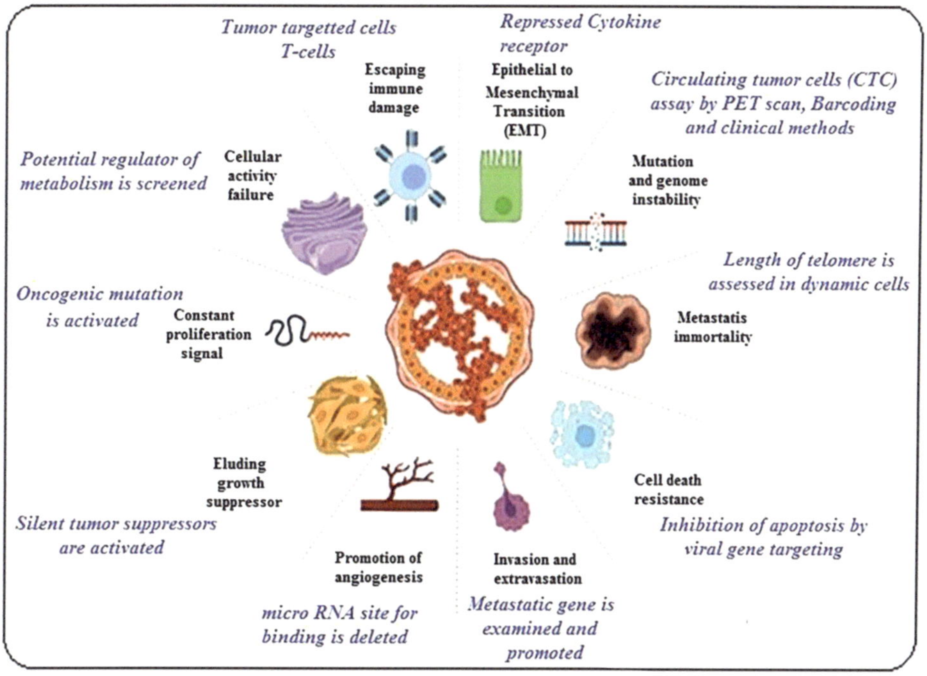

Figure 1.
Illustration depicting hallmarks of cancer.

ethical issues, the consumption of this nature's gift is kept to a minimum [81]. Now on the take for medical fields, the compound cannot be directly taken or consumed intravenously, there is a proper procedure to use up [66]. The bioactive compound, which shows potential as an anti-cancer agent, needs to be fully accepted by the body to get maximum benefits [82]. Here comes the role of genetic engineering, which further modifies and makes microalgae and other special compounds hostile to the human body [82]. The complete transformation is a required process for the nucleus and chloroplast. However, the studies point out the biological synthesis of lipids, carotenoids, and photosynthesis. *Chlamydomonas reinhardtii* is the supreme class for modification and engineering processes [83]. The lipids and carotenoids have the utmost qualities, which were discussed with several examples in previous sections [84]. The latest update regarding this is the use of the CRISPR gene editing tool, but it might not work on all microalgae as it also has some of its limitations like delivery by the large size of proteins, activity of nuclease activity, etc. In this section, the study of RBCM-Algae is taken [85]. The microalgae are isolated from *Chlorella vulgaris*, which is engineered with red blood cells from human [86]. The purpose is to get a layer of blood on the membrane to serve some respectful purposes [87]. This engineered element promotes oxygen content and tumor tissues because the tumor grows better in hypoxia conditions and increasing the levels would make it sensitive [88]. Then it is subjected to irradiation, which many times is successful to remove the tumor completely [89]. But this is not always the case. It is effective against breast cancer [90].

4.2 Effectiveness over other therapies

The traditional protocol to treat tumor lines is radiation and chemotherapy [9]. This set does not usually affect the patient's condition in various aspects but numerous other factors in the body of the patient either benefit worsen day by day [11]. The survivor must be able to fight the tumor, but he won't be able to fight the opportunist diseases which might take over [14]. These aspects only regress the condition of the body, which then many times leads to a relapse of the malignant tumor and then the person falls into cancer syndromes like cachexia [91]. The delay in detection of the tumor, carelessness, cost of the treatment, the ability of the cells to divide uncontrollably, metastasize to a secondary site, etc. point out the severity of the disease [91]. Research on the new drugs and different drug delivery systems is being done to provide maximum comfort and relief to the patient. The major side effect observed in the patient is that immunity is chiefly affected [9]. It is seen that the condition takes a toll toll on health [1]. The already in-market drugs, which are known as anti-malignant drugs, inhibit or cease the escalation of the outgrowth. It stimulates the action of the immune system and, along with it, triggers the killer cell activity naturally and initiates the defense mechanism of the host [9]. As it is known cancer multiplies in the hypoxia condition [3]. Cancer cells are immune to oxygen species which are reactive or scavenging capacity of the fundamental unit of cell stirs programmed cell death called apoptosis [92]. The standard chemotherapy eliminates tumor tissues, liver, nephron, nausea, anorexia, and mucous membrane inflammation [91]. The sector of nutraceuticals composed of antioxidant supplements is supposed to cover up for the side effects [9]. These are generally plant-based herbal medications, and the antioxidants polyphenols, carotenoids, etc. [25]. The chemicals produced by plants, which are called phytochemicals, work as antioxidants to perform defense against chemotherapy in normal cells by exerting pressure on radiation or other forms of treatment via oxidative stress [9].

4.3 Reactive oxygen species and its role in metastasis

As explained earlier, the electrons which are not paired up usually have one or two-electron atoms in the shell, which is the outermost shell [93]. It is formed in the body in the form of products like reactive species of oxygen, nitrogen, sulfur, etc. It is formulated by the cytosol [94]. One aspect of it is to get the form of a radical one with unpaired electrons, and the other one without unpaired electrons of these species [95]. It is also a form of secondary messenger [93]. The general question arising here is why oxidative stress occurs, and the easiest way of approaching the solution is that there is a misbalanced proportion in the generation and detoxification of the species [31]. The activation of all phases of them is by several transcription factors like NF-κB, activator proteins like in AP, and p53, where most mutations take place [93]. It works in both directions; that is, it works in the tumorigenic's favor as well as against the tumorigenic via signaling pathway of cancer MAPK/AP-1/NF- κB and it also triggers inflammation cytokines, chemokine, etc. [93]. When there is an escalation of reactive species, there is an elevation in cancer cell division. Tumor cells require higher levels of reactive oxygen species as cancer cells prefer to proliferate in hypoxia conditions without even cell death. Reactive species have a vital role to play in metastasis and angiogenesis [93]. It also provides resistance to chemotherapy [32]. The category of transporter protein P-glycoproteins which is thereby also known as a multi-drug resistant, that work on moving out of particles "efflux" of cancer-inhibiting drugs from malignant cells [31].

4.4 An experiment with *Spirulina* and cancer cell lines

The simplest experiment was performed with the help of *Spirulina* [28]. It was the key ingredient in the protocol. The easy methodology adopted was devised after considering cancer patients. The patients taking four cycles of chemotherapy were randomized into groups and categorized as "control" and "patient consuming the medication" [35]. In the first two cycles, the *Spirulina* was made to be consumed in 3 capsules of 100 mg each, 3 times a day [28]. The patients were taking their medications as normal, as prescribed. It influenced many things like it inhibits tumor growth interferon and interleukins like the *TNF-alpha* [35]. It works against HIV and increases the growth of CD4 cells and works on liver inflammation. The reduction of breast malignant is 87 to 13 percent [28, 39]. When the blood test was taken, the account of WBC and NEUTR. It was better with patients treated with *spirulina* and low for control [39]. The hemoglobin levels were approximately the same, whereas *myelosuppression* was lower in the treated patients rather than in the control group. IgM and CD8 were higher in treated patients [28]. The fact is that *Spirulina* slows down the *MAPK pathway.* When CD8 needs to be more numbered, as when it is exhausted, it increases apoptosis [39]. This was effective against *HepG2, MCF-7, AG49, HT29* which are liver, breast, alveolar basal epithelial cell, and colon cancer cell lines respectively [28]. A few limitations were also reported like the population size was comparatively small, and the effects after treatment were not observed [28]. There was a lack of evidence for treatment applied to the *neoplasm*. Observation and research were not performed at the molecular level [35]. Vitamin extract from the algale variant is effective for metastasis treatment and its effectiveness is presented in **Table 3**. Coenzyme Q: Ubiquinone in combination with α-lipoic acid and cisplatin ceasees colon cancer cell line HCT116 [29].

Vitamin	Algae source and potential	Cancer	Reference
• Pro vitamin A	*Tetraselmis suecica* • Retinoland rhodopsin	• Skin, Breast, Oral, Lung, Hepatic, Prostatic	[30, 96]
• Cobalamin	Porphyra sp. • DNA repair, Methylation of histone, Folate elevation regress breast malignancy.	• Lung cancer	[89, 97]
• Ascorbic acid	*Chaetocerosmuelleri* Regulate hypoxia condition via HIF1α and protects from opportunist tuberculosis infection.	• Antioxidant effect in treatment	[29, 98]
• Vitamin D	*S. costatum* • blocks cell cycle via CDK - p21, p27, preventing cancer, anti-neuro degenerative and alter insulin growth factor *IGF-1*	• colorectal, breast, pancreatic and prostate	[29, 98]
• Tocopherol	*P. lutheri* • per-oxidation of lipid, damage by oxidation	• Pancreatic cancer	[29, 98]

Table 3.
Vitamin extract from algal variant effective for metastasis treatment and effectiveness.

5. Challenges and limitations

5.1 Life expectancy post-treatment

The cost of cancer treatment is always reported to be costly, and it is observed that the majority of the families couldn't even afford the treatment [41]. Developed countries tend to even afford it to some extent, but developing countries struggle a lot. Poor countries cannot even think of fighting cancer [9]. The treatment is not even accessible to many people. Going into depth and breaking some of the myths, the cost-bearing factor is subdivided by the stage of a cancer diagnosis [39]. It is based on how severe the malignancy is and what stage it is in transition to [41].

A person who gets diagnosed at the very first or early stage can fully recover from it and the billing amount is very less but if the stage is the second, third, or fourth stage, the bill amount tends to increase [28]. The body of the patient is often not even capable of surviving the chemo and radiation it becomes hollow [39]. The body loses its virtues and qualities, and eventually, the fight to defeat cancer fails and a high number of deaths have been reported in recent years [43]. The company for cancer drugs has increased several folds and the point to highlight here is that the treatment is costly due to the commercialization and industrialization of the medications [41]. People tend to believe the brand names of medications as safest but in fact, both the substances are the same [45].

The life expectancy factor comes into play when the treatment is done, and the patient is free from any kind of medication related to cancer [28]. The condition of the patient improves to some extent in the first or second stage, but it hasn't proven to be very beneficial for other stages or severe cases like leukemia [9]. The therapy if includes microalgae will increase the cost as well. The ethical issues need to be kept in check [8]. The availability and incorporation into the therapy will require new technology as well as a new system to sustain its shelf life as well [9]. The only way to have cheap treatment is to find out the fittest technology and techniques and fulfill the requirements as well as to have multiple units in countries so there is no stress for the production of medication [35]. Proper analysis and molecular studies need to be done for future updates [9].

5.2 Limiting factors of the treatment

Each treatment has its pros and cons, and so this also has the same [41]. The limitations reported were very few, and the research is being done to have a deeper knowledge of this aspect [45]. The high intake of nutraceuticals can be toxic, which can cause side effects rather than improvising the condition [9]. There should be a proper balance between the protection of healthy and tumor cells called prophylaxis and inhibition of the growth of only cancer cells known as therapeutic doses [31]. When the concentration of antioxidants increases, the low concentrations of free radicals escalate [3]. The supplements which are derived from plants and herbs are observed to have a higher risk of pharmacokinetic interaction with agents of chemotherapy [92]. The synergic effect is by concentration, which decides whether it is beneficial or harmful. The mixture of antioxidants increases that effect in several phases and is more important than a single one [9]. A limiting factor is an aspect that lowers the possibility of achieving the best. Here, comes the most important aid to getting the treatment, which is the cost [28]. The amount we pay includes goods and services tax (GST), hospital amenities and services, and imported medicines [35]. The medicines are very expensive because only developed countries have the services and industry to manufacture their medicines and could afford these medicines to some extent. But a large proportion of the world starves for it [41]. The technology and pharma industries take up their roles, which builds up their money graphs. Medicines and treatment are costly due to technology and less accessible to most people. The elite classes of society could afford them easily [43]. The point is the world doesn't revolve around a small proportion. Many people use up all their savings, but very few can win the battle [28]. Supportive aids like antiemetics and other medications are already at higher rates. For instance, in the case of colorectal cancer, the use of cisplatin and fluorouracil types of drugs will increase the cost, and then the synthesis will increase it. Chemotherapy, diagnostics, and all add up to it [29].

5.3 Cultivation issues with microalgae

The cultivation of microalgae is required on a large scale considering its potential benefits for developing the target medicine and other applications. The need for arable land, light, oxygen, pH and other climatic conditions is required [12]. They can grow in various places, but they will possess various properties and potential [12]. If a specific industry requires a specific compound extract that is only found in a particular area, then these conditions are mandatory to be fulfilled [28]. Now, considering this, the laboratory where attempts are made to cultivate them, but they show setbacks. The setbacks include not being able to provide all conditions, chemicals to some extent, and obviously, the technology which makes it even harder to process [9]. The ability to protect it from contamination is a tough task as well [70]. Proper instruments are not there to aid in this. Nature is incredible as the laboratories are trying hard. To take a step further large bioreactors have opted, but they are not getting beneficial results [12].

5.4 Lack of technology and awareness

Cancer treatment requires more sophisticated technology. The invention is the mother of all innovations, so the technology has upgraded itself a lot. The biotechnology industries have grown so much because of the tools and medication [47]. The technology is an expensive brand, which costs a lot. It takes a lot for a technology to be launched and reach out to the people [50]. The technicians working on the prevention and treatment of cancer sometimes become clueless because proper protocol and complete research on a disease like cancer are still incomplete. The disease is an elaborative and extensive field that requires many aspects of treatment but is still not researched [51]. Technology and awareness are missing; therefore, the requirement is for a higher level that could diagnose cancer at the earliest stage, and surgery and chemotherapy should have alternative ways as well [36].

6. Recent advances and technology

The innovations take technology and techniques to sustain them. The betterment lies in the up-gradation of the recent technology and creating innovations in the current timeframe [12]. The more progressive the medication gets, the more effective it will become but more supplements are required to overcome the side effects [9]. Till now, no treatment has been found which is free of side effects [5]. The role of *Spirulina* is always vital and valued, even in space technology. The food additives like *Chlorella vulgaris* and *Spirulina pacifica* [28] their biomass production is used in several areas of biotechnology because of the bioactive molecules and composition that are beneficial. The primary and secondary metabolic products have found the place where they fits perfectly [19].

The bioactive elements have to cope with environmental stress, which stands out as a challenge. Some of the variant features are known and discovered and tested, while many are anonymous to the field till now [19]. The food industry, repair mechanisms, defense, etc. fields are very much flourishing but alterations and modifications cannot harm them [14]. The best way to use microalgae as a drug delivery system is fine, but combining it with nanotechnology can do wonders [98]. Nanotechnology promises a better future because of its compact size and dimensions and its adaptability to various metals and compounds, making it industrially and biologically acceptable and compatible [96]. The nanotubes or nanostructure can be successfully engineered and incorporated with other treatments [99]. Already, nanotechnology is working with various diseases and treatments. This combination can lead to better innovations and the betterment of humanity without even hampering nature [98].

At present, Nanotechnology is the leading technology that has evolved with time and is considered to be the best one to cure malignancy at a nanoscale level [99]. The biopolymer synthesized from carbon, platinum, and gold colloids can be taken up to cure the disease [100]. The nanorods prove to be beneficial for small tumors [9] as well as various sizes and morphologies of nanoparticles can be directly incorporated into the treatment [98]. The biological synthesis of the nanoparticle is not a complex phenomenon because no reducing, as well as capping agents, are involved. Instead, the bioactive compounds present in microalgae themselves act as strong reducing and capping agents leading to the synthesis of nanoparticles [100]. Sometimes both reducing, as well as capping agents, could also be added as a reactant to form the

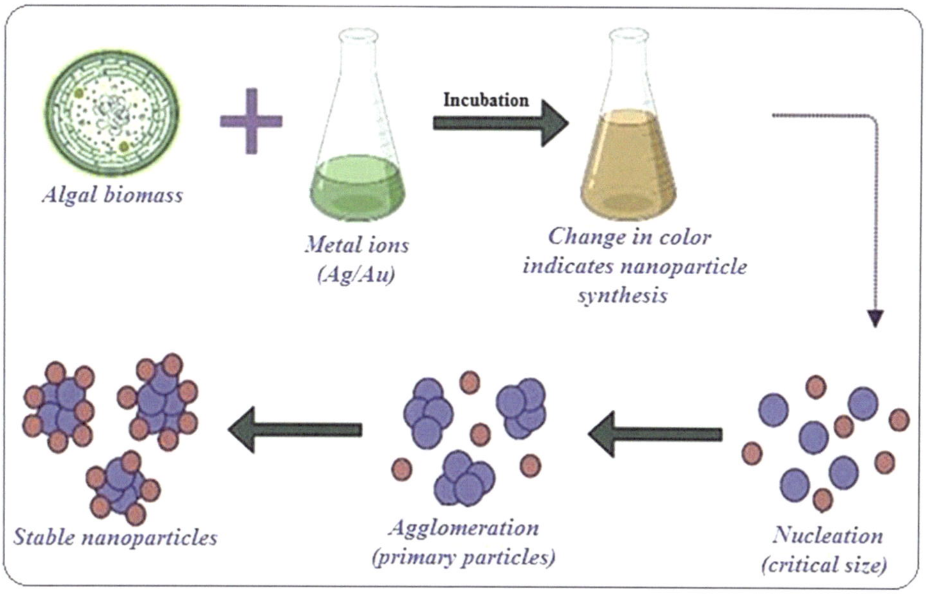

Figure 2.
Schematic representation of nanoparticle synthesis from algae.

desired and most efficacious products. The reducing agent can be sodium citrate, sodium borohydride, or ascorbic acid, and the capping agent could be polymers like polyvinyl chloride, thiols, or citrate [98, 97]. The passivating agent is to prevent aggregation and promotes nucleation. The therapeutic molecule can also be kept inside the molecule synthesis so that it could target the tumor [28]. The schematic representation of nanoparticle synthesis from microalgae extracts is presented in **Figure 2**.

Microalgae have various biotechnological applications. There is a need to have supplements and capsules that could be consumed to prevent cancer and are also proved to be human-friendly [9]. The algal extracts could directly be incorporated into them just like we intake probiotics [101]. Rapid advancement in pharmaceutical and diagnostic technology allow rapid blood tests could detect malignancy on an early stage [96]. A recent study shows promising results with this concept, it is proven to be effective against a few cancer cells lines [9]. There should be an advancement in technology to cultivate microalgae on a large scale without exploitation of nature, considering the ethical issue at top priority [9, 51]. People should be made aware of the benefits to humanity and the environment. Microalgae could be a gift and a curse to many issues be it environment, pharmaceutical, biotechnology, chemical industries, etc. [102]. This is a nature's gift that is meant to cherished and protected [73, 103].

7. Conclusion

Microalgae are rich in industrial and human consumption needs and requirements. The present study is an attempt to highlight the importance and potential of bioactive compounds isolated from microalgae as an anti-cancer agent. The aim of presenting a suitable treatment with unique features and lesser side effects using nanotechnology and other technologies was discussed concerning the species of microalgae. The incorporation of bioactive compounds can be useful when the proper dosage required by the patient is known. The concentration of bioactive compounds added to the final formulation as anti-cancerous compounds should be added carefully in order to get the desired results. The antioxidants found in the microalgae prove to be anti-inflammatory, anti-cancer, anti-diabetic, and skin protectors as well. The microalgae have evolved to such an extent that they can thrive in extreme conditions, but still, they are a major oxygen supplier to the environment. Extensive and deep investigations need to be carried out to explore more varieties of microalgae so that they can be taken down for more experiments with cancer cell lines. A technology that could cultivate microalgae in the lab needs to be developed so that ethical issues are controlled and its proper utilization could lead to significant benefits. The role of nanotechnology can be important as the smaller the scale, the bigger the possibility of treating the disease. The aim is to assess and analyze the root cause, which has to be inhibited and removed to ensure a better lifestyle. A significant view is that technology is required for the cultivation of microalgae in a short period and diagnostic tools to identify tumor growth in the body.

Acknowledgements

This work was supported by the Department of Biotechnology (Grant No: BT/PR 15650/AAQ/3/815/2016).

Conflict of interest

The authors declare no conflict of interest.

Author details

Abhishek Saxena, Aditi Raj and Archana Tiwari
Diatom Research Laboratory, Amity Institute of Biotechnology, Amity University, Noida, India

*Address all correspondence to: panarchana@gmail.com

References

[1] Fitzmaurice C, Dicker D, Pain A, Hamavid H, Moradi-Lakeh M, Macintyre MF, et al. The global burden of cancer 2013. JAMA Oncology. 2015;**1**(4):505-527

[2] Torpy JM, Lynm C, Glass RM. Cancer: The basics. JAMA. 2010;**304**(14):1628-1628

[3] Jain KK. Personalised medicine for cancer: From drug development into clinical practice. Expert Opinion on Pharmacotherapy. 2005;**6**(9):1463-1476

[4] Cragg GM, Newman DJ. Plants as a source of anti-cancer agents. Journal of Ethnopharmacology. 2005;**100**(1-2):72-79

[5] Bertram JS. The molecular biology of cancer. Molecular Aspects of Medicine. 2000;**21**(6):167-223

[6] Welch DR, Hurst DR. Defining the hallmarks of metastasis. Cancer Research. 2019;**79**(12):3011-3027

[7] Newman DJ, Cragg GM. Natural products as sources of new drugs over the 30 years from 1981 to 2010. Journal of Natural Products. 2012;**75**(3):311-335

[8] Singh J, Saxena RC. Chapter 2 - An Introduction to Microalgae: Diversity and Significance. Kim S.-K, editors. Handbook of Marine Microalgae. Academic Press. 2015. p. 11-24. ISBN: 9780128007761. https://doi.org/10.1016/B978-0-12-800776-1.00002-9

[9] Abd El-Hack ME, Abdelnour S, Alagawany M, Abdo M, Sakr MA, Khafaga AF, et al. Microalgae in modern cancer therapy: Current Knowledge. Biomedicine & Pharmacotherapy. 2019;**111**:42-50

[10] Raja R, Hemaiswarya S, Kumar NA, Sridhar S, Rengasamy R. A perspective on the biotechnological potential of microalgae. Critical Reviews in Microbiology. 2008;**34**(2):77-88

[11] Prabakaran G, Moovendhan M, Arumugam A, Matharasi A, Dineshkumar R, Sampathkumar P. Quantitative analysis of phytochemical profile in marine microalgae Chlorella vulgaris. International Journal of Pharm and Biological Science. 2018;**8**(2):562-565

[12] Tiwari A, Melchor-Martínez EM, Saxena A, Kapoor N, Singh KJ, Saldarriaga-HernándezS,etal.Therapeutic attributes and applied aspects of biological macromolecules (polypeptides, fucoxanthin, sterols, fatty acids, polysaccharides, and polyphenols) from diatoms—A review. International Journal of Biological Macromolecules. 2021;**171**:398-413

[13] Martínez Andrade KA, Lauritano C, Romano G, Ianora A. Marine microalgae with anti-cancer properties. Marine Drugs. 2018;**16**(5):165

[14] Das P, Aziz SS, Obbard JP. Two phase microalgae growth in the open system for enhanced lipid productivity. Renewable Energy. 2011;**36**(9):2524-2528

[15] Ibrahim K, Ramli R, Abdul Rashid AH, Mohd Yusof YA. Antimicrobial property of water and ethanol extract *Chlorella vulgaris*: A value-added advantage for a new wound dressing material. International Medical Journal. 2015;**22**(5): 399-401

[16] Cha KH, Koo SY, Lee DU. Antiproliferative effects of carotenoids extracted from Chlorella ellipsoidea and Chlorella vulgaris on human

colon cancer cells. Journal of Agricultural and Food Chemistry. 2008;**56**(22):10521-10526

[17] Saadaoui I, Rasheed R, Abdulrahman N, Bounnit T, Cherif M, Al Jabri H, et al. Algae-derived bioactive compounds with anti-lung cancer potential. Marine Drugs. 2020;**18**(4):197

[18] Somasekharan SP, El-Naggar A, Sorensen PH, Wang Y, Cheng H. An aqueous extract of marine microalgae exhibits antimetastatic activity through preferential killing of suspended cancer cells and anticolony forming activity. Evidence-Based Complementary and Alternative Medicine. 2016;**2016**:9730654

[19] Wang HM, Pan JL, Chen CY, Chiu CC, Yang MH, Chang HW, et al. Identification of anti-lung cancer extract from *Chlorella vulgaris* CC by antioxidant property using supercritical carbon dioxide extraction. Process Biochemistry. 2020;**45**(12):1865-1872

[20] Newman DJ, Cragg GM. Natural products from marine invertebrates and microbes as modulators of antitumor targets. Current Drug Targets. 2006;**7**(3):279-304

[21] Swapnil S, Bruno B, Udhaya R, Krishna K, Sandhya S, Waman P. Bioactive compounds derived from microalgae showing antimicrobial activities. Journal of Aquaculture Research and Development. 2014;**5**(3):1-4

[22] Brooks DR, McLennan DA, McLennan DA. Phylogeny, Ecology, and Behavior: A Research Program in Comparative Biology. Chicago: University of Chicago Press; 1991

[23] Landsberg JH. The effects of harmful algal blooms on aquatic organisms. Reviews in Fisheries Science. 2002;**10**(2):113-390

[24] Mahapatra GP, Raman S, Nayak S, Gouda S, Das G, Patra JK. Metagenomics approaches in discovery and development of new bioactive compounds from marine actinomycetes. Current Microbiology. 2020;**77**(4):645-656

[25] Mondal A, Bose S, Banerjee S, Patra JK, Malik J, Mandal SK, et al. Marine cyanobacteria and microalgae metabolites—A rich source of potential anticancer drugs. Marine Drugs. 2020;**18**(9):476

[26] Saunders CM. The care of the terminal stages of cancer. Annals of the Royal College of Surgeons of England. 1967;**41**(Suppl):162-169

[27] Jackson SE, Chester JD. Personalised cancer medicine. International Journal of Cancer. 2015;**137**(2):262-266

[28] Yeh AC, Ramaswamy S. Mechanisms of cancer cell dormancy—Another hallmark of cancer? Cancer Research. 2015;**75**(23):5014-5022

[29] Manlusoc J, Hsieh CL, Hsieh CY, Salac E, Lee YT, Tsai PW. Pharmacologic application potentials of sulfated polysaccharide from marine algae. Polymers. 2019;**11**(7):1163

[30] Ferdous UT, ZNB Y. Medicinal prospects of antioxidants from algal sources in cancer therapy. Frontiers in Pharmacology. 2021;**12**:593116. DOI: 10.3389/fphar.2021.593116

[31] Lehne G. P-glycoprotein as a drug target in the treatment of multidrug resistant cancer. Current Drug Targets. 2000;**1**(1):85-99

[32] Yokoyama C, Sueyoshi Y, Ema M, Mori Y, Takaishi K, Hisatomi H. Induction of oxidative stress by anticancer drugs in the presence and

absence of cells. Oncology Letters. 2017;**14**(5):6066-6070

[33] Moussavou G, Kwak DH, Obiang-Obonou BW, Maranguy CA, Dinzouna-Boutamba SD, Lee DH, et al. Anticancer effects of different seaweeds on human colon and breast cancers. Marine Drugs. 2014;**12**(9):4898-4911

[34] Paul S, Kundu R. Antiproliferative activity of methanolic extracts from two green algae, Enteromorpha intestinalis and Rizocloniumriparium on HeLa cells. DARU Journal of Pharmaceutical Sciences. 2013;**21**(1):1-12

[35] Yang TY, Chang GC, Hsu SL, Huang YR, Chiu LY, Sheu GT. Effect of folic acid and vitamin B12 on pemetrexed antifolate chemotherapy in nutrient lung cancer cells. BioMed Research International, 2013; 389046.

[36] Charmsaz S, Prencipe M, Kiely M, Pidgeon GP, Collins DM. Innovative technologies changing cancer treatment. Cancers. 2018;**10**(6):208

[37] Ropellato J, Carvalho MM, Ferreira LG, Noseda MD, Zuconelli CR, Gonçalves AG, et al. Sulfatedheterorhamnans from the green seaweed Gayralia oxysperma: Partial depolymerization, chemical structure and antitumor activity. Carbohydrate Polymers. 2015;**117**:476-485

[38] Pugazhendhi A, Prabhu R, Muruganantham K, Shanmuganathan R, Natarajan S. Anticancer, antimicrobial and photocatalytic activities of green synthesized magnesium oxide nanoparticles (MgONPs) using aqueous extract of Sargassum wightii. Journal of Photochemistry and Photobiology B: Biology. 2019;**190**:86-97

[39] Meropol NJ, Schrag D, Smith TJ, Mulvey TM, Langdon RM Jr, Blum D, et al. American Society of Clinical Oncology guidance statement: The cost of cancer care. Journal of Clinical Oncology. 2009;**27**(23):3868-3874

[40] Bakunina I, Chadova O, Malyarenko O, Ermakova S. The effect of fucoidan from the brown alga Fucus evanescence on the activity of α-N-acetylgalactosaminidase of human colon carcinoma cells. Marine Drugs. 2018;**16**(5):155

[41] Ouhtit A, Ismail MF, Othman A, Fernando A, Abdraboh ME, El-Kott AF, et al. Chemoprevention of rat mammary carcinogenesis by spirulina. The American Journal of Pathology. 2014;**184**(1):296-303

[42] Mei C, Zhou S, Zhu L, Ming J, Zeng F, Xu R. Antitumor effects of laminaria extract fucoxanthin on lung cancer. Marine Drugs. 2017;**15**(2):39

[43] Understanding Cancer Prognosis. National Cancer Institute. https://www.cancer.gov/about-cancer/diagnosis-staging/prognosis [Accessed January 9, 2018]

[44] Genin M, Clement F, Fattaccioli A, et al. M1 and M2 macrophages derived from THP-1 cells differentially modulate the response of cancer cells to etoposide. BMC Cancer. 2015;**15**:577

[45] Markman M, Luce R. Impact of the cost of cancer treatment: An internet-based survey. Journal of Oncology Practice. 2010;**6**(2):69-73

[46] Jose GM, Raghavankutty M, Kurup GM. Sulfated polysaccharides from Padina tetrastromatica induce apoptosis in HeLa cells through ROS triggered mitochondrial pathway. Process Biochemistry. 2018;**68**:197-204

[47] Kim P. Cost of cancer care: The patient perspective. Journal of Clinical Oncology. 2007;**25**(2):228-232

[48] Shao P, Liu J, Chen X, Fang Z, Sun P. Structural features and antitumor activity of a purified polysaccharide extracted from *Sargassum horneri*. International Journal of Biological Macromolecules. 2015;**73**:124-130

[49] Ghannam A, Murad H, Jazzara M, Odeh A, Allaf AW. Isolation, structural characterization, and antiproliferative activity of phycocolloids from the red seaweed Laurencia papillosa on MCF-7 human breast cancer cells. International Journal of Biological Macromolecules. 2018;**2018**(108):916-926

[50] Bernard PS, Wittwer CT. Real-time PCR technology for cancer diagnostics. Clinical Chemistry. 2002;**48**(8):1178-1185

[51] Ahlbrandt J, Lablans M, Glocker K, Stahl-Toyota S, Maier-Hein K, Maier-Hein L, et al. Modern information technology for cancer research: What's in IT for me? An overview of technologies and approaches. Oncology. 2020;**98**(6):363-369

[52] Talero E, García-Mauriño S, Ávila-Román J, Rodríguez-Luna A, Alcaide A, Motilva V. Bioactive compounds isolated from microalgae in chronic inflammation and cancer. Marine Drugs. 2015;**13**(10):6152-6209

[53] Oftedal L, Selheim F, Wahlsten M, Sivonen K, Døskeland SO, Herfindal L. Marine benthic cyanobacteria contain apoptosis-inducing activity synergizing with daunorubicin to kill leukemia cells, but not cardiomyocytes. Marine Drugs. 2010;**8**(10):2659-2672

[54] Hau AM, Greenwood JA, Löhr CV, Serrill JD, Proteau PJ, Ganley IG, et al. Coibamide A induces mTOR-independent autophagy and cell death in human glioblastoma cells. PLoS One. 2013;**8**(6):e65250

[55] Gilmore TD. Introduction to NF-κ B: Players, pathways, perspectives. Oncogene. 2006;**25**(51):6680-6684

[56] Carmichael WW. Cyanobacteria secondary metabolites—The cyanotoxins. Journal of Applied Bacteriology. 1992;**72**(P6):445-459

[57] Siddiq A, Dembitsky V. Acetylenic anticancer agents. Anti-Cancer Agents in Medicinal Chemistry (Formerly Current Medicinal Chemistry-Anti-Cancer Agents). 2008;**8**(2):132-170

[58] Hosokawa M, Wanezaki S, Miyauchi K, Kurihara H, Kohno H, Kawabata J, et al. Apoptosis-inducing effect of fucoxanthin on human leukemia cell line HL-60. Food Science and Technology Research. 1999;**5**(3):243-246

[59] Banker R, Carmeli S. Tenuecyclamides A– D, cyclic hexapeptides from the Cyanobacterium Nostoc s pongiaeforme var. tenue. Journal of Natural Products. 1998;**61**(10):1248-1251

[60] Davidson BS. New dimensions in natural products research: Cultured marine microorganisms. Current Opinion in Biotechnology. 1995;**6**(3):284-291

[61] Fabregas J, Herrero C. Vitamin content of four marine microalgae. Potential use as source of vitamins in nutrition. Journal of Industrial Microbiology and Biotechnology. 1990;**5**(4):259-263

[62] Palozza P, Torelli C, Boninsegna A, Simone R, Catalano A, Mele MC, et al. Growth-inhibitory effects of the astaxanthin-rich alga Haematococcuspluvialis in human colon cancer cells. Cancer Letters. 2009;**283**(1):108-117

[63] Gügi B, Costaouec T, Le Burel C, Lerouge P, Helbert W, Bardor M.

Diatom-specific oligosaccharide and polysaccharide structures help to unravel biosynthetic capabilities in diatoms. Marine Drugs. 2015;**13**:5993-6018

[64] Lee JY, Jung WH, Lim EH, Joo MK, Lee BJ, Park JJ, et al. Anticancer effects of astaxanthin and alpha tocopherol in esophageal cancer cell lines. The Korean Journal of Helicobacter and Upper Gastrointestinal Research. 2011;**11**(3):170-175

[65] Rickards RW, Rothschild JM, Willis AC, de Chazal NM, Kirk J, Kirk K, et al. Calothrixins A and B, novel pentacyclic metabolites from Calothrix cyanobacteria with potent activity against malaria parasites and human cancer cells. Tetrahedron. 1999;**55**(47):13513-13520

[66] Benemann J. Microalgae for biofuels and animal feeds. Energies. 2013;**6**(11):5869-5886

[67] Takahashi K, Hosokawa M, Kasajima H, Hatanaka K, Kudo K, Shimoyama N, et al. Anticancer effects of fucoxanthin and fucoxanthinol on colorectal cancer cell lines and colorectal cancer tissues. Oncology Letters. 2015;**10**(3):1463-1467

[68] Peng J, Yuan JP, Wu CF, Wang JH. Fucoxanthin, a marine carotenoid presents in brown seaweeds and diatoms: Metabolism and bioactivities relevant to human health. Marine Drugs. 2011;**9**(10):1806-1828

[69] Fleischauer AT, Simonsen N, Arab L. Antioxidant supplements and risk of breast cancer recurrence and breast cancer-related mortality among postmenopausal women. Nutrition and Cancer. 2003;**46**(1):15-22

[70] Umemura K, Yanase K, Suzuki M, Okutani K, Yamori T, Andoh T. Inhibition of DNA topoisomerases I and II, and growth inhibition of human cancer cell lines by a marine microalgal polysaccharide. Biochemical Pharmacology. 2003;**66**(3):481-487

[71] De Jesus Raposo MF, De Morais AMB, De Morais RMSC. Marine polysaccharides from algae with potential biomedical applications. Marine Drugs. 2015;**13**:2967-3028

[72] Kusaikin MI, Ermakova SP, Shevchenko NM, Isakov VV, Gorshkov AG, Vereshchagin AL, et al. Structural characteristics and antitumor activity of a new chrysolaminaran from the diatom alga *Synedra acus*. Chemistry of Natural Compounds. 2010;**46**:1-4

[73] Skjermo J, Størseth TR, Hansen K, Handå A, Øie G. Evaluation of β-(1 → 3, 1 → 6)-glucans and High-M alginate used as immunostimulatory dietary supplement during first feeding and weaning of Atlantic cod (*Gadus morhua* L.). Aquacult. 2006;**261**:1088-1101

[74] Damonte EB, Matulewicz MC, Cerezo AS. Sulfated seaweed polysaccharides as antiviral agents. Current Medicinal Chemistry. 2004;**11**(18):2399-2419

[75] Carte BK. Biomedical potential of marine natural products. Bioscience. 1996;**46**(4):271-286

[76] Hartmann R, Meisel H. Food-derived peptides with biological activity: From research to food applications. Current Opinion in Biotechnology. 2007;**18**(2):163-169

[77] Sheih IC, Fang TJ, Wu TK, Lin PH. Anticancer and antioxidant activities of the peptide fraction from algae protein waste. Journal of Agricultural and Food Chemistry. 2010;**58**(2):1202-1207

[78] Fennell BJ, Carolan S, Pettit GR, Bell A. Effects of the antimitotic natural product dolastatin 10, and related peptides, on the human malarial parasite Plasmodium falciparum. Journal of Antimicrobial Chemotherapy. 2003;**51**(4):833-841

[79] Kobayashi M, Natsume T, Tamaoki S, Watanabe JI, Asano H, Mikami T, et al. Antitumor activity of TZT-1027, a Novel Doiastatin 10 Derivative. Japanese Journal of Cancer Research. 1997;**88**(3):316-327

[80] Bajpai VK, Shukla S, Kang SM, Hwang SK, Song X, Huh YS, et al. Developments of cyanobacteria for nano-marine drugs: Relevance of nanoformulations in cancer therapies. Marine Drugs. 2018;**16**(6):179

[81] Ghosh A, Khanra S, Mondal M, Halder G, Tiwari ON, Saini S, et al. Progress toward isolation of strains and genetically engineered strains of microalgae for production of biofuel and other value-added chemicals: A review. Energy Conversion and Management. 2016;**113**:104-118

[82] Radakovits R, Jinkerson RE, Darzins A, Posewitz MC. Genetic engineering of algae for enhanced biofuel production. Eukaryotic Cell. 2010;**9**(4):486-501

[83] Zeng X, Danquah MK, Chen XD, Lu Y. Microalgae bioengineering: From CO_2 fixation to biofuel production. Renewable and Sustainable Energy Reviews. 2011;**15**(6):3252-3260

[84] Shabbir MAB, Hao H, Shabbir MZ, Hussain HI, Iqbal Z, Ahmed S, et al. Survival and evolution of CRiSPR–Cas system in prokaryotes and its applications. Frontiers in Immunology. 2016;7:375

[85] Qiao Y, Yang F, Xie T, Du Z, Zhong Z, Qi Y, et al. Engineered algae: A novel oxygen-generating system for effective treatment of hypoxic cancer. Science Advances. 2020;**6**(21):5996. DOI: 10.1126/sciadv.aba5996

[86] Song BH, Lee DH, Kim BC, Ku SH, Park EJ, Kwon IH, et al. Photodynamic therapy using chlorophyll-a in the treatment of acne vulgaris: A randomized, single-blind, split-face study. Journal of the American Academy of Dermatology. 2014;**71**(4):764-771

[87] Zaid AAA, Hammad DM, Sharaf EM. Antioxidant and anticancer activity of Spirulina platensis water extracts. International Journal of Pharmacology. 2015;**11**:846-851. DOI: 10.3923/ijp.2015.846.851

[88] Cassidy J, Clarke S, Díaz-Rubio E, Scheithauer W, Figer A, Wong R, et al. Randomized phase III study of capecitabine plus oxaliplatin compared with fluorouracil/folinic acid plus oxaliplatin as first-line therapy for metastatic colorectal cancer. Journal of Clinical Oncology: Official Journal of the American Society of Clinical Oncology. 2008;**26**(12):2006-2012

[89] Gwiazda KS, Grier AE, Sahni J, Burleigh SM, Martin U, Yang JG, et al. High efficiency CRISPR/Cas9-mediated gene editing in primary human T-cells using mutant adenoviral E4orf6/E1b55k "helper" proteins. Molecular Therapy. 2016;**24**(9):1570-1580

[90] Beusterien K, Grinspan J, Kuchuk I, Mazzarello S, Dent S, Gertler S, et al. Use of conjoint analysis to assess breast cancer patient preferences for chemotherapy side effects. The Oncologist. 2014;**19**(2):127-134

[91] Dhanapal R, Saraswathi T, Govind RN. Cancer cachexia. Journal of Oral and Maxillofacial Pathology : JOMFP. 2011;**15**(3):257-260

[92] Harris IS, DeNicola GM. The complex interplay between antioxidants and ROS in cancer. Trends in Cell Biology. 2020;**30**(6):440-451

[93] Aggarwal V, Tuli HS, Varol A, Thakral F, Yerer MB, Sak K, et al. Role of reactive oxygen species in cancer progression: Molecular mechanisms and recent advancements. Biomolecules. 2019;**9**(11):735

[94] Morgan MJ, Liu ZG. Crosstalk of reactive oxygen species and NF-κBsignaling. Cell Research. 2011;**21**(1):103-115

[95] Harraan D. Aging: A theory based on free radical and radiation chemistry. Journal of Gerontology. 1956;**11**(3):298-300

[96] Reilly RM. Carbon nanotubes: Potential benefits and risks of nanotechnology in nuclear medicine. Journal of Nuclear Medicine. 2007;**48**(7):1039-1042

[97] Yu Z, Gao L, Chen K, et al. Nanoparticles: A new approach to upgrade cancer diagnosis and treatment. Nanoscale Research Letters. 2021;**16**:88

[98] Jin C, Wang K, Oppong-Gyebi A, Hu J. Application of nanotechnology in cancer diagnosis and therapy: A mini-review. International Journal of Medical Sciences. 2020;**17**(18):2964-2973

[99] Bhattacharya J, Shang B, Su CK, Goldman DP. Technological advances in cancer and future spending by the elderly. Health Affairs (Project Hope). 2005;**24**(2):W5R53-W5R66

[100] Awasthi R, Roseblade A, Hansbro PM, Rathbone MJ, Dua K, Bebawy M. Nanoparticles in cancer treatment: Opportunities and obstacles. Current Drug Targets. 2018;**19**(14):1696-1709

[101] Nagarajan R. Nanoparticles: Building blocks for nanotechnology. 2008; 10.1021/bk-2008-0996.ch001

[102] Rizwan M, Mujtaba G, Memon SA, Lee K, Rashid N. Exploring the potential of microalgae for new biotechnology applications and beyond: A review. Renewable and Sustainable Energy Reviews. 2018;**92**:394-404

[103] Pulz O, Gross W. Valuable products from biotechnology of microalgae. Applied Microbiology and Biotechnology. 2004;**65**(6):635-648

Chapter 17

Algal Biorefinery: A Synergetic Sustainable Solution to Wastewater Treatment and Biofuel Production

Gulshan Kumar Sharma, Shakeel Ahmad Khan, Amit Kumar, Ittyamkandath Rashmi, Fayaz Ahmad Malla and Gopal Lal Meena

Abstract

In the recent years, due to heavy surge in the price of petrochemical products, researchers are getting interest towards renewable bioenergy resources such as algal-based biomass. In order to meet a world energy demand, current bioeconomy challenges and to produce valuable products, intensive and integrated research on algal biorefinery is highly required. Even though several research carried out study for the conversion of algae biomass to biofuel, but none of these proved economically viable. Hence, range of value added product (biodiesel, biochar, fertilizer, etc.) must be produced subsequently from algae. The utilization of microalgae for biomass production is better than agricultural crops as microalgae do not required fresh water for its growth, it can readily grow on wastewater throughout the year. Generation of wastewater is severe concern throughout the world and discharge of wastewater without proper treatment in to water bodies causes water pollution. Microalgae bear vast potential in significantly deescalating pollutant load (nitrate, TDS, ammonium, phosphate, organic load) from wastewater. The harvested algal biomass after remediation has significance role in producing biofuels and by-products in a sustainable way. In this chapter, emphasis would be given on role of algae in wastewater treatment and its biorefinary approach for sustainable energy development.

Keywords: microalgae, wastewater, phycoemediation, biorefinery

1. Introduction

Globally, the severe problem humanity is facing today is the availability of fresh water, wastewater generation and energy supply. As the continuous use of fossil fuel are depleting day by day the natural stock of fossil fuels. This natural energy reserve may end in next 45–50 years. This depletion is posing stress to continue the various anthropogenic activities i.e. industry, agriculture, production of precious chemicals for food, and pharmaceutical which directly affects the global economics due to shortage of energy sources. In these circumstances, this is the high time to identify and develop alternative, cost-effective, efficient renewable energy resources for enhancing the sustainability of anthropogenic activities. Algal biomass could be utilized to generate extensively energy support as algae bear high productivity of biomass [1]. The more concentration for diversification of agro-ecosystem from food to fuel is also fulfilled by the algal biofuel as these living being are also not required agricultural land due to their aquatic nature. As per the report of the Central Pollution Control Board of India [2], 71853 MLD [million liters per day] wastewater (considering both sewage and industrial discharge) is discharged into the water bodies of India and out of which only 37% get treated. In these days, pollution of natural resources especially water is also at alarming point and the climate change making it more serious. Thus, minimum contamination/waste of water and reuse of the contaminated/waste/used water is also the highly required. The presently available technologies of wastewater treatment are not only costly but also generate huge amount of sludge. The generated sludge after wastewater treatment essentially need to be treated and disposed, these two requirements further increases the financial effectiveness of the any technology [3, 4]. The algae mediated wastewater treatment is an environmentally sustainable and efficient approach and can be integrated with secondary wastewater treatment process. Algae are the small, mostly aquatic, photosynthetic (converts sunlight into the oil form stored energy) organisms, currently, getting more attention due to their capabilities to address the different environmental issues including energy. Microalgae have been noted for their enormous potential to remediate waste water i.e. Phyco-remediation. Phycoremediation is the utilization of alga culture to remove/biotransformation of pollutants, nutrients, xenobiotic from waste water. Phycoremediation can handle more than one environmental problem such as pH correction, BOD, COD and TDS removal simultaneously over the chemical methods. Phycoremediation consider highly eco-friendly as did not cause secondary pollution. Presently, biodiesel production utilizing microalga is not economically sound due to its cost. Thus, algal biorefinery concept can serve an important option to minimize the microalgal biofuel cost. Algal biorefinery is the analogous concept to present petroleum refinery as petroleum refinery produces multiples products including fuels from petroleum. Algal biorefinery is having potential to increase the values of the products obtained from the biomass feed stocks. Algal biorefinery can be integrated among biomass conversion process, fuels (low value, but high volume), intermediate compounds (low volume, but high value) and value added chemicals along with electricity generation through advanced technologies such as combined heat and power (CHP) technology. Microalgae are having high capacity to convert the solar energy to chemical energy per unit land than terrestrial phototrophs due to their high productive rate. Thus, microalgae can address the increasing energy demands as well as growing environmental issues such as climate change. Beside this, microalgae having some advantages as feedstock for value added product generation. The microalgae are capable of synthesize huge quantity of lipids (20–50% dry cell weight). The growth of algae is very fast compared to terrestrial plants (double the biomass within 20–25 days), so can be used for bioremediation [4]. Algae do not require arable land and fresh water for the growth. Algal biomass can also contribute significantly to reduce the enhanced atmospheric carbon. Keeping this view, the present chapter is focused towards the utilization of algae in wastewater treatment, biofuel, biofertilizer production, CO_2 sequestration, bioremediation and challenges with future perspective through algal biorefinery interventions.

2. Wastewater characteristics and their treatment

Due to industrialization, population expansion and modern life style the wastewater generation is increasing day by day. The discharged of wastewater without proper treatment in to waterbodies are continuously overloading the fresh water bodies and minimizing their self-cleaning capacity. This overloading of waste into the freshwater bodies, disturbing the nutrient recycling process along with the disturbance in biogeochemical cycles i.e. nitrogen, carbon and water cycles through physical (evaporation, precipitation etc.), ecological (eutrophication, bio-magnification etc.) and biological process (photosynthesis, nitrogen fixation, respiration etc.). Availability of fresh water/irrigation water for fulfillment of daily requirement of the human society is also decreasing in the present changing environment [5]. To meet out the current demand of water for various anthropogenic activities, this is very necessary to increase the water reuse potential. This can be achieved through proper treatment of the wastewater generated during different activities. Wastewater is generally composed of water and wastes originate from commercial, industrial, home and institution activities. Wastewater at the point of origin contains high organic load, colorants, pesticides, heavy metals, hydrocarbons, numerous pathogens, toxic compounds and nutrients. The minimum treatment of this generated wastewater is required and recommended before disposing into the environment. The mechanistic understanding of the environmental effects, influencing factors, controls and effective utilization of the treatment process is essential to design the treatment process and its operation. There are three methods which are used to treat wastewater. These methods are physical, chemical and biological methods. The treatment of waste water is general divided into three systems of treatment i.e. primary, secondary and tertiary based on the capacity to remove the different contaminants. The general wastewater treatment process details are depicted in **Figure 1**.

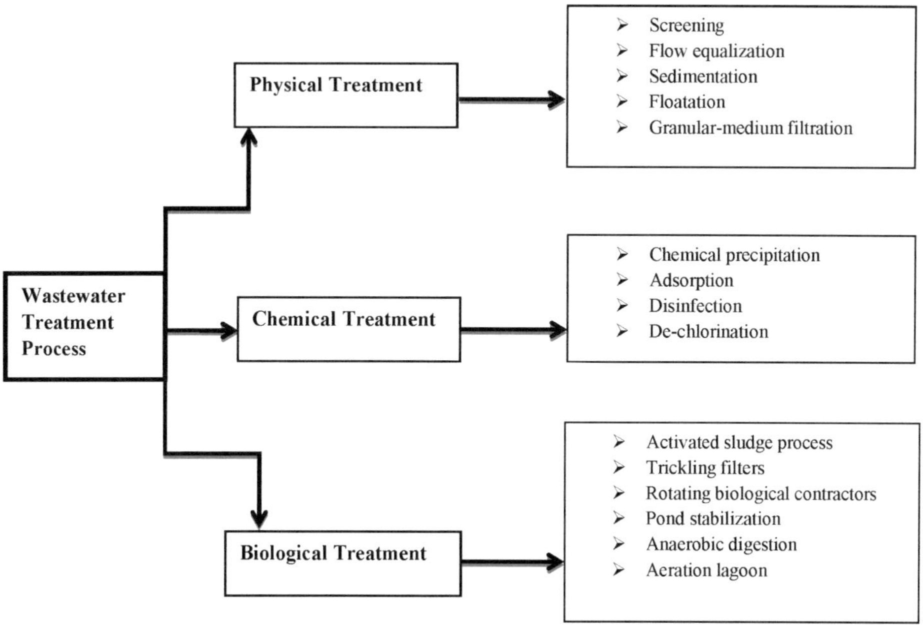

Figure 1.
General wastewater treatment process.

3. Algae mediated wastewater treatment

Algae are autotrophic organism however; there are some other algae, which are heterotrophy or mixotrophy in nature. The dominant mode of microalgae metabolism in wastewater is heterotrophic (approximately 50%) in nature. The heterotrophic microalgae metabolize organic components in wastes, and convert them into organic biomass along with inorganic components. Microalgae contribute approximately 50% of global primary production (GPP) i.e. most efficient convertor of solar energy to chemical energy and act as producer of aquatic food chain [6]. The quantity and quality of bioactive compounds of microalgae is based on the ambient environmental, ecological factors and taxonomic position. The removal of this generated algal biomass results in the purification of wastewater as their removal decreases the biological oxygen demand especially in case of their death to minimizing the chance of back release of nutrients in the ecosystem.

In 1960s Oswald and Gotta, [7] reported the potential role of microalgae for the removal of pollution load from the tertiary wastewater treatment by algae. Phycoremediation is the removal/biotransformation of pollutants such as nutrients, xenobiotics from wastewater and CO_2 from air. Thus, phycoremediation can be used for to extract nutrient from municipal wastewater/effluents which are rich in organic matter; to complete removal/transformation and degradation of xenobiotic compounds utilizing as biosorbent; to treatment of acidic wastewaters; to sequestrate CO_2; and to detect toxic compounds using algae-based biosensors. There are various studies which recommends the removal of nitrogen and phosphorous from wastewater to protect the waterbodies from eutrophication [5, 8–12]. The controlled growth of algae in wastewater leads to reduction of contamination load on natural resources and can also be enhance reuse efficiency. The utilization of algae in the treatment of different waste such as agro-based industrial wastes, sewage, industrial wastes (metal finishing, paper, and textile) and even landfill leachate [10, 12–15]. Waste mitigation potential of an algal species entirely depends on the algal productivity, nutrient and pollutant removal efficiency, and cost of biomass harvest [16–18]. In addition to removal of pollutant load from wastewater, algae make available oxygen (O_2) to bacteria (heterotrophic aerobic) for mineralization of pollutants and CO_2 produces by bacterial catabolism is subsequently consumed by the photosynthetic activity of algae (**Figure 2**). The photosynthetic process of algae is reduces the pollutant volatilization through mechanical aeration and contribute to reduce the cost of operation directly. Thus, the dual purpose utilization of microalgae in biorefinery approaches is

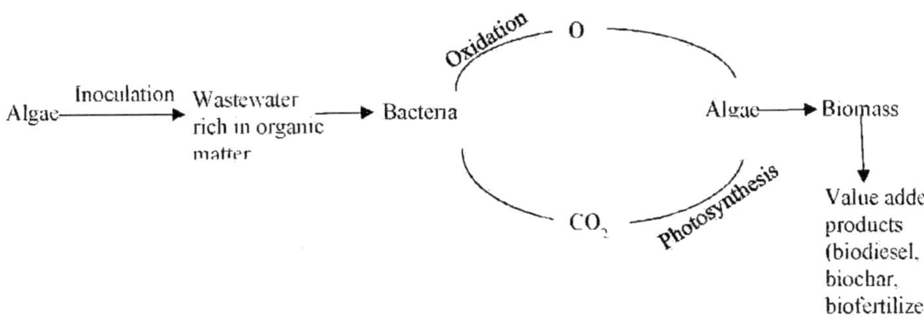

Figure 2.
Principle of photosynthetic oxygenation in BOD removal.

S. No	Microalgae	Waste	Culture system	Biomass productivity	Reduction of pollutants	References
1.	*Chlorella minutissima*	Primary treated sewage wastewater	Race way ponds	0.44 ± 0.04 g/L	Reduction of TDS, P, NH_4^+, NO_3^-, BOD and COD by 94.3%, 67.4%, 48.2%, 88.8%, 93.2% and 80.5%, respectively	[13]
2.	*Chlorella pyrenoidosa*	Sewage treatment plant wastewater and synthetic wastewater	Bioreactor	—	SWW*: Reduction of Nitrate (99%), phosphate (77%), and COD (61%) from SWW; STPW: Reduction of nitrate (99%), phosphate (94%), and COD (87%)	[23]
3.	*Chlorophyta, cyanobacteria, Acinetobacter, and Pseudomonas*	Polyacrylamide (PAM)-containing wastewater	RAB reactors	—	PAM**, COD, TOC, and TN removed by 64.1 ± 2.0, 58 ± 1.5, 34.5 ± 1.5, and 85 ± 6.0%, respectively.	[9]
4.	*C. minutissima, Scenedesmus spp N. muscorum* and *Consortium*	Sewage wastewater	20 L capacity of plastic bottles	0.4 g/L dry biomass of *Chlorella*	*Chlorella* reduces NH_4^+-N (92%), NO_3^-N (87%), PO_4^{3-} -P (85%), and reduces TDS (96%), BOD (90%) and COD (81%). *Scenedesmus spp* removed 72% TDS and 92% NH_4^+-N. Out of selected *C. minutissima* performed better	[14]
5.	*Scenedesmus obliquus*	Wastewater treatment plant	25 L glass containers	0.88 ± 0.04 g/L	Removal efficiencies of 71.2 ± 3.5% COD, 81.9 ± 3.8% NH_4^+, ~100.0% NO_3^-, and 94.1 ± 4.7% PO_4^{3-}.	[24]
6.	*S. obliquus*	Paddy-soaked wastewater	Polybags (PB), photobioreactors (PBR) and race way ponds (RWP)	340 ± 2 mg/L/d	Reduction of Ammonical-N (96%), phosphates (97%)	[25]
7.	*Scenedesmus* sp. ISTGA1	Municipal sewage wastewater	Bioreactor	1.81 g/L	Reduction in BOD and COD by 86.74% and 88.82% respectively	[26]
8.	*C. pyrenoidosa, Scenedesmus abundans and Anabaena ambigua*	primary treated sewage wastewater	Bioreactor	—	52–88% reduction in the nutrient concentration	[27]

S. No	Microalgae	Waste	Culture system	Biomass productivity	Reduction of pollutants	References
9.	*C. minutissima*	Primary treated sewage wastewater and tertiary treated Common Effluent Treatment Plant (CETP)	5 L capacity laboratory grade plastic tray	Sewage wastewater 0.75 g/L while it was 0.43 g/L in CETP wastewater	Reduction from wastewater within 12 Days, TDS (90–98%), N (70–80%), P (60–70%), K (45–50%)	[10]
10.	*Isochrysis*	Modified f/2 medium with Palm Oil Mill Effluent (POME) and inorganic fertilizer	Photobioreactor (PBR) (1 L) versus outdoor cultures (glass aquarium, 9 L)	Biomass- PBR, 69 mg m − 2 day−1, Outdoor, 92 mg m − 2 day −1	NO_3^-, PBR – 38.1%, Outdoor- 46.4%, PO_4^{3-}, PBR – 86.6%, Outdoor- 83.3%	[28]
11.	*Chlorella minutissima, Scendesmus, Nostoc muscorum*	Sewage wastewater	Poly house in plastic trays	0.79 ± 0.02 and 0.78 ± 0.01 g/Lin *Scenedesmus* and *C. minutissima*, respectively	≥ 90% reduction in TDS, BOD and Ammonium-N.	[8]

**SWW-Sewage wastewater; **PAM-polyacrylamide.*

Table 1.
Selective examples of the microalgae for wastewater treatment along with system utilized biomass productivity and targeted pollutants.

providing high sustainable solution to the long standing environmental concerns than any other equivalent approaches. The important products of the algal biorefinery are biomanure, biodiesel, ethanol, pharmaceutical, fish feed, biohydrogen and several other valuable products [19–22]. The role of various microalgae in the remediation of wastewater is given in **Table 1**.

4. Algal bio-refinery based production

Algal biorefinery approach aims to promote harvesting of several value-added products from the algae feedstock, towards economic and environmental effectivity of algal based technology. There are plenty of the research efforts has been made to harness the algal biomass to produce biofuel, biofertilizer, biodiesel, pharmaceutical products as well as for wastewater remediation. The industrial production of the biofuel through algal biomass utilizing different photobioreactors could be possible to burden off the current energy demand. The basic process of biodiesel production through bio-refinery is explained in **Figure 3**. Additionally, the nutrient qualities such as carbohydrates, proteins, nitrogen, phosphorous, potassium and other nutrients of the algal biomass can also be utilized as nutrient feed for animal as well as for fish etc. The algal dry biomass composition contains up to 46% Carbon (C), 10% Nitrogen (N) and 1% Phosphates (P) and 1 kilogram (kg) of dry algal biomass utilizes up to 1.7 kg carbon dioxide (CO2) (Hu et al., 2008). The algal biomass after remediation of wastewater having vast potential as biofertilizer. The N,P and K content in microalgae algae biomass varies from 7 to 9%, 1–2% and 0.1–1%, respectively [5, 13]. The algal biomass can also be used as biofertilizer in agro-ecosystems. The algal biomass not only provides essential nutrients to the agricultural crops but also significantly contribute to improve the soil carbon and soil fertility [5, 13]. Sharma et al. [5] in a study conducted on the impact of algae biomass as manure on the nitrate leaching reported that microalgal manure are slow releasing in nature and less leaching of nitrate as compared to chemical fertilizer was observed, so application of microalgae biomass

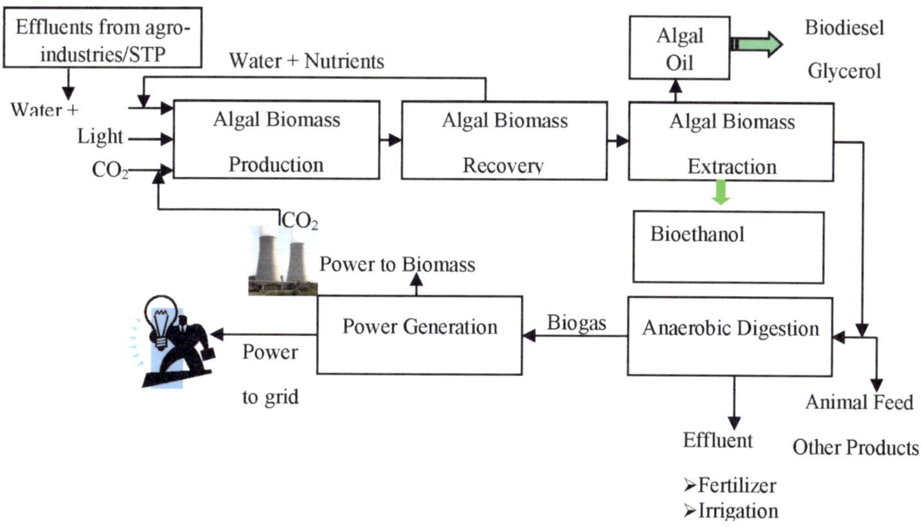

Figure 3.
A conceptual process (Biorefinery based) for producing microalgae biofuels for better economy.

also results in reduction of nitrate leaching from agricultural fields as compared to chemical fertilizer, hence prevent eutrophication of the water bodies. Thus based on the available literature and recent research it can be concluded and put forward that algal biorefinery is one of the most promising cutting-edge economic alternative of existing traditional technologies to cater the environment through direct reduction of the primary and/or secondary pollutants as well as sustainable solution for essential required developmental process.

5. Microalgae as potential source of biofuels

Microalgae has a potential to deliver renewable energy resources such as biofuels. Due to the problem of global warming (burning of fossil fuels) and day-to-day surge in petroleum, prices the role of microalgae has been rethink by various domains for using as a source of clean energy. The reason behind microalgal biomass as suitable feedstock for biofuel as the algae has high biomass productivity, high lipid content and high photosynthetic efficiency than terrestrial plant. Algae is considered as third generation biofuel and have advantage over first and second generation biofuel in terms of readily available, ability to grow throughout the year, water consumption is very less, can grow on wastewater, ability to grow under harsh condition, and high biomass production. The oil content of algae compared to first and second generation biofuel is given in **Table 2**. In this section we will discussed the important product obtained from algae as biodiesel, biofertilizer and biochar.

5.1 Biodiesel

The recent research developments in microalgae reveals that microalgal biomass is one of the promising sources of biodiesel, which partially could met the demand of transportation sector. Using microalgae to produce biodiesel will not compromise production of food, fodder and other products derived from crops. The microalgae

Source	Oil (Liter/ hectare)
Algae	1,00,000
Oil Palm	1413
Coconut	2684
Jatropha	741
Rapeseed/Canola	1187
Peanut	1057
Sunflower	954
Safflower	776
Soybeans	636
Hemp	364
Corn	172

(Demirbas and Demirbas, [29]; Ahmad et al. [30]).

Table 2.
Comparison of algae with different crops for biofuel.

species such as *Kirchneriella lunaris, Ankistrodesmus fusiformis, Chlamydocapsa bacillus*, and *Ankistrodesmus falcatus* are prominent species for biodiesel production as they contain high polyunsaturated FAME [31]. The comparison of various oil yielding crops is given in **Table 2**. Thus, considering the potential of the algal based biodiesel production, it can be concluded that the biodiesel can be used to displace fossil diesel partially/completely. The oil percentage in various algae are in the range of 20–50% (**Table 3**) and increase in oil content can be achieved >80% by weight of dry biomass in microalgae. The oil productivity of microalgae is the mass of oil produced per unit volume of the microalgal broth per day, depends on the algal growth rate and the oil content of the biomass.

5.2 Bio-fertilizer

The microalgae can assimilate excess N&P from the wastewater and convert it into the valuable biomass which has potential as a manure for the agricultural crops. Various researches have reported that %N content in the dry algae biomass is significantly higher than the available organic manure (cow dung, farmyard manure etc.) [10, 13, 36]. The NPK content of dry algae biomass ranged from 3 to 7%, 0.5–2% and 0.4–0.8%, respectively [13, 14, 36, 37]. The algal based fertilizers are composed of high OC which support to increase the moisture retention capacity and nutrient bioavailability than chemical fertilizers and other organic inputs such as farm yard manure [38]. Algal bio-fertilizer being rich in carbohydrates, soluble protein contents and other important plant organic nutrients, ensure higher vegetative yield [39, 40]. The algal-biofertilizer input also enhance the microflora of the soils along with the availability of inorganic nutrients [13]. Renuka et al. [41] confirms that the microalgae-based biofertilizer decreases the nutrient losses as nutrients are slowly release into the soil and available to the crop in longer periods than the synthetic fertilizers. In a leaching experiment conducted by Sharma et al. [5] the application algae biomass (*C. minutiisma*) after harvesting from sewage wastewater results in reduction of nitrate leaching in spinach crops as compared to application of chemical fertilizer, hence prevent eutrophication in water bodies. The immobilization and mineralization of any fertilizer depends on its C:N ratio. If the C:N ratio of any fertilizer is more than 20, it

Microalgae	Oil content (%)	References
Botryococcus terribilis	49	[32]
Chlorella vulgaris	41–58	[33]
Chlorella emersonni	23–63	[33]
Chlamydomonas sp.	22.7	[33]
Desmodesmus sp.	6.5–9.1	[34]
Chlorella sp.	28–53	[34]
Scenedesmus sp.	17–24	[33]
Scenedesmus obliqus	30–50	[34]
Nannochloris sp.	31–68	[35]
Chlorella salina	11	[32]

Table 3.
Oil content of microalgae.

promotes immobilization and therefore not advisable for application in soil. The C:N ratio of phycoremediated algae manure is around 9.16, hence its application promotes mineralization in the soil [13]. In addition, algae fertilizer also reported to reduce nitrate leaching from the agricultural fields than synthetic fertilizer [5, 21]. Therefore, it can be summarized that phycoremediation of sewage wastewater with biofertilizer production is a resource conservation approach and recycling of wastewater as well as nutrient for improvement in crop quality.

5.3 Biochar

Algae biomass is potential feedstock for various value added products. Since last decades, interest has been raised in production of biochar from microalgae biomass. As biochar is rich in organic carbon, so its application enhances carbon sequestration and improving the soil quality [42–44]. Generally, carbon, hydrogen, nitrogen and sulfur content in biochar is 48.45, 1.78, 1.47, and 0.78 (wt%) and it varies with the feedstock [45]. The microalgae derived biochar (*Chlorella vulgaris* FSP-E) is slightly alkaline in nature having carbon, hydrogen, nitrogen, oxygen and sulfur content (% dry wt) is 61.32,3.55, 9.76, 11.92, and 0.02% [46]. Similarly, Chaiwong et al. [47] reported volatile matter 16.8%, carbon 62.4%, and nitrogen 2.1% in spirogyra microalgae derived biochar. Generally, compared to lignocellulosic biochar, algae derived biochar have low organic carbon content and CEC, but having high nitrogen, P, K, Ca and Mg content [48]. Due to its high nutrient content and ion exchange capacity, algae biochar can be utilized for agricultural inputs and adsorbents in wastewater remediation [42]. Being an alkaline in nature, algae biochar could be used as amendment in acidic soil. Due to high biosorption capacity of associated with the large amount of functional group, microalgae biochar results enhancing the efficiency for the removal of organic contaminants [49]. Producing algae biochar also results in sequestration of atmospheric carbon dioxide, hence prevent global warming. Biochar is the carbon-enriched (coke) obtained after pyrolysis under temperatures (600–700°C) and under anaerobic conditions. The produce yield from pyrolysis is related to parameters, such as temperature, heating rate, and residence time [50]. The yield of biochar increased with decrease in pyrolysis temperature, and with increase in the duration. Chen et al. [45] showed that the yield of biochar algae in terrified microalgae residue at the temperature ranged from 200 to 300°C with a residence time of 15–60 min. Similarly, the yield of 50.8–95.7% in microalgae *Chlamydomonas* sp. JSC4 under the temperature of 200–300°C for 15–60 min [51]. Hence, it can be concluded that, production of algal biochar is expected to contribute to a further sustainable environment in the future.

5.4 Carbon dioxide sequestration

Global climate is a challenging issue, and reason behind is increasing concentration of greenhouse gases in atmosphere. Currently, CO_2 concentration in the atmospheres is around 400 ppm and it may reach to 750 ppm by the end of century [52]. CO_2 is well known greenhouse gases contributing climate change and global warming. The industrialization, and population expansion is the main cause of greenhouse gases emission. Several technologies has develop for capturing CO_2, although biological capture of CO_2 is a potential and attraction alternative. The algae mediated CO_2 fixation coupled with wastewater treatment is gaining attention as compared to terrestrial plants [53]. The microalgae that are effective in CO_2 sequestration generally belongs to *Chlorococcum*, *Chlorella*, *Scenedesmus* and *Euglena* genus. The carbon dioxide sequestration potential of microalgae is around 10–50 times higher than terrestrial plants [54]. The nutrients content in wastewater (N & P) can be utilized by microalgae for source of food and resulting biomass could be utilized for biofuel, biofertilzer, biochar and value added products. Microalgae can be grown in photobioreactor by carbon dioxide from the point sources such as industry, cement kiln, thermal power plant etc. Tang et al. [55] conducted a study on the impact of CO_2 concentration on biomass productivity of algae *Chlorella pyrenoidosa* in a photobioreactor and found that at 10% CO2 concentration, biomass production was highest (1.8 g/L). However the process is cumbersome and faced problem in down streaming process (harvesting). Open pond system and closed PBR are generally suggested for the purpose of growing algae. Open pond system/raceway ponds are cost effective, but significant amount of CO_2 loss to the atmosphere as compared to closed PBR. The CO_2 sequestration with remediation of wastewater thorough algae is cost efficient, sustainable, and recycling approach.

6. Microalgal biomass production

6.1 Open ponds

Cultivation of algae in open pond is oldest and simple practice in which algae are cultivated in similar condition as external environment. This type of system was first introduced in 1950s [56, 57]. Open pond consists of close loop system for circulation which is around 0.3 m depth with a paddlewheel constructed as clay, or plastic-lined ponds. Paddlewheel is used for circulation of water and for proper aeration. Open pond system is still used for large scale production of algae in outdoor condition. With time various designs has emerged for open-pond systems, but three designs (race-way ponds, circular ponds, and unstirred ponds) succeeded for mass multiplication of algae (**Figure 4**).

6.2 Closed photobiorector

As name suggest, growing of algae in closed system. Closed photobioreactor can produce higher biomass than closed system, but it is not cost effective. The most common cultivation technology in closed system is the photobioreactor (PBR) (**Figure 5**). Typically, closed reactors include tubular and flat bioreactors. The system consists of glass or plastic, although glass PBR is frequently used for large scale production [58]. In closed system, glass tubes are arranged normally in vertical, helical of in horizontal

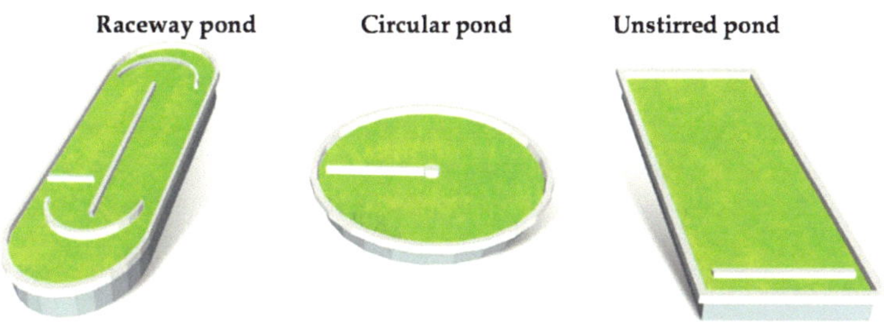

Figure 4.
Open pond system for algae biomass production.

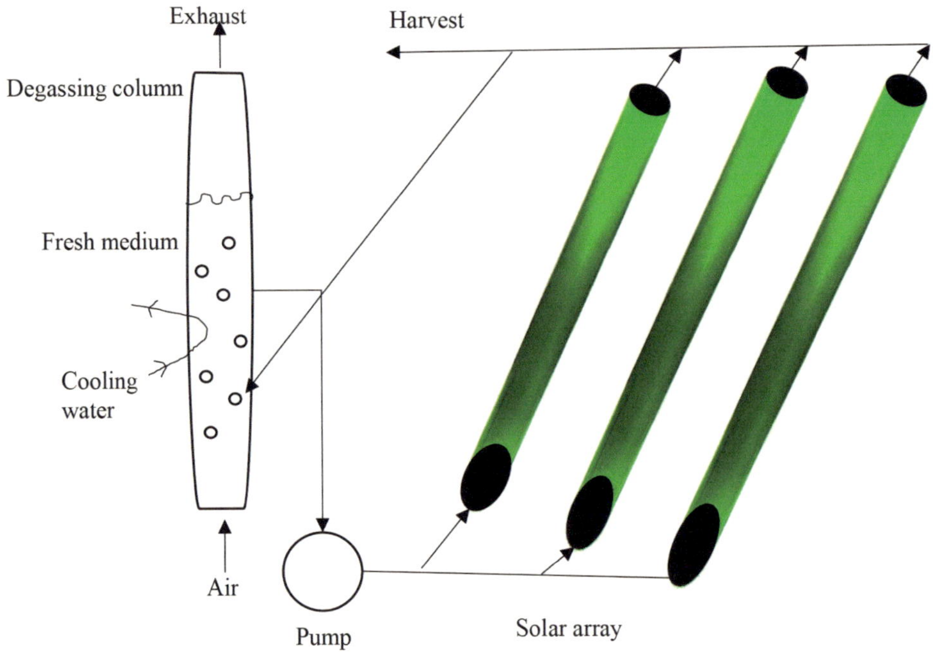

Figure 5.
Tubular photobioreactor with parallel tubes.

Method of cultivation	Advantages	Disadvantages
Open system	Cost effective, easy to maintain	Biomass production is low, low light use efficiency, high risk of contamination of other microorganism, not suitable for all sensitive microalgae species
Closed system	High biomass yield, high sunlight use efficiency, less space is required, can be suitable, highly suitable for monoculture and sensitive species	High cost in construction as well as in maintenance including cleaning of reactor

Yaakob et al. [60].

Table 4.
Comparison between open and closed photobioreactor system.

manner and mechanical pumps fixed to allow CO_2 and O_2 exchange. Closed PBR has advantage over open system as it harnesses more sunlight and hence enhance productivity (from 20 to 40 g/m /d) in short span of time, although it is costly due to complexity in structure [59]. The advantages and disadvantages between open pond and closed bioreactor is given in **Table 4**.

7. Challenges and future perspective

Despite of the importance of algae mediated wastewater treatment and further production of several bioproducts form harvested algae biomass such as (fuel, feed, food, ferilizer), some challenges are also associated with algae technology. Various microbial contaminants can also be act as inhibitors to algae growth. The pH and organic impurities such as i.e. lignin and tannins present in wastewater can affects the algal growth negatively and the concentration of heavy metals above the permissible levels can unfit the products for subsequent utilization of the pharmaceutical products. The microbe (bacteria, protozoa) present in wastewater may affect the growth of algae, there pre-treatment methods such as autoclaving, filtration is not feasible at large scale production. Therefore, advanced technologies are required for the removal of pathogens particularly for the commercial scale production. The major problem associated with conventional wastewater treatment process is generation of sludge. The algae mediated wastewater treatment process overcomes the problem of sludge generation as sludge contains only algae biomass [61]. Different types of wastewater has different composition in terms of pollution load like TDS, heavy metals contents, dissolved oxygen, so the selection of microalgae and its strain should be according to the source of the wastewater, resistibility to the pollution load, easily accessible and achieve the goal of preferred outcome. The harvesting methods of microalgae from wastewater are tedious, costly and laborious too, particularly for the unicellular microalgae. But with scientific development, biotechnological approach and emergence of advanced technology, the problem of microalgae harvesting would be elucidated. Genetic modifications of microalgae hold a great potential for biofuel production from commercialization point of view. However, there are certain challenges that need to be overcome for its large scale production. Hence, more and more studies are required to unfold the enzymatic pathway of lipid/ biofuel production to understand the mechanism involved in the process. To date, several metabolic engineering processes have been developed for enhanced production of algal biofuel, high carbohydrate and lipid content in algal biomass and improving the photosynthetic efficiency of algal species through the cellular expression or down regulation of various genes encoding a specific enzyme [62–64]. The complex nature of fatty acid biosynthetic pathway and lack of molecular transformation techniques for most of the oleaginous microalgae is cumbersome for genetic engineering process. Moreover, enhanced lipid production through genetic manipulation are not fully evolved and recent advancement in the genetic engineering methodology and techniques still promising to reach the desired goal. For the commercial purpose, mass multiplication of microalgae is required. The growth of microalgae is governed by the temperature, seasonal variations and climatic conditions. The laboratory facility with controlled condition of temperature, humidity and invariable seasonal variation is required for the mass multiplication. By viewing the importance of microalgae in the wastewater treatment, production of various valuable products and its combination with the other emerging technologies would definitely overcome the current challenges and cost in near future.

8. Conclusion

Algae are considered as a third generation biofuel, having high oil content than terrestrial crops. In the present scenario, biorefinery approach of microalgae is a promising approach towards reducing the cost of operation of decontamination as well as fuel production. Algae can easily grow on wastewater, which further preserving the resources (arable land and fresh water) for other purposes. In spite of producing various value added products from harvested algae biomass, it can act as a potential agent for wastewater remediation. Microalgae biomass production after wastewater remediation, could be a suitable fertilizer option. The microalgae biomass production reduces the organic load, and TDS, in wastewater which may further utilized as ferti-irrigation, hence reduces the burden on utilization of fresh water in a green circular economy. In addition, production of microalgae biochar which is rich in organic carbon, further enhances carbon sequestration and improving the soil quality and productivity. In this way, algae mediated wastewater treatment integrated with biochar, biodiesel and biofertilizer production from algae biomass is a recycling and resource conservation practice.

Acknowledgements

The authors are thankful to the ICAR-Indian Institute of Soil and Water Conservation, Dehradun for providing support.

Conflict of interest

The authors declare that they have no known competing financial interests or personal relationships that could have appeared to influence the work reported in this book chapter.

Notes/thanks/other declarations

Place any other declarations, such as "Notes", "Thanks", etc. in before the References section. Assign the appropriate heading. Do NOT put your short biography in this section. It will be removed.

Author details

Gulshan Kumar Sharma[1*], Shakeel Ahmad Khan[2], Amit Kumar[3], Ittyamkandath Rashmi[1], Fayaz Ahmad Malla[4] and Gopal Lal Meena[1]

1 ICAR- Indian Institute of Soil and Water Conservation, Research Centre, Kota, Rajasthan, India

2 Division of Environmental Sciences, ICAR- Indian Agricultural Research Institute, New Delhi, India

3 Central Muga Eri Research and Training Institute, Jorhat, Assam, India

4 Higher Education Department, Government Degree College Tral, Jammu and Kashmir, India

*Address all correspondence to: gulshansharma2222@gmail.com

References

[1] Yadav KK, Krishnan S, Gupta N, Prasad S, Amin AA, Cabral-Pinto MMS, et al. Review on evaluation of renewable bioenergy potential for sustainable development: Bright future in energy practice in India. ACS-Sustainable chemistry and Engineering. 2021. DOI: 10.1021/acssuschemeng.1c03114

[2] CPCB [Central Pollution Control Board of India]. National Inventory of Sewage Treatment Plants, March 2021. Available from: https://cpcb.nic.in/status-of-stps/. [Accessed 27 January 2022].

[3] Bosnic M, Buljan J, Daniels RP. Pollutants in Tannery Effluents. US/RAS/92/120, United Nations Industrial Development Organization, Regional Programme for Pollution Control in the Tanning Industry in South-east Asia. 2000.

[4] Kabdasli I, Tunay O, Cetin MS, Olmez T. Assessment of magnesium ammonium phosphate precipitation for the treatment of leather tanning industry wastewaters. Water Science Technology. 2002;**46**:231e239

[5] Sharma GK, Khan SA, Srivastava M, Bhattacharyya R, Sharma A, Gupta N, et al. Phycoremediated N-fertilization approaches on reducing environmental impacts of agricultural nitrate leaching. Journal of Cleaner Production. 2022;**345**: 131120. DOI: 10.1016/j.jclepro.2022.131120

[6] Masojidek J, Prasil O. The development of microalgal biotechnology in the Czech Republic. Journal of Industrial Microbiology Biotechnology. 2010;**37**(12):1307-1317

[7] Oswald WJ, Gotaas HB. Photosynthesis in sewage treatment. Transactions of the American Society of Civil Engineers. 1957;**122**(1):73-97

[8] Sharma GK, Khan SA, Fayaz MA, Gupta N. Nutrient sequestration and phycoremediation of sewage wastewater by selective microalgae. Green Farming. 2014;**5**:623-626

[9] Zhang H, Xin L, An Z, Liu Z, Tang C, Zhao X. Treatment of polyacrylamide-polluted wastewater using a revolving algae biofilm reactor: Pollutant removal performance and microbial community characterization. Bioresource Technology. 2021;**332**:125132

[10] Malla FA, Khan SA, Khan R, Sharma GK, Gupta N, Abraham G. Phycoremediation potential of Chlorella minutissima on primary and tertiary treated wastewater for nutrient removal and biodiesel production. Ecological Engineering. 2015;**75**:343-349

[11] Khan SA, Malla FA, Rashmi Malav LC, Gupta N, Kumar A. Potential of wastewater treating Chlorella minutissima for methane enrichment and CO_2 sequestration of biogas and producing lipids. Energy. 2018;**150**:153-168

[12] Dinnebier HCF, Matthiensen A, Michelon W, Tápparo DC, Fonseca TG, Favretto R, et al. Phycoremediation and biomass production from high strong swine wastewater for biogas generation improvement: An integrated bioprocess. Bioresource Technology. 2021;**332**:125111

[13] Sharma GK, Khan SA, Shrivastava M, Bhattacharyya R, Sharma A, Gupta DK, et al. Circular economy fertilization: Phycoremediated algal biomass as biofertilizers for sustainable crop production. Journal of Environmental Management. 2021;**287**:112295

[14] Khan SA, Sharma GK, Malla FA, Kumar A, Rashmi GN. Biorefinery potential of algal biomass

grown in sewage wastewater for phycoremediation, biodiesel and manure production. Journal of Cleaner Production. 2019;**211**:1412-1419

[15] Caprita FC, Ene A, Cantaragiu CA. Valorification of Ulva rigida algae in pulp and paper industry for improved paper characteristics and wastewater heavy metal filtration. Sustainability. 2021;**13**:10763

[16] Olguin EJ. Phycoremediation: Key issues for cost-effective nutrient removal processes. Biotechnology Advance. 2003;**22**(1-2):81-91

[17] Mohsenpoura SF, Hennige S, Willoughby N, Adeloye A, Gutierrez T. Integrating micro-algae into wastewater treatment. A review. Science of the Total Environment. 2021;**752**:142168

[18] Al-Jabri H, Das P, Khan S, Thaher M, AbdulQuadir M. Treatment of wastewaters by microalgae and the potential applications of the produced biomass—A review. Water. 2021;**13**:27

[19] Khan SA, Hussain MZ, Prasad S, Barnerjee UC. Prospects of biodiesel production from microalgae in India. Journal of Renewable and Sustainable Energy Reviews. 2009;**13**(9):2361-2372

[20] Sturm BSM, Lamer SL. An energy evaluation of coupling nutrient removal from wastewater with algal biomass production. Applied Energy. 2011;**88**:3499-3506

[21] Sharma GK, Khan SA, Shrivastava M, Gupta N, Kumar S, Malav LC, et al. Bioremediation of sewage wastewater through microalgae (*Chlorella minutissima*). Indian Journal of Agricultural Sciences. 2020;**90** (10):2024-2028

[22] Kusmayadi A, Leong YK, Yen HW, Huang CY, Chang JS. Microalgae as sustainable food and feed sources for animals and humans – Biotechnological and environmental aspects. Chemosphere. 2021;**271**:129800

[23] Kumari P, Varma AK, Ravi S, Thakur LS, Mondal P. Phycoremediation of wastewater by Chlorella pyrenoidosa and utilization of its biomass for biogas production. Journal of Environmental Chemical Engineering. 2021; **9**:104974

[24] Ansari FA, Ravindran B, Gupta SK, Nasr M, Rawat I, Bux F. Techno-economic estimation of wastewater phycoremediation and environmental benefits using *Scenedesmus obliquus* microalgae. Journal of Environmental Management. 2019;**240**:293-302

[25] Umamaheswari J, Shanthakumar S. Phycoremediation of paddy-soaked wastewater by indigenous microalgae in open and closed culture system. Journal of Environmental Management. 2019;**243**:435-443

[26] Tripathi R, Gupta A, Thakur IS. An integrated approach for phycoremediation of wastewater and sustainable biodiesel production by green microalgae, Scenedesmus sp. ISTGA1. Renewable Energy. 2019;**135**:617-625

[27] Brar A, Kumar M, Singh RP, Vivekanand V, Pareek N. Phycoremediation coupled biomethane production employing sewage wastewater: Energy balance and feasibility analysis. Bioresource Technology. 2020;**308**:123292

[28] Vairappan CS, Yen AM. Palm oil mill effluent (POME) cultured marine microalgae as supplementary diet for rotifer culture. Journal of Applied Phycology. 2008;**20**:603-608

[29] Demirbas A, Fatih DM. Importance of algae oil as a source of biodiesel. Energy Conversation Management. 2011;**52**:163-170

[30] Ahmad AL, Mat Yasin NH, Derek CJC, Lim JK. A review-microalgae as a sustainable energy source for biodiesel production. Renew Sustainable Energy Review. 2011;**15**:584-593

[31] Nascimento IA, Marques SSI, Cabanelas ITD, et al.. Screening microalgae strains for biodiesel production: Lipid productivity and estimation of fuel quality based on fatty acids profiles as selective criteria. Bioenergy Research. 2013;**6**:1-13

[32] Menegazzo LM, Fonseca GG. Biomass recovery and lipid extraction processes for microalgae biofuels production: A review. Renew. Sustainable Energy Review. 2019;**107**:87-107

[33] Deshmukh S, Kumar R, Bala K. Microalgae biodiesel: A review on oil extraction, fatty acid composition, properties and effect on engine performance and emissions. Fuel Processing Technology. 2019;**191**:232-247

[34] Ferreira GF, Rios Pinto LF, Filho RM, Fregolente LV. A review on lipid production from microalgae: Association between cultivation using waste streams and fatty acid profiles. Renewable Sustainable Energy Review. 2019;**109**:448-466

[35] Chisti Y. Biodiesel from microalgae. Biotechnology Advances. 2007;**25**:294-306

[36] Mulbry W, Kondrad S, Pizarro C. Biofertilizers from algal treatment of dairy and swine manure effluents: Characterization of algal biomass as a slow release fertilizer. International Journal of Vegetable. 2007;**12**(4):107-125

[37] Nayak M, Swain DK, Sen R. Strategic valorization of de-oiled microalgal biomass waste as biofertilizer for sustainable and improved agriculture of rice (*Oryza sativa* L.) crop. Science of Total Environment. 2019;**682**:475-484

[38] Kumar G, Bawaja P. Biofertilizer: A tool for sustainable agriculture in changing environment. In: Ansari MW, Kumar S, Kaula BC, Wattal RK, editors. Introduction to Challenges and Strategies to Improve Crop Productivity in Changing Environment. Dwarka: Public Pvt. Ltd; 2018. pp. 83-92

[39] Shaaban MM. Green microalgae water extract as foliar feeding to wheat plants. Pakistan Journal of Biological Sciences. 2001;**4**(6):28-632

[40] Abd El Moniem EA, Abd-Allah ASE. Effect of green algae cells extract as foliar spray on vegetative growth, yield and berries quality of superior grapevines. Journal of American European Agricultural and Environmental Science. 2008;**4**(4):427-433

[41] Renuka N, Guldhe A, Prasanna R, Singh P, Bux F. Microalgae as multi-functional options in modern agriculture: Current trends, prospects and challenges. Biotechnology Advances. 2018;**36**:1255-1273

[42] Sharma GK, Jena RK, Hota S, Kumar A, Ray P, Fagodiya RK, et al. Recent Development in Bioremediation of Soil pollutants through biochar for environmental sustainability. In: Singh J, Singh C, editors. Biochar Applications in Agriculture and Environment Management. Cham: Springer; 2020a. pp. 122-140

[43] Malyan SK, Kumar SS, Fagodiya RK, Ghosh P, Kumar A, Singh R, et al. Biochar for environmental sustainability in the energy-water-agroecosystem

nexus. Renewable and Sustainable Energy Reviews. 2021;**149**:111379

[44] Gupta DK, Gupta CK, Dubey R, Fagodiya RK, Sharma GK, Keerthika A, et al. Role of biochar in carbon sequestration and greenhouse gas mitigation. Biochar Application Agricultural and Environmental Management. 2020:141-165. DOI:10.1007/978-3-030-40997-5_7

[45] Chen Z, Li K, Pu L. The performance of phosphorus (P)-doped activated carbon as a catalyst in air-cathode microbial fuel cells. Bioresource Technology. 2014;**170**:379-384

[46] Yu KL, Show PL, Ong HC, Ling TC, Chen W, Amran M. Biochar production from microalgae cultivation through pyrolysis as a sustainable carbon sequestration and biorefinery approach. Clean Technology Environmental Policy. 2018;**20**:2047-2055

[47] Chaiwong K, Kiatsiriroat T, Vorayos N, Thararax C. Biochar production from freshwater algae by slow pyrolysis. Maejo International Journal of Science and Technology. 2012;**6**(2):186-195

[48] Jindo K, Mizumoto H, Sawada Y, Sanchez-Monedero MA, Sonoki T. Physical and chemical characterization of biochars derived from different agricultural residues. Bio Geosciences. 2014;**11**(23):6613-6621

[49] Zheng H, Guo W, Li S, Chen Y, Wu Q, Feng X, et al. Adsorption of p-nitrophenols (PNP) on microalgal biochar: Analysis of high adsorption capacity and mechanism. Bioresource Technology. 2017;**244**(2):1456-1464

[50] Hodgson E, Lewys-James A, Ravella SR, Thomas-Jones S, Perkins W, Gallagher J. Optimisation of slow-pyrolysis process conditions to maximise char yield and heavy metal adsorption of biochar produced from different feedstocks. Bioresource Technology. 2016;**214**:574-581

[51] Yu KL, Lau BF, Show PL, Ong HC, Ling TC, Chen WH, et al. Recent developments on algal biochar production and characterization. Bioresource Technology. 2017;**246**:2-11

[52] International Panel of Climate Change [IPCC]. Synthesis Report. Contribution of Working Groups I, II and III to the Fifth Assessment Report of the Intergovernmental Panel on Climate Change [Core Writing Team, eds R. K. Pachauri and L. A. Meyer (Geneva: IPCC), 2014.

[53] Razzak SA, Ali SAM, Hossain MM, deLasa H. Biological CO_2 fixation with production of microalgae in wastewater – A review. Renewable and Sustainable Energy Reviews. 2017;**76**:379-390

[54] Langley NM, Harrison STL, Van Hille RP. A critical evaluation of CO_2 supplementation to algal systems by direct injection. Biochemical Engineering Journal. 2012;**68**:70-75

[55] Tang D, Han W, Li P, Miao X, Zhong J. CO_2 biofixation and fatty acid composition of Scenedesmus obliquus and Chlorella pyrenoidosa in response to different CO_2 levels. Bioresources Technology. 2011;**102**:3071-3076

[56] Meier RL. Biological cycles in the transformation of solar energy into useful fuels. In: Solar Energy Research. Madison, WI: University Wisconsin Press; 1955. pp. 179-183

[57] Golueke CG, Oswald WJ. Biological conversion of light energy to the chemical energy of methane. Applied Microbiology. 1959;7(4):219-227

[58] Brennan L, Owende P. Biofuels from microalgae-a review of technologies for production, processing and extractions of biofuels and coproducts. Renewable Sustainable Energy Review. 2010;**14**:557-577

[59] Buhrer H. Light within algal cultures; Implications from light intensities with lens. Aquatic Science. 2000;**62**:91-103

[60] Yaakob Z, Kamarudin KF, Rajkumar R, Takriff MS, Badar SN. The current methods for the biomass production of the microalgae from wastewaters: An overview. World Applied Sciences Journal. 2014;**31**(10):1744-1758

[61] Bhatia SK, Mehariya S, Bhatia RK, Kumar M, Pugazhendhi A, Awasthi MK, et al. Wastewater based microalgal biorefinery for bioenergy production: Progress and challenges. Science of Total Environment. 2021;**751**:141599

[62] Li K, Liu S, Liu X. An overview of algae bioethanol production. International Journal of Energy Research. 2014;**38**(8):965-977

[63] Rastogi RP, Pandey A, Larroche C, Datta M. Algal Green Energy – R&D and technological perspectives for biodiesel production. Renewable and Sustainable Energy Reviews. 2018;**82**(3):2946-2969

[64] Ho SH, Chen CNN, Lai YY, Lu W, Chang JS. Exploring the high lipid production potential of a thermotolerant microalga using statistical optimization and semi-continuous cultivation. Bioresource Technology. 2014;**163**:128-135